# Studies in Systems, Decision and Control

Volume 464

The series "Studies in Systems, Decision and Control" (SSDC) covers both new developments and advances, as well as the state of the art, in the various areas of broadly perceived systems, decision making and control–quickly, up to date and with a high quality. The intent is to cover the theory, applications, and perspectives on the state of the art and future developments relevant to systems, decision making, control, complex processes and related areas, as embedded in the fields of engineering, computer science, physics, economics, social and life sciences, as well as the paradigms and methodologies behind them. The series contains monographs, textbooks, lecture notes and edited volumes in systems, decision making and control spanning the areas of Cyber-Physical Systems, Autonomous Systems, Sensor Networks, Control Systems, Energy Systems, Automotive Systems, Biological Systems, Vehicular Networking and Connected Vehicles, Aerospace Systems, Automation, Manufacturing, Smart Grids, Nonlinear Systems, Power Systems, Robotics, Social Systems, Economic Systems and other. Of particular value to both the contributors and the readership are the short publication timeframe and the world-wide distribution and exposure which enable both a wide and rapid dissemination of research output.

Indexed by SCOPUS, DBLP, WTI Frankfurt eG, zbMATH, SCImago.

All books published in the series are submitted for consideration in Web of Science.

Orestes Llanes-Santiago
Editor

# Proceedings of 19th Latin American Control Congress (LACC 2022)

Springer

*Editor*
Orestes Llanes-Santiago
Universidad Tecnológica de la Habana José
Antonio Echeverría, Cujae
Marianao, La Habana, Cuba

ISSN 2198-4182 ISSN 2198-4190 (electronic)
Studies in Systems, Decision and Control
ISBN 978-3-031-26360-6 ISBN 978-3-031-26361-3 (eBook)
https://doi.org/10.1007/978-3-031-26361-3



This Springer imprint is published by the registered company Springer Nature Switzerland AG
The registered company address is: Gewerbestrasse 11, 6330 Cham, Switzerland

# Preface

The 19th. Latin American Congress of Automatic Control was scheduled to be held in December 2020 in Havana, Cuba. For this reason, the International Program Committee and the National Organizing Committee began the work of organizing it in June 2019.

To form the International Program Committee, 157 outstanding international researchers in the different areas related to Automatic Control were invited. Finally, 123 of them answered affirmatively, with which 23 countries were represented.

In December 2019, the first call for papers was launched and in March 2020 the Covid-19 pandemic spread throughout the world, forcing isolation between and within most countries. This forced the holding of the Congress to be postponed until 2022, when it was decided to be held virtually. Also, due to a lack of funding which was exacerbated by the pandemic, a large number of professors and researchers were not enable to attend in person.

A total of 53 papers were sent to Congress, which shows how Latin American professors and researchers continued their research work despite the difficult conditions imposed by the pandemic. Of the submitted papers, 30 were accepted for presentation based on rigorous analysis of scientific novelty and their quality. For their presentation in this book, the accepted papers were divided into 5 major areas that appear in the book in the following order:

- Robust and Nonlinear Control
- Fault Diagnosis in Industrial Systems
- Robotic and Autonomous Systems
- Modeling, Identification and Delayed Systems
- Low-Cost Systems and Biomedical Applications

The organization of any Congress involves many people who must be recognized. The Organizing Committee of the 19th Latin American Automatic Control Congress (LACC 2022) would like to firstly thank Prof. Janusz Kacprzyk, Series Editor of Springer Nature Publisher, and Dr. Thomas Ditzinger, Editorial Director of Interdisciplinary and Applied Sciences and Engineering of Springer Nature Publisher, for their support to the Latin American Automatic Control community with the publication of the proceedings of LACC 2022 in the Book series Studies in Systems, Decision and Control. Also, we want to thank all the colleagues who constitute the International Program Committee members and other outstanding reviewers for the professional and serious work done in reviewing the articles sent to LACC 2022. Finally, we want to thank the Latin American Automatic Control Community for their support in holding the Congress.

La Habana, Cuba
November 2022

Orestes Llanes-Santiago
Adrián Rodríguez-Ramos

# Contents

**Robust and Nonlinear Control**

**Robust Admissibilization of Discrete-Time Descriptor LPV Systems Based-on $\mathcal{H}_2$-Control** ........ 3
Addison Ríos-Bolívar

**Fault Tolerance in Feedback Control Loops** ........ 13
Jose A. Carrazana and Efrain Alcorta Garcia

**Robust PID Control with Anti-Windup Compensation for ULDS** ........ 23
Addison Ríos-Bolívar and Orestes Llanes-Santiago

**Observer-Based Tracking Control for PMSM Rooted on Port-Hamiltonian Systems Structural Properties** ........ 35
F. Ramos-García, C. Granados-Salazar, M. Rojas, and G. Espinosa-Pérez

**Sliding Mode Control Based on a Generalized Reduced Fractional Order Model: Theoretical Approach** ........ 47
Antonio Di Teodoro, Diego Ochoa-Tocachi, Hanna Aboukheir, and Oscar Camacho

**Tuning and Optimization Software for Non-linear PID Controllers with Guaranteed Robustness** ........ 59
Mercedes Ramírez Mendoza, Liset Mayo Martí, Guillermo González Yero, and Pedro Albertos Pérez

**Sliding Mode Control Based on a Nonlinear PID Surface for Chemical Processes** ........ 71
Mateo Vásquez, Joseline Yanascual, Marco Herrera, Francisco Rossomando, and Oscar Camacho

## Fault Diagnosis in Industrial Processes

**Exploratory Data Analysis on the Poincaré Features of the Electrical Torque Oriented to the Severity Diagnosis of a Gearbox Tooth Breakage** ..... 85
Mariela Cerrada, María-Victoria Mejía, Renato Ortega, Diego Cabrera, and Rene-Vinicio Sánchez

**A Proposal of Training Data Selection for Remaining Job Cycle Time Prediction in Industrial Monitoring: A Real Case Study** ..... 97
Alejandro García-Celis, Adriana Villalón-Falcón, Randy Piñero-Aguilar, Fernando Suárez-Concepción, Alberto Prieto-Moreno, and Orestes Llanes-Santiago

**Energy-Efficient Wavelet Transform Implementation for Fault Diagnosis** ..... 109
Iliover Vega-Gonzalez and Jorge Ramírez-Beltrán

**Poincaré Images Extracted from Vibration Signals are Useful Features for Fault Classification in a Reciprocating Compressor** ..... 119
Ruben Medina, Mariela Cerrada, Diego Cabrera, and René Vinicio Sánchez

**A Novel Approach for Detection and Location of Cyber-Attacks in Water Distribution Networks** ..... 129
Claudia Rodríguez-Martínez, Marcos Quiñones-Grueiro, Antônio J. Silva Neto, and Orestes Llanes-Santiago

**Predictive Diagnosis of Isolated Photovoltaic Systems Based on PWM & MPPT Charge Regulators** ..... 141
Emilio García-Moreno, Eduardo Quiles-Cucarella, Antonio Correcher-Salvador, and Francisco Morant-Anglada

## Robotic and Autonomous Systems

**Path Following Control of Autonomous Tractor Using Non-linear Guidance Law** ..... 155
Alexander Rodríguez Conte, Luis E. Hernández Morales, and Luis Hernández Santana

**Integral Line-of-Sight Controller for 3-D Path Following in Presence of Marine Currents** ..... 165
Yunier Valeriano Medina, Omar Milián Morón, Leidys Miranda Jiménez, and Anailys Hernández Julián

**Battery Health-Aware MPC Planning for Autonomous Racing Vehicles** ..... 177
Sergio E. Samada, Vicenç Puig, and Fatiha Nejjari

**Leader-Follower Formation Tracking Strategies Using Heterogeneous Robots for Monitoring and Remediation Applications** ........ 189
Jesus Santiaguillo-Salinas, Hiram N. Garcia-Lozano, and Rafael F. Gonzalez-Zarate

**Modeling, Identification and Delayed Systems**

**Model-Based Efficient Selection of Centrifugal Pumps for Simple Branch Pumping Systems** ........ 203
Cláudio C. Pellegrini, Jacqueline Pedrera-Yanes, Orestes Llanes-Santiago, and Guillermo Vilalta-Alonso

**Reinforcement Learning with Probabilistic Boolean Networks in Smart Grid Models** ........ 215
Pedro Juan Rivera Torres, Carlos Gershenson García, María Fernanda Sánchez Puig, Mario Franco, and Samir Kanaan Izquierdo

**Selection of Model Structure Using PSO Method** ........ 225
Carlos Alberto García Vázquez and Ana Isabel González Santos

**Neuro-Adaptive PID Helicopter Controller Based on Atomic Functions** ........ 239
O. F. Garcia–Castro, L. E. Ramos–Velasco, E. Escamilla–Hernández, R. Garcia–Rodriguez, M. A. Vega–Navarrete, C. R. Domínguez–Mayorga, and L. N. Oliva–Moreno

**Predictor Based Control Scheme for High-Order Delayed Systems with One Unstable Pole** ........ 251
Alán Ricardo Pineda Higuera, Luis Alberto Barragan Bonilla, María Margarita Vargas Mondragón, Basilio del Muro Cuéllar, Juan Francisco Márquez Rubio, and Gonzalo Isaac Duchén Sánchez

**Observer Based Control of a Scalar Neutral Delayed System** ........ 263
J. F. Marquez Rubio, A. Urquiza Castro, R. J. Vazquez Guerra, B. del Muro Cuéllar, and E. Aranda Bricaire

**Low Cost Systems and Biomedical Applications**

**ESP32 Based Low-Power and Low-Cost Wireless Sensor Network** ........ 275
Alejandro Perdomo-Campos, Iliover Vega-González, and Jorge Ramírez-Beltrán

**Low-Cost Portable System for the Estimation of Air Quality** ........ 287
Erik Hernandez-Rodriguez, David Kairúz-Cabrera, Alain Martinez, Rosa Amalia González-Rivero, and Olivier Schalm

**"José Martí" International Airport Runway Lights System Control and Monitoring** ........ 299
Dayra Ivette Saurí Vázquez and Ariel Omar Cepero Díaz

**Automatic System to Characterization of Creatinine by Optic Measurement '*On Line*' in Wastes Samples During Hemodialysis** ...... 309
Abel Calle-Herranz, Carmenchu Regueiro-Busoch, Carmen Brígida Busoch-Morlán, and Angel Regueiro-Gómez

**Automatic System to Continuous and Differential Acquisition of Blood Pressure During Tilt Test** ........ 319
C. Regueiro-Busoch, Y. J. Almeida-Pichardo, L. A. Noa-Lloren, C. B. Busoch-Morlán, and A. Regueiro-Gómez

# Robust and Nonlinear Control

# Robust Admissibilization of Discrete-Time Descriptor LPV Systems Based-on $\mathscr{H}_2$-Control

**Addison Ríos-Bolívar**

**Abstract** The main contribution of this work is centered in the robust admissibilization of discrete-time linear descriptor systems (DS) with variant parameters (LPV). The technique is based on the synthesis of robust controllers that allow to obtain performance index in $\mathscr{H}_2$ for closed-loop system. The DSs are considered to have parametric uncertainties of polytopic type and external perturbations, as described by:

$$E(\rho)x_{k+1} = A(\rho)x_k + B_\omega(\rho)\omega_k + B(\rho)u_k; \quad y_k = C(\rho)x_k + D(\rho)\omega_k$$

where $\rho$ is the parametric variation. From the characterization of the $\mathscr{H}_2$-norm as a Linear Matrix Inequality (LMI), the necessary and sufficient admissibility conditions in strict linear matrix inequality form are derived, which allow to design the robust controller by state feedback or static output feedback (SOF) using computational tools. The obtained results are extensions of the derivatives of the problem of the robust admissibility based on the $\mathscr{H}_\infty$-control of discrete-time linear DS type LPV.

## 1 Introduction

Since its introduction in 1977 [1], descriptor systems (DS) have been one of the main research fields within control theory, [2, 3]. Unlike their regular counterparts in state space, a DS allows a representation that incorporates algebraic constraints in their physical variables.

On the other hand, the context of linear parameter variable (LPV) systems refers to linear dynamical systems whose state-space representations depend on exogenous non-stationary parameters [4]. LPV systems are a generalization of LTV systems [5].

A. Ríos-Bolívar (✉)
Universidad de Los Andes, Facultad de Ingeniería, Departamento de Sistemas de Control, Mérida 5101, Venezuela
e-mail: ilich@ula.ve



O. Llanes-Santiago (ed.), *Proceedings of 19th Latin American Control Congress (LACC 2022)*, Studies in Systems, Decision and Control 464,
https://doi.org/10.1007/978-3-031-26361-3_1

When there are combined the modeling of physical systems with uncertain parameters, there arise dynamic systems representing uncertain DS. As is well known, for modeling many applications and technical processes, only approximate models are available, so that the analysis of DSs subject to uncertainties has been a very active research line, [6, 7]. In that order of ideas, one of the main research topics has been the robust admissibility of DS, which has been focused on the state feedback control and $\mathscr{H}_\infty$-norm, [8–10]. In order to extend these results to the case of $\mathscr{H}_2$ control, in this paper the robust admissibilization of discrete-time linear descriptor systems (DS) with variant parameters (LPV) is studied. A technique for robust admissibility is presented, which consists of the design of robust controllers based on the characterization of the $\mathscr{H}_2$-norm as LMIs, [11, 12].

## 2 Descriptor LPV System

Let be a descriptor LPV system

$$E(\rho)x_{k+1} = A(\rho)x_k + B(\rho)u_k, \qquad y_k = C(\rho)x_k \tag{1}$$

where $x \in \mathbb{R}^n$ is the vector of descriptor variable (instead of state vector), $\mathbb{E} \in \mathbb{R}^{m\times n}$, with $\text{rank}(\mathbb{E}) = r$ and $0 < r \leq n$ for all $\rho$, which is called the descriptor matrix. $A \in \mathbb{R}^{m\times n}$, $B \in \mathbb{R}^{m\times q}$, $C \in \mathbb{R}^{p\times n}$. $\rho_k \in \mathbb{R}^l$ is the varying parameter vector

$$\rho_k \triangleq \left[\rho_{1_k},\ \rho_{2_k},\ \cdots,\ \rho_{l_k}\right]^T \tag{2}$$

The descriptor matrix $E(\rho)$ is assumed to be rank-invariant, that is, $\text{rank}(\mathbb{E}) = r$ and $0 < r \leq n$ for all admissible uncertainties $\rho$. This last asseveration is important for the robust admissibilization, when $E$ is an uncertain matrix, [13].

The range of each varying parameter $\rho_i$, $i = 1, 2, \ldots, l$, is given by

$$\rho_{i_k} \in [\underline{\rho}_i, \overline{\rho}_i], \quad \underline{\rho}_i < \overline{\rho}_i, \quad i = 1, 2, \ldots, l \tag{3}$$

Equation (3) represents a convex hull whose number of the vertices is $N = 2^l$. It corresponds to the operating region of the LPV system (1) and is also called the parameter box. The vertex set of the parameter box (3) is defined by

$$\Omega \triangleq \left\{\rho_k \in \mathbb{R}^l :\quad \rho_{i_k} = \underline{\rho}_i \ \text{or}\ \overline{\rho}_i, \quad i = 1, 2, \ldots, l\right\} \tag{4}$$

The descriptor LPV system (1) can be transformed into a polytopic model which is constructed by linearly combining LTI models in the vertex set $\Omega$ (4):

$$\sum_{j=1}^{N}\alpha_j(\rho)E_jx_{k+1}=\sum_{j=1}^{N}\alpha_j(\rho)A_jx_k+\sum_{j=1}^{N}\alpha_j(\rho)B_ju_k,\quad y_k=\sum_{j=1}^{N}\alpha_j(\rho)C_jx_k \tag{5}$$

Thus,

$$\begin{bmatrix}E_j & A_j & B_j & C_j\end{bmatrix}\triangleq\begin{bmatrix}E(\rho^j) & A(\rho^j) & B(\rho^j) & C(\rho^j)\end{bmatrix} \tag{6}$$

$$\rho^j\in\Omega,\ \ j=1,2,\ldots,N.\ \ \text{With}\ \sum_{j=1}^{N}\alpha_j=1,\quad \alpha_j\geq 0$$

In [14] a general framework for to transform the Descriptor System (DS) (1) in form of descriptor polytopic models (5) is given.

**Definition 1** For (1) with $m=n$, if $\forall\rho$, $p(\mathbf{z})=\det(\mathbf{z}E(\rho)-A(\rho))$ satisfies that $p(\mathbf{z})\neq 0$, it is said that the pair $(E(\rho), A(\rho))$ is *regular*. Otherwise, it is called *singular*.

**Definition 2** Consider the system (1) $(m=n)$, with $\kappa=\deg(\det(\mathbf{z}E(\rho)-A(\rho))$. If $\kappa=r$ is said that the DS is of *impulse-free* or causal.

For the non-trivial case $E(\rho)\neq 0$, the property of impulse-free it implies regularity.

**Definition 3** Consider the system (1). It is said that the DS is *admissible* if it is regular, impulse-free (causal) and stable.

In which it follows, the system (1) is considered regular and impulse-free, consequently, the system admissibility corresponds to its stability.

Let consider the open-loop stability of system (1), in [15] a novel condition is proposed:

**Theorem 1** (Stability of open-loop: [15]) *The system* (1) *is admissible if and only if there exists a positive definite matrix* $Q\in\mathbb{R}^{n\times n}$ *and a symmetric matrix* $S\in\mathbb{R}^{(n-r)\times(n-r)}$ *such that*

$$A_j^T\left(Q-\overline{E}_j^TS\overline{E}_j\right)A_j-E_j^TQE_j<0,\qquad j=1,2,\ldots,N \tag{7}$$

*with* $\overline{E}_j$ *denotes a matrix with the properties of* $N(\overline{E}_j)=R(E_j)$, *i.e.* $\overline{E}_jE_j=0$, *and* $\overline{E}_j\,\overline{E}_j^T>0$. *There,* $N(\cdot)$ *is called the nullspace of the matrix and* $R(\cdot)$ *defines the range space.*

***Proof*** See [15]. □

The matrix $\overline{E}\in\mathbb{R}^{(n-r)\times n}$, which is of full column ranks, it is composed of base of $\ker E$.

Using the Schur complement, then:

**Lemma 1** *The system* (1) *is admissible if and only if there exists a positive definite matrix* $Q \in \mathbb{R}^{n\times n}$, *a symmetric matrix* $S \in \mathbb{R}^{(n-r)\times(n-r)}$ *and a matrix* $G \in \mathbb{R}^{n\times n}$ *such that*

$$\begin{bmatrix} -E_j^T Q E_j & A_j^T G \\ G^T A_j & -G - G^T + Q - \overline{E}_j^T S \overline{E}_j \end{bmatrix} < 0, \qquad j = 1, 2, \ldots, N \tag{8}$$

A new extended and improved condition has been presented in [16], which is described in the following theorem:

**Theorem 2** (Stability of open-loop: [16]) *The system* (1) *is admissible if and only if the following equivalent statements hold.*

*(i) There exist matrices* $Q > 0$ *and* $S = S^T$ *satisfying the LMI* (7).
*(ii) There exist matrices* $Q > 0$ *and* $S = S^T$ *satisfying the LMI*

$$A_j \left( Q - \overline{\overline{E}}_j S \overline{E}_j^T \right) A_j^T - E_j Q E_j^T < 0, \qquad j = 1, 2, \ldots, N \tag{9}$$

*with* $\overline{E}_j^T E_j^T = 0$ *and* $\overline{E}_j^T \overline{\overline{E}}_j > 0$.
*(iv) There exist matrices* $Q > 0$, $S = S^T$, $F$ *and* $G$ *satisfying,* $\forall\ j = 1, 2, \ldots, N$, *the following LMI*

$$\begin{bmatrix} -E_j^T Q E_j + A_j^T F^T + F A_j & -F + A_j^T G^T \\ -F^T + G A_j & -G - G^T + Q - \overline{E}_j^T S \overline{E}_j \end{bmatrix} < 0 \tag{10}$$

*(v) There exist matrices* $Q > 0$, $S = S^T$, $F$ *and* $G$ *satisfying,* $\forall\ j = 1, 2, \ldots, N$, *the following LMI*

$$\begin{bmatrix} -E_j Q E_j^T + A_j F^T + F A_j^T & -F + A_j G^T \\ -F^T + G A_j^T & -G - G^T + Q - \overline{\overline{E}}_j S \overline{E}_j^T \end{bmatrix} < 0 \tag{11}$$

***Proof*** See [16]. □

## 2.1 Robust Stabilization by State Feedback Control

In order to design appropriate controllers, it is necessary to evaluate the characterization of the controllability condition for DS.

**Definition 4** A linear DS is said controllable if, for any time $k > 0$, $x_0 \in \mathbb{R}^n$ and $x_f \in \mathbb{R}^n$, there exits a control $u$ such that $x_k = x_f$

**Lemma 2** (Controllability of Descriptor LPV System) *Consider system* (1), *which is said controllable if and only if, for all* $\rho$*:*

$$rank\,[zE(\rho) - A(\rho)\ \ B(\rho)] = n,\ \ \forall z \in \mathbb{C},\ \ z\,\textit{finite and}\ \ rank\,[E(\rho)\ \ B(\rho)] = n \tag{12}$$

***Proof*** See [17, 18]. □

In (12), the first condition talks about the controllability of the finite dynamic modes. The second condition is related to the controllability of the impulsive modes.

The Definition 3 and the Lemma 2 allow to establish conditions for the control of DS in the sense of its stabilization [16, 19].

Let consider that system (1) is controllable for all $\rho$, thus, the robust admissibilization of system (1) is considered by means of a state feedback control. Then, consider the control $u_k$ given by

$$u_k = \mathbf{K}x_k, \tag{13}$$

then, the system in cloed-loop is given by

$$E(\rho)x_{k+1} = (A(\rho) + B(\rho)\mathbf{K})\,x_k \tag{14}$$

From the Theorem 2 on the equivalent dual system, the following result is determined:

**Lemma 3** *The system* (1) *is robustly admissible via a state feedback controller* (13)*, if and only if there exists a positive definite matrix* $Q \in \mathbb{R}^{n\times n}$*, a symmetric matrix* $S \in \mathbb{R}^{(n-r)\times(n-r)}$ *and a matrix* $G \in \mathbb{R}^{n\times n}$*, a matrix* $F \in \mathbb{R}^{n\times n}$ *and the matrices* $X \in \mathbb{R}^{q\times n}$*,* $Y \in \mathbb{R}^{q\times n}$*, such that,* $\forall\ j = 1, 2, \ldots, N$*:*

$$\begin{bmatrix} -E_jQE_j^T + A_jF^T + B_jX + FA_j^T + X^TB_j^T & -F + A_jG^T + B_jY \\ -F^T + GA_j^T + Y^TB_j^T & -G - G^T + Q - \overline{\overline{E}}_jS\overline{E}_j^T \end{bmatrix} < 0, \tag{15}$$

*where the feedback gain is given by*

$$\boldsymbol{K} = Y(G^T)^{-1} \tag{16}$$

***Proof*** Applying Lemma 1 with respect to the dynamic matrix of the closed-loop $A_j + B_j\mathbf{K}$, then the change of variables $Y = \mathbf{K}G^T$ and $X = \mathbf{K}F^T$ must be used. □

# 3 Main Results

## 3.1 LMI Characterization of the $\mathscr{H}_2$-Norm for Descriptor LVP System

A first strict LMI condition for $\mathscr{H}_2$ control of LTI descriptor systems is presented in [20]. Let consider the descriptor LPV system type polytopic:

$$E(\rho)x_{k+1} = A(\rho)x_k + B_\omega(\rho)\omega_k + B(\rho)u_k, \quad y_k = C(\rho)x_k + D(\rho)\omega_k \tag{17}$$

where $\omega_k \in \mathbb{R}^d$ are perturbation signals. The matrices $B_\omega(\rho)$, $D(\rho)$ have appropriated dimensions. Then, the transfer function $H_{\omega y}(\rho, \mathbf{z}) = C(\rho)(\mathbf{z}E(\rho) - A(\rho))^{-1}B_\omega(\rho) + D(\rho)$ is defined. Then, $\|H_{\omega y}\|_2^2 = \mathrm{tr}[D^T(\rho)D(\rho) + B_\omega^T(\rho)QB_\omega(\rho)]$, such that $Q = Q^T > 0$ and $A^T(\rho)QA(\rho) - E^T(\rho)QE(\rho) + C^T(\rho)C(\rho) < 0$. Thus, the following result can be established:

**Theorem 3** *The discrete-time descriptor LPV system (17) is admissible and satisfies $\|H_{\omega y}\|_2^2 < \mu$, if and only if there exists a positive definite $Q \in \mathbb{R}^{n\times n}$, a symmetric matrix $S \in \mathbb{R}^{(n-r)\times(n-r)}$ and the matrices $W \in \mathbb{R}^{d\times d}$ with tr($W$)$< \mu$, $G$ satisfying, $\forall j = 1, 2, \ldots N$, the following LMI of the equivalent statements:*

*(i)*

$$\begin{bmatrix} E_j^TQE_j & A_j^TQ & C_j^T \\ QA_j & Q & 0 \\ C_j & 0 & \mathbb{I} \end{bmatrix} > 0, \quad \begin{bmatrix} W & B_{\omega j}^TQ & D_j^T \\ QB_{\omega j} & Q & 0 \\ D_j & 0 & \mathbb{I} \end{bmatrix} > 0 \tag{18}$$

*(ii)*

$$\begin{bmatrix} E_jQE_j^T & A_jQ & B_{\omega j} \\ QA_j^T & Q & 0 \\ B_{\omega j}^T & 0 & \mathbb{I} \end{bmatrix} > 0, \quad \begin{bmatrix} W & C_jQ & D_j \\ QC_j^T & Q & 0 \\ D_j^T & 0 & \mathbb{I} \end{bmatrix} > 0 \tag{19}$$

*(iii)*

$$\begin{bmatrix} E_jQE_j^T & A_jG & B_{\omega j} \\ G^TA_j^T & G + G^T - Q + \overline{\overline{E}}_jS\overline{E}_j^T & 0 \\ B_{\omega j}^T & 0 & \mathbb{I} \end{bmatrix} > 0 \quad \begin{bmatrix} W & C_jG & D \\ G^TC_j^T & G + G^T - Q + \overline{\overline{E}}_jS\overline{E}_j^T & 0 \\ D_j^T & 0 & \mathbb{I} \end{bmatrix} > 0 \tag{20}$$

***Proof*** See [20]. □

In order to obtain strict and less conservative inequalities, the following is obtained:

**Lemma 4** *The discrete-time descriptor LPV system (17) is admissible and satisfies $\|H_{\omega y}\|_2^2 < \mu$, if and only if:*

*(i) there exists a positive definite $Q \in \mathbb{R}^{n\times n}$, a symmetric matrix $S \in \mathbb{R}^{(n-r)\times(n-r)}$ and the matrices $W \in \mathbb{R}^{d\times d}$ with tr($W$)$< \mu$, and $G$ satisfying, $\forall j = 1, 2, \ldots N$, the following LMI*

$$\begin{bmatrix} E_j^T Q E_j & A_j^T Q & C_j^T \\ QA_j & Q & 0 \\ C_j & 0 & \mathbb{I} \end{bmatrix} > 0, \qquad \begin{bmatrix} W & B_{\omega j}^T Q & D_j^T \\ QB_{\omega j} & Q & 0 \\ D_j & 0 & \mathbb{I} \end{bmatrix} > 0 \tag{21}$$

*(ii) there exists a positive definite $Q \in \mathbb{R}^{n\times n}$, a symmetric matrix $S \in \mathbb{R}^{(n-r)\times(n-r)}$ and the matrices $W \in \mathbb{R}^{d\times d}$ with tr($W$)< $\mu$, $G$ and $F$ satisfying, $\forall j = 1, 2, \ldots N$, the following LMI*

$$\begin{bmatrix} E_j^T Q E_j - FA_j - A_j^T F^T & -F + A_j^T G & C_j^T \\ -F^T + G^T A_j & G + G^T - Q + \overline{E}_j^T S \overline{E}_j & 0 \\ C_j & 0 & \mathbb{I} \end{bmatrix} > 0 \tag{22}$$

$$\begin{bmatrix} W & B_{\omega j}^T G & D^T \\ G^T B_{\omega j} & G + G^T - Q + \overline{E}_j^T S \overline{E}_j & 0 \\ D_j & 0 & \mathbb{I} \end{bmatrix} > 0 \tag{23}$$

*(iii) there exists a positive definite $Q \in \mathbb{R}^{n\times n}$, a symmetric matrix $S \in \mathbb{R}^{(n-r)\times(n-r)}$ and the matrices $W \in \mathbb{R}^{p\times p}$ with tr($W$)< $\mu$, $G$ and $F$ satisfying, $\forall j = 1, 2, \ldots N$, the following LMI*

$$\begin{bmatrix} E_j Q E_j^T - FA_j^T - A_j F^T & -F + A_j G & B_{\omega j} \\ -F^T + G^T A_j^T & G + G^T - Q + \overline{\overline{E}}_j S \overline{E}_j^T & 0 \\ B_{\omega j}^T & 0 & \mathbb{I} \end{bmatrix} > 0 \tag{24}$$

$$\begin{bmatrix} W & C_j G & D \\ G^T C_j^T & G + G^T - Q + \overline{\overline{E}}_j S \overline{E}_j^T & 0 \\ D_j^T & 0 & \mathbb{I} \end{bmatrix} > 0 \tag{25}$$

***Proof*** The proof is based on establishing equivalence between the LMI (22) and and its corresponding in (18), [15, 16, 21]. Thus, consider the matrix $R = R^T = (Q - \overline{E}_j^T S \overline{E}_j) > 0$, with $\overline{E}_j E_j = 0$ and $\overline{E}_j \; \overline{E}_j^T > 0$, such that $\|H_{\omega y}\|_2^2 = \text{tr}[D^T(\rho)D(\rho) + B^T(\rho)RB(\rho)]$, and $E^T(\rho)RE(\rho) - A^T(\rho)RA(\rho) - C^T(\rho)C(\rho) > 0$.

*Sufficiency*: Assuming that the condition (22) is satisfied, then multiplying by the left by $\begin{bmatrix}\mathbb{I} & -A_j^T & C_j^T\end{bmatrix}$ and the right by $\begin{bmatrix}\mathbb{I} & -A_j^T & C_j^T\end{bmatrix}^T$, the following inequality is obtained:

$$E_j^T Q E_j - A_j^T R A_j - C_j^T C_j > 0, \tag{26}$$

which represents the corresponding LMI extended characterization of (18), that is exactly the necessary and sufficient condition for descriptor LPV system (17) to be admissible with a sub-optimal performance in $\mathscr{H}_2$.
*Necessity*: Assuming that the condition (26) is satisfied, then there exist a matrix $G$ such that $G + G^T - R > 0$, therefore

$$\begin{bmatrix} E_j^T Q E_j - A_j^T R A_j - C_j^T C_j & 0 \\ 0 & G + G^T - R \end{bmatrix} > 0, \tag{27}$$

which is equivalent to

$$\begin{bmatrix} E_j^T Q E_j - A_j^T R A_j & 0 & C_j^T \\ 0 & G + G^T - R & 0 \\ C_j & 0 & \mathbb{I} \end{bmatrix} > 0. \tag{28}$$

In effect, (27) is obtained pre-multiplying (28) by $\mathfrak{T} = \begin{bmatrix} \mathbb{I} & 0 & -C_j^T \\ 0 & \mathbb{I} & 0 \end{bmatrix}$ and post-multiplying it by $\mathfrak{T}^T$. Consequently, in order to obtain (22), the LMI (28) is multiplied by the right by $\mathfrak{U} = \begin{bmatrix} \mathbb{I} & 0 & 0 \\ A_j & \mathbb{I} & 0 \\ 0 & 0 & \mathbb{I} \end{bmatrix}$ and by the left by $\mathfrak{U}^T$. Thus, by choosing $F = A_j^T R - A_j^T G^T$, the condition (22) is obtained. Finally, (iii) is the dual equivalent of (ii).

□

### 3.2 $\mathscr{H}_2$-Control for Descriptor LVP System

From of the LMI characterization of the $\mathscr{H}_2$-norm, we are interesting in to design a control law by state feedback in order that the closed-loop system be robustly admissible.

**Lemma 5** *The discrete-time descriptor LPV system (17) is admissible and satisfies $\|H_{\omega y}\|_2^2 < \mu$ via a state feedback controller (13), if and only if: there exists a positive definite $Q = Q^T \in \mathbb{R}^{n\times n}$, a symmetric matrix $S \in \mathbb{R}^{(n-r)\times(n-r)}$ and the matrices $W \in \mathbb{R}^{p\times p}$ with tr($W$)< $\mu$, $G$ and $Y$ satisfying, $\forall j = 1, 2, \ldots N$, the following LMI*

$$\begin{bmatrix} E_j Q E_j^T & A_j G + B_j Y & B_{\omega j} \\ G^T A_j^T + Y^T B_j^T & G + G^T - Q + \overline{\overline{E}}_j S \overline{E}_j^T & 0 \\ B_{\omega j}^T & 0 & \mathbb{I} \end{bmatrix} > 0 \tag{29}$$

$$\begin{bmatrix} W & C_j G & D \\ G^T C_j^T & G + G^T - Q + \overline{\overline{E}}_j S \overline{E}_j^T & 0 \\ D_j^T & 0 & \mathbb{I} \end{bmatrix} > 0 \tag{30}$$

*where the feedback gain is given by*

$$K = Y G^{-1} \tag{31}$$

***Proof*** Applying item (iii) of Theorem 3, using the dynamic matrix of the closed-loop $A_j + B_j\mathbf{K}$, then the change of variable $Y = \mathbf{K}G$ must be used. □

## 3.3 Example

Let consider the descriptor polytopic system

$$E = \begin{bmatrix} 1 & 0 & 0 \\ 0 & 0 & 0 \\ 2 & 0 & 1 \end{bmatrix}, \quad A = \begin{bmatrix} -.25 & 0.1\rho & 0.01\rho \\ -0.5 & 0.5 - 0.01\rho & 2 - 0.01\rho \\ 0.75 & -1 + 0.005\rho & -1.5 + 0.005\rho \end{bmatrix}, \quad B = \begin{bmatrix} 0 \\ 0 \\ 1 \end{bmatrix}, \quad B_\omega = \begin{bmatrix} 0 & 0 \\ 0.1 & 0 \\ 0.2 & 0.1 \end{bmatrix}$$

$$C = \begin{bmatrix} 2 & 0 & 2 \end{bmatrix}, \quad D_\omega = \begin{bmatrix} 0.01 & -0.5 \end{bmatrix}$$

The matrix A is assumed to be uncertain with $\rho \in [-1, 1]$. Considering $\overline{E}^T = \begin{bmatrix} 0 & 1 & 0 \end{bmatrix}$ and $\mu = 1$, then

$$G = 1.0e + 04 \times \begin{bmatrix} 0.0665 & -0.1966 & 0.0196 \\ -0.0215 & 2.6149 & -0.0070 \\ 0.0221 & -0.7053 & 0.0072 \end{bmatrix}$$

$$Y = 1.0e + 04 \times \begin{bmatrix} -0.0759 & 1.8041 & -0.0221 \end{bmatrix}, \qquad S = 7.9698e + 08$$

Therefore, the feedback gain is

$$\mathbf{K} = \begin{bmatrix} -1.3242 & 0.9999 & 1.5183 \end{bmatrix}$$

In order to evaluate the robust admissibility, Fig. 1 shows the closed-loop pole distribution with respect to the parametric variation and its projection in the complex plane. There is evident that the closed loop system is robustly admissible.

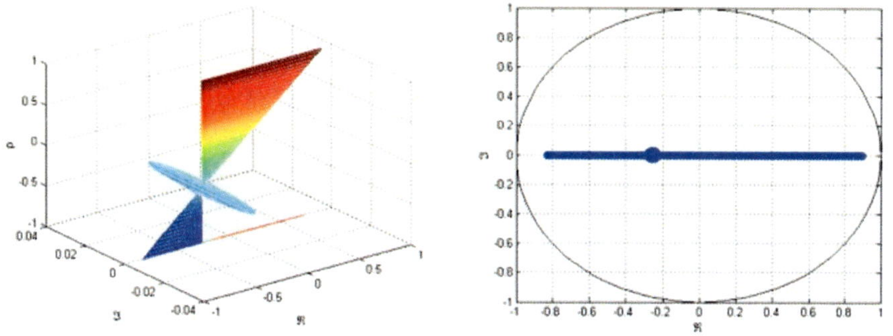

**Fig. 1** The pole distribution and its projection for the closed-loop system

## 4 Conclusions

A method for the robust admisibilization of discrete-time LPV-type descriptor systems has been presented. The technique is based on the synthesis of robust controllers from satisfying performance index in $\mathscr{H}_2$. In order to avoid computational difficulties, an extended characterization of the $\mathscr{H}_2$-norm has been presented, which allows obtaining strict and less conservative linear matrix inequalities. Thus, the necessary and sufficient admissibility conditions in strict linear matrix inequality have also been derived. The results obtained can be extended to robust admissibility by considering the location of poles in LMI regions.

## References


1. Luenberger, D.G.: IEEE Trans. Automat. Control. **22**(3), 312 (1977)
2. Feng, Y., Yagoubi, M.: Robust Control of Linear Descriptor Systems, vol. 194. Springer (2017)
3. Marir, S., Chadli, M.: IEEE/CAA J. Autom. Sin. **6**(3), 685 (2019)
4. Shamma, J.S.: On overview of LPV systems. Control of Linear Parameter Varying Systems with Applications, pp. 3–26. Springer (2012)
5. Duan, G., Yu, H.H.: LMIs in Control Systems: Analysis, Design and Applications. CRC Press, Center for Control Theory and Guidance Technology, China (2013)
6. Chadli, M., Shi, P., Feng, Z., Lam, J.: Asian J. Control. **19**(6), 2192 (2017)
7. Zhang, B., Xu, S., Ma, Q., Zhang, Z.: Autom. **104**, 1 (2019)
8. González, A., Estrada-Manzo, V., Guerra, T.M.: Int. J. Syst. Sci. **48**(15), 3215 (2017)
9. Autom. **91**, 136 (2018)
10. Rodríguez, C., Barbosa, K.A., Coutinho, D.: IFAC-PapersOnLine **51**(25), 78 (2018)
11. Pipeleersa, G., Demeulenaere, B., Sweversa, J., Vandenbergheb, L.: Syst. & Control. Lett. **58**(7), 510 (2009)
12. Hilhorst, G., Pipeleers, G., Oliveira, R., Peres, P., Swevers, J.: In: Proceeding 19th IFAC World Congress, vol. 1, pp. 9307–9312. IFAC, Cape Town, South Africa (2014)
13. Mao, W.J.: IET Control. Theory & Appl. **6**(17), 2676 (2012)
14. Fujimori, A.: Trans. Jpn. Soc. Aeronaut. Space Sci. **47**(156), 138 (2004)
15. Zhang, G., Xia, Y., Shi, P.: Autom. **44**(3), 886 (2008)
16. Chadli, M., Darouach, M.: Autom. **48**(2), 449 (2012)
17. Sjöberg, J.: Descriptor Systems and Control Theory. Linköping University Electronic Press (2005)
18. Scholz, L.: TU Berlin (WS2014/15) (2015)
19. Chaabane, M., Tadeo, F., Mehdi, D., Souissi, M.: Math. Probl. Eng. **2011**(1), 1 (2011)
20. Ikeda, M., Lee, T., Uezat, E.: In: Proceeding $39^{th}$ IEEE Conference on Decision and Control, vol. 1, pp. 601–604. IEEE (2000)
21. Chadli, M., Darouach, M.: IEEE Trans. Autom. Control. **59**(2), 494 (2013)

# Fault Tolerance in Feedback Control Loops

Jose A. Carrazana and Efrain Alcorta Garcia

**Abstract** Fault tolerant control (FTC) seeks to avoid the effects of performance deterioration of a system when a change of one or more parameters outside the tolerance limits, that is, when a fault occurs. The starting point is that every feedback control loop has a certain degree of fault tolerance (like in passive fault tolerant control). So that, when the effects of a fault exceed certain limits (i.e., when a fault is no longer considered tolerated by the nominal control) is when additional control action is required (active fault tolerant control), i.e., a corrective action is applied to compensate the effect of the fault in question on a system. This paper focuses on discussing the limits of nominal operation (in the context of a parameter change). in order to define also the intolerance to a certain fault. Frequently, these limits are taken for granted, although they are not always clearly determined, and yet they play a central role in the operation of fault tolerant control schemes. Here we propose the use of interval observers to delimit the limits of acceptable operation, even in the presence of faults. This indicates that the control is tolerating the fault in question. When the system variables exceed the obtained delimitations, we decree that the corresponding fault is not being tolerated by the system control. This gives rise to the use of some scheme that returns the system variables within acceptable limits. This last part is not part of the present work. A variable load speed control system is used to demonstrate the concepts introduced.

## 1 Introduction

Fault tolerance in control loops can be understood in different ways, however in this paper it is understood as a way to prevent the effects of a fault from spreading throughout a system while recovering, as far as possible, the nominal performance. It is clear that the concept of fault tolerance described above is not useful in all

J. A. Carrazana · E. A. Garcia (✉)
Universidad Autónoma de Nuevo León, Facultad de Ingeniería Mecánica y Eléctrica, San Nicolás de los Garza, Mexico
e-mail: efrain.alcortagr@uanl.edu.mx

 
O. Llanes-Santiago (ed.), *Proceedings of 19th Latin American Control Congress (LACC 2022)*, Studies in Systems, Decision and Control 464,
https://doi.org/10.1007/978-3-031-26361-3_2

possible failure cases, however, we consider only the set of failures, (we will call them admissible) for which the definition of tolerance makes sense.

In the community of diagnosis and fault tolerance there are many efforts to make the control loop can respond adequately to the effect of faults include: adaptive control laws, with robust controllers, reconfiguring the control, etc.. However, most of the efforts to make fault tolerance modify the nominal control loop. Another aspect is the explicit requirement to know the size of the fault to be able to compensate its effect, see for example [2], or using some non-linear techniques, such as [6], [8] as well as statistical tools, see for example [3].

In the last decade, a fault-tolerant scheme called cascade approach, see [7], [1], was proposed. This scheme maintains nominal control and does not require explicit knowledge of the fault size. The effect of disturbances is also corrected. The mentioned scheme requires to be able to discriminate between failures that can be tolerated by the nominal control and those that require supplementary control action [5]. This activity has been previously recognized by the team working around Prof. Staroswieci, see for example the paper: [9]. In the work cited above, a tolerance measure is proposed based on a performance index similar to those used in optimal control. This works well when a fault cancel the control canal.

Since the nominal control presents a certain degree of fault tolerance (passive fault tolerant control) and corrective action is only required when the fault level produces deviations outside the tolerable limits of system operation. In this work, interval observers, such as those considered in [4], are used to delimit the acceptable operating limits, even in the presence of faults. This indicates that the control is tolerating the fault in question. When the system variables exceed the obtained delimitations, we decree that the corresponding fault is not tolerated by the control. This gives us the opportunity to implement some additional control strategy. This last part is not part of the present work. A variable load speed control system is used to demonstrate the proposed scheme.

In the next section we present some preliminaries that are useful for subsequent development. Section 3 presents the use of interval observers to determine if a fault is being tolerated by the nominal control. Section 4 presents a description of the system under study. Section 5 presents an example of application. Section 6 presents conclusions.

## 2 Preliminaries

The objective of Fault Tolerant Control (FTC) is to eliminate the effect of faults occurring in the functional units of the control system, as well as to prevent the effect of the fault from propagating and to maintain the performance of the faulted system as close to nominal as possible, thus maintaining safe and continuous operation in control applications. The material in this section has been taken mainly from [3].

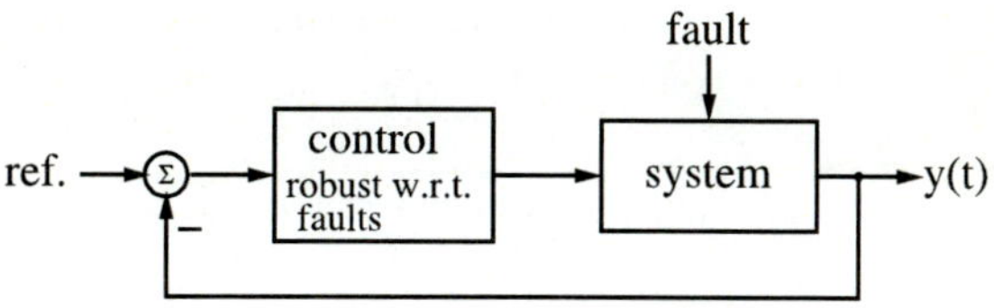


**Fig. 1** Scheme with passive fault tolerant control

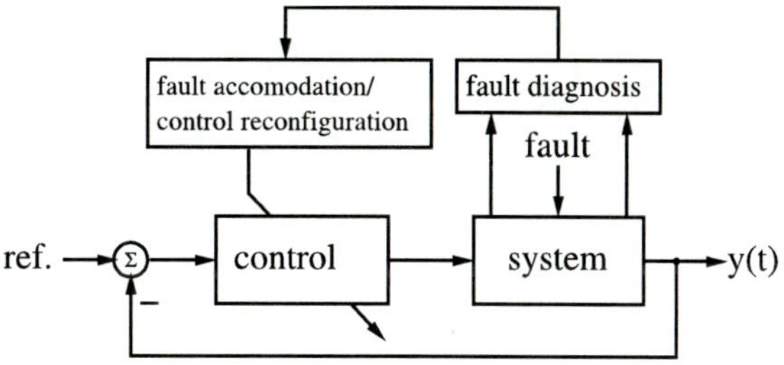


**Fig. 2** Active fault tolerant control scheme

The different strategies for the implementation of fault-tolerant control can be classified into two general approaches.

**Pasive Fault Tolerant control**. This focuses on using the property of back-fed systems to cope with disturbances or changes in system dynamics to mitigate the effects of faults. Practically, it consists of a robust design of the back-fed control system to make it immune to certain faults. See Fig. 1.

**Active Fault Tolerant Control**. This approach consists of using information from the fault diagnosis system to activate the control reconfiguration mechanism designed to compensate for the effect of faults. Active fault-tolerant control is more costly to implement as it requires monitoring and diagnostic capabilities, but offers a better solution for specific faults. See Fig. 2.

## *2.1 A Fault Modeling*

Let a nominal system, i.e. fault free system, given by:

$$
\begin{aligned}
\dot{x}_n(t) &= A_n x_n(t) + B_n u(t) \\
y_n(t) &= C_n x_n(t)
\end{aligned}
$$

Where the sub-index $n$ shows that the corresponding system matrices are fault free (nominal)

**Fault**: No admissible deviation of at least one characteristic property or parameter of a system from its acceptable, usual or standard condition.

Failures affecting a system are often represented by a variation of the system parameters. Thus, in the presence of a fault, the system model can be written as:

$$\dot{x}_f(t) = A_f x_f(t) + B_f u(t)$$
$$y_f(t) = C_f x_f(t)$$

where the new faulty system matrices are defined as:

$$A_f = A_n + \Delta A.$$

$$B_f = B_n + \Delta B.$$

$$C_f = C_n + \Delta C.$$

y $\Delta A$, $\Delta B$, $\Delta C$ correspond to the system parameters deviations with respect to the nominal ones.

Different types of faults can occur, from typical faults that change only a small part of the plant to a complete loss of operation of one or more components. Therefore, from the fault model, three types of failures are distinguished to be considered: sensor, actuator and component faults.

## 3 When a Fault is Tolerated by a Control System?

As previously mentioned, active fault-tolerant control requires the ability to discriminate between faults that can be tolerated by the nominal control of the system from those that need to be compensated by an action in addition to the nominal one. [5]. Inspired by the work of [9], the following definitions and procedures are proposed to achieve the desired discrimination.

**Definition 1** Consider a closed-loop system and a fault $f_i(t)$. This fault is called **tolerated** (such that the system is tolerant to this fault) if in the presence of this fault the following is satisfied:

(1) The feedback system remain stable.
(2) The system output remains inside of a region around the nominal output system.
(3) The nominal feedback control law $u(t)$ remain bounded and realizable.

△

**Definition 2** Let be $f_i(t)$ a fault which is not tolerated is called **recoverable** if there exists a supplementary control efforce such that in the resulting control loop the fault $f_i(t)$ becomes tolerated.

△

According to definition 1, it is necessary to be able to measure deviations of the output from nominal in order to determine whether the fault in question can be considered tolerated. For this purpose, it is proposed to use a tool known as an output interval observer, based on a state interval observer. For this purpose, the results described mainly in [4] and the references contained therein are used.

Note that the class of interval observers considered here are based on the concept of cooperative systems:

**Cooperative control systems**: a system in which the matriz **A** is a Metzler matrix.

**Metzler Matrix**: is a matrix with no negative diagonal elements (bigger or equal to zero).

Based on these concepts it is possible to define two observers, one that can estimate the upper bound of the output and another that can estimate the lower bound. The range of the coordinates can be set by a percentage of variation of some elements of the system matrices. The design is not trivial and requires satisfying some technical requirements [4].

The structure of the observers as well as the design for the particular case of the variable load speed control system is discussed in Sect. 5.

## 4 Variable Load Speed Control System DR300

As a case study, the speed control system of a variable load motor is considered., DR300 from the firma **AMIRA**. See Fig. 3.

The plant is represented by a permanently excited DC motor where the input signal (armature current) is provided by a current control loop. The sensors for the output signal (speed) are a tachogenerator and an incremental encoder. The free end of the motor shaft is firmly coupled to the shaft of a second identical motor. See Fig. 4.

By determining the voltage terminal ($U_A$) as input, the load ($M_L$) as disturbance, and the motor shaft speed ($M_L$) as output, we obtain the block diagram in Fig. 5. When the DC motor, as described above, is connected to a supply voltage, the speed of the motor can be easily adjusted to the desired easily adjusted to the desired values (Fig. 6).

But this structure is simply an open-loop control because changes in load can result in speed changes since the armature voltage is not automatically adjusted to the desired values. While the fundamental characteristic of a closed-loop control is

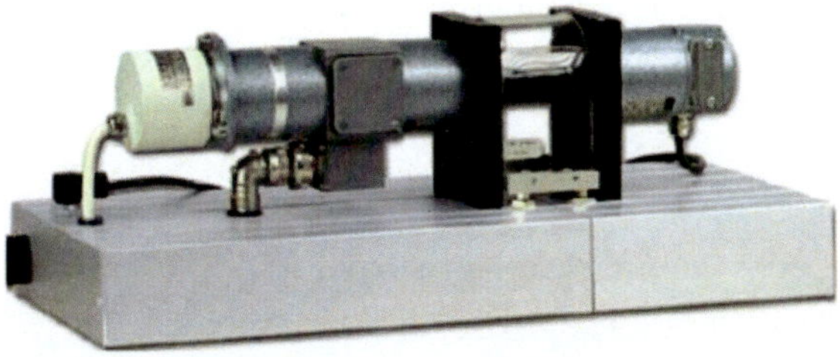

**Fig. 3** Variable load speed control system DR300 used in the study case

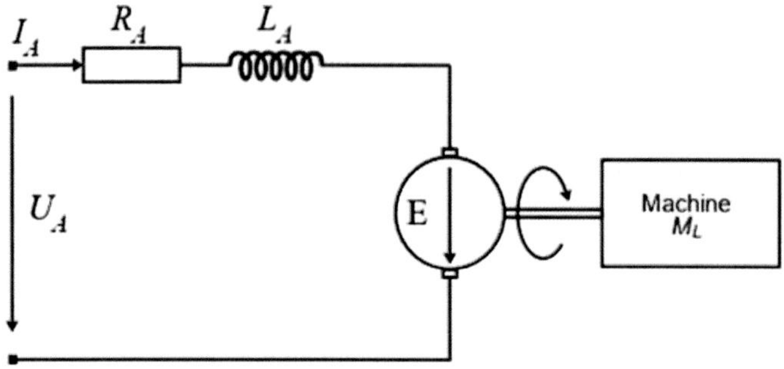


**Fig. 4** Schematic description of the electromechanic system

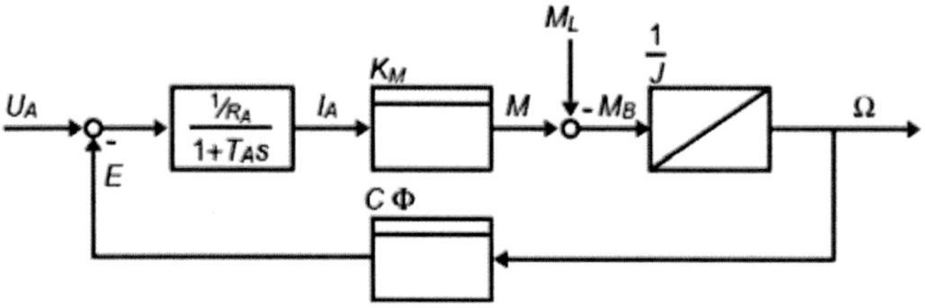


**Fig. 5** Block diagram of the loaded motor

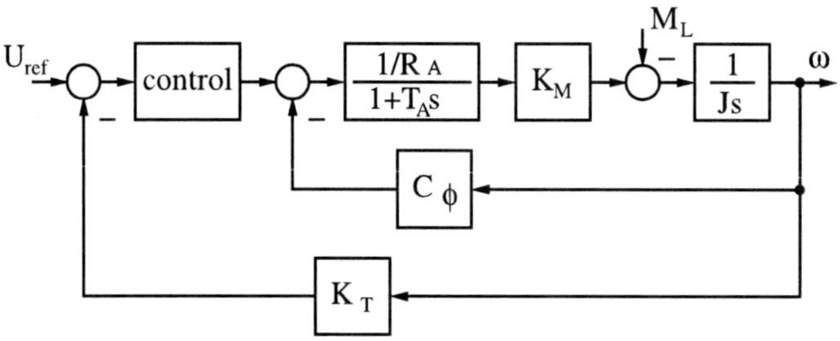


**Fig. 6** Block diagram of the speed control loop

that the reference variable (set point) does not change even in the event of perturbing external influences (load changes). Therefore, it is required that the controlled variable Ω be measured continuously and compared to the reference variable. In the event of deviations, the control voltage $U_s$ of the actuator must be adjusted by a suitable controller (Table 1).

**Table 1** Values for the parameters

| Parameter | Symbol | Value | Units |
|---|---|---|---|
| Total inertia | J | $80.45 * 10^-6$ | kg*m$^2$ |
| Voltage constant | CΦ | $6.27 * 10^-3$ | V/Rpm |
| Motor constant | $K_M$ | 0.06 | Nm/A |
| Armature inductance | $L_A$ | 0.003 | H |
| Resistance | $R_A$ | 3.13 | Ohm |
| Tacho Voltage | $K_T$ | $2.5 * 10^-3$ | V/Rpm |
| Time constant | $T_A$ | | |

## 5 Study Case: Variable Load Speed Control System

In the example below, actuator failures are assumed and the parameter values are summarized in the table below are summarized in the following table.

The simulations were performed using Python, where the speed control loop shown in Fig. 5, which represent the nominal speed control.

The path of the nominal speed output is bounded by an upper and a lower bound that delimit the region in which the nominal control is tolerant to certain faults. These coordinates are obtained using interval observers using the following equations.

The first pair of equations belongs to the lower bound of the interval observer.

$$\dot{\underline{x}}(t) = (\underline{A} - L\underline{C})\underline{x}(t) + (\underline{B} - L\underline{D})u(t) + L(y(t) + \underline{\epsilon})$$
$$\underline{y}(t) = \underline{Cx}(t) + \underline{D}u(t)$$

While the second pair represents the upper dimension.

$$\dot{\overline{x}}(t) = (\overline{A} - L\overline{C})\overline{x}(t) + (\overline{B} - L\overline{D})u(t) + L(y(t) + \overline{\epsilon})$$
$$\overline{y}(t) = \overline{C}\overline{x}(t) + \overline{D}u(t)$$

The results of the simulations are shown below, Fig. 7 shows the nominal velocity trajectory when the system is fault free as well as the upper and lower bounds that delimit the threshold at which the nominal system is fault tolerant. The design of the gain L was performed in a straightforward manner, since in the case of our example no intervals in $A$ are considered, i.e., $\bar{A} = \underline{A}$ as with the $C$ matrix ($\bar{C} = \underline{C}$). The difference in the two observers in our example is in the $B$ matrix.

Figure 10 shows the effect of a 10% actuator fault, showing that the nominal speed control is tolerant to failures of this magnitude and the response is within the allowable threshold (Fig. 8).

In the Fig. 9 it can be seen how the nominal speed trajectory goes out of the established limits, which means that the nominal control is not tolerant to failures in the actuator greater than or equal to 15%.

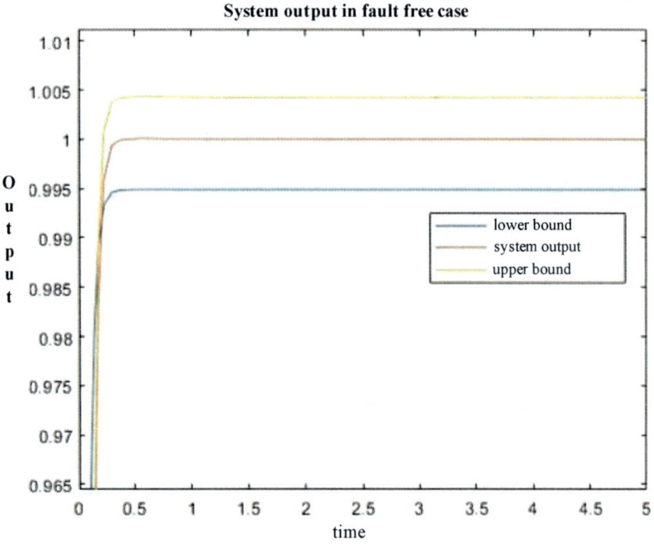


**Fig. 7** Nominal output trajectory (without faults)

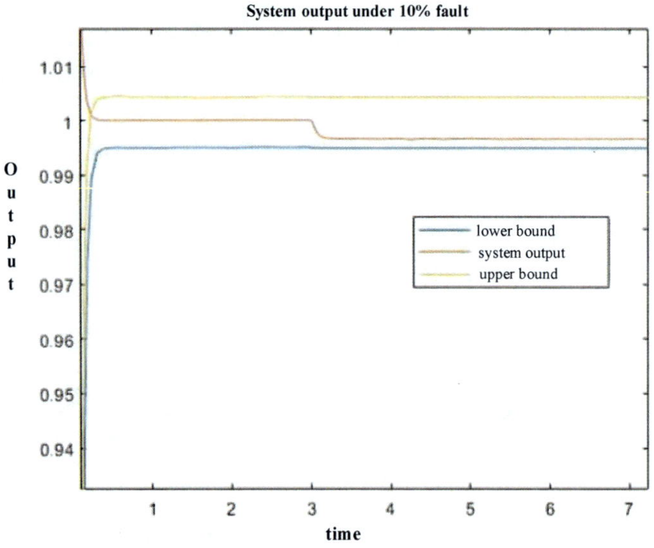


**Fig. 8** Output trajectory with a 10%

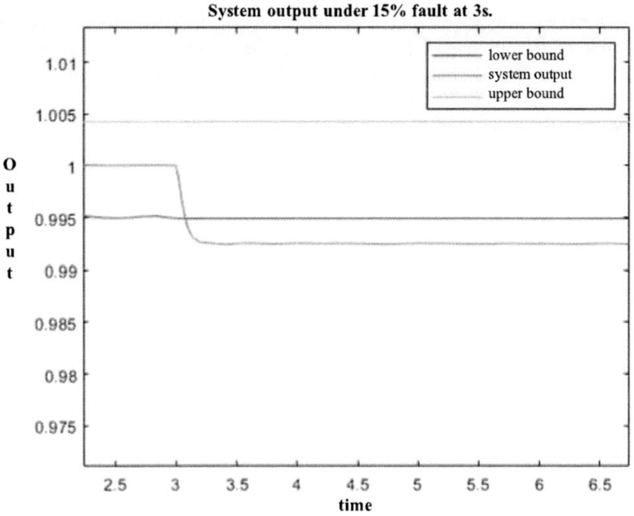


**Fig. 9** Output trajectory with a 15% fault

## 6 Conclusion

In the present work it is proposed to use output interval observers to decide when a fault is tolerated or not by a control loop. Note that the fact that a fault is tolerable is associated with a specific control as well as with the acceptable deviation specification. This latter fact is related to the specific application and may change from user to user. The proposed scheme with the associated definitions is implemented in a variable load speed control system. As shown in the results, the proposed scheme allows to make the discriminations of the different fault levels studied. In future work, a software solution for the design of interval observers is being built.

## References

1. Acosta-Santana, P., Rodriguez-Alfaro, L.H., Alcorta-Garcia, E.: A cascade structure approach to controler configuration. In: Proceedings 2nd International Conference On Control and Fault-Tolerant Systems SysTol'13 (2013)
2. Blanke, M., Kinnaert, M., Lunze, J., Staroswiecki, M.: Diagnosis and Fault-Tolerant Control. Springer (2016)
3. Ding, S.X.: Advanced Methods for Fault Diagnosis and Fault-Tolerant Control. Springer (2020)
4. Efimov, D., Raïssi, T.: Design of interval observers for uncertain dynamical systems. Autom. Remote. Control./Avtom. I Telemekhanika, MAIK Nauk./Interperiodica. **77**(2), 191–225 (2016)
5. Gonzalez Martinez, J.L., Alcorta-Garcia, E.: Fault-tolerant supplementary control synthesis for dc motor with variable load. In: Memorias del 2021 Congreso Nacional de Control Automático1, 3-15 de Octubre, Guanajuato, México (2021)
6. Patan, K.: Robust and Fault-Tolerant Control: Neural-Network-Based Solutions. Springer (2019)

7. Rodriguez-Alfaro, L., Alcorta-Garcia, E., Espinosa-Pérez, G.: Un enfoque cascada para control activo tolerante a fallas en msip. Congreso Nacional de Control Automático (2013)
8. Shen, Q., Jiang, B., Shi, P.: Fault Diagnosis and Fault-Tolerant Control Based on Adaptive Control Approach. Springer (2017)
9. Staroswiecki, M., Berdjag, D.: A general fault tolerant linear quadratic control strategy under actuator outages. Int. J. Syst. Sci. **41**(8), 971–985 (2010)

# Robust PID Control with Anti-Windup Compensation for ULDS

**Addison Ríos-Bolívar and Orestes Llanes-Santiago**

**Abstract** This paper presents an approach for the synthesis of robust PID controllers with compensation Anti-Windup (AW) for Uncertain Linear Descriptor Systems (ULDS). In a first stage, the ULDS are considered to pose parametric uncertainties of polytopic type and disturbances, as parametric variation and actuator saturation. From a condition of existence of a linear injective application, representing the generalized inverse matrix of $\mathbb{E}$, the original ULDS is transformed to a linear parameter variable (LPV) system. Then, the synthesis of the robust multivariable PID controller by static output feedback (SOF) is obtained considering performance indices in $\mathscr{H}_\infty$, described as linear matrix inequalities, LMIs. The PID controllers are designed without taking into account the physical restrictions of the actuators. In a second stage, taking into account the actuator saturation, an AW gain compensation is designed, which consider a performance index that involves the difference between the actuator output signal and the controller output and its effect on the control signal, then, the closed loop performance and the changes in the saturation limits of the actuators are evaluated. The minimization of the $\mathscr{H}_\infty$ norm is considered, which are described as a convex optimization problem of the involved transfer functions, in conjunction with the specification of closed loop performance in the absence of saturation. Performance of the proposed synthesis technique is illustrated by numerical example.

A. Ríos-Bolívar (✉)
Universidad de Los Andes, Facultad de Ingeniería, Departamento de Sistemas de Control, Mérida 5101, Venezuela
e-mail: ilich@ula.ve

O. Llanes-Santiago
Universidad Tecnológica de La Habana José Antonio Echeverría, CUJAE, Marianao, La Habana, Cuba
e-mail: orestes@tesla.cujae.edu.cu



O. Llanes-Santiago (ed.), *Proceedings of 19th Latin American Control Congress (LACC 2022)*, Studies in Systems, Decision and Control 464,
https://doi.org/10.1007/978-3-031-26361-3_3

## 1 Introduction

Since its introduction in 1977 [13], descriptor systems (DS) have been one of the main research fields within control theory [5, 9, 14]. Unlike their regular counterparts in state space, a DS allows a representation that incorporates algebraic constraints in their physical variables.

On the other hand, the context of linear parameter variable (LPV) systems refers to linear dynamical systems whose state-space representations depend on exogenous non-stationary parameters [17]. LPV systems are a generalization of LTV systems [3, 6].

When there are combined the modeling of physical systems with uncertain parameters, there arise dynamic systems representing uncertain DS. As is well known, for modeling many applications and technical processes, only approximate models are available, so that the analysis of DSs subject to uncertainties has been a very active research line, [2, 7].

By another hand, the main objective of any control system design is its practical implantation. When the control system synthesis for real applications is treated, always the flexibility and simplicity of their implementation must be studied. In this context, the static output feedback allows all the kindness for an implementation without many requirements. In that same context, the PID controllers have been extensively studied throughout the history of automatic control and are widely used in industrial control, [18, 20, 21].

From a practical point of view, it is common to find control schemes of complex industrial processes that must satisfy multiple objectives and must operate in different control modes, [1, 12]. Because of the physical limitations (saturation) and commutation of the controllers, the process control input will be different from the output of these controllers. This situation, known as *windup* is an inconsistency among the control input to the process and the internal states of the controller. The problem of control system analysis and synthesis for LTI systems subject to plant input limitations and substitutions can be handled by means of *compensation*, which consists of two stages: the first stage considers the control system design without taking into account the physical constraints of the actuators. The second stage is related to the design of some compensation through an additional feedback to minimize the negative effects of the actuators saturation. In general, this technique is studied as Anti-Windup compensation problem (AW) [19, 23].

In this paper, a design method of control by PID with robust AW compensation for ULSD is presented. The method consists of designing multivariable PID controllers without taking into account the actuator saturation, considering a class particular of ULDS, those that admit a linear transformation in order to transform the original ULDS to an LPV system. After, a compensation gain is designed in order to diminish the adverse effects of the saturation. This compensation gain is designed taking into account the closed loop performance and the changes in the saturation limits of the actuators. The robust performance property is analyzed on the basis of the closed loop transfer of the actuator output related to the difference between the control signal effective to the process and the controller output, which is considered as a disturbance signal. Thus, the minimization of the $\mathscr{H}_\infty$ norm is considered here, which are described as a convex optimization problem of the transfer functions, in conjunction with the specification of closed loop performance in the absence of saturation. The uncertainty parameters are supposed to belong to a polytope. The stabilization problem is formulated in the context of linear matrix inequalities (LMI).

## 2 Descriptor Systems and LPV Systems

### 2.1 Descriptor Systems

A DS is dynamically defined by

$$\mathbb{E}\dot{z}(t) = \mathcal{F}z(t) + Bu(t), \qquad h(t) = \mathcal{C}z(t) \tag{1}$$

where $z(t) \in \mathbb{R}^n$ is the vector of descriptor variable (instead of state vector), $\mathbb{E} \in \mathbb{R}^{m\times n}$, with rank$(\mathbb{E}) = r$ and $0 < r \leq n$, which is called the descriptor matrix. $\mathcal{F} \in \mathbb{R}^{m\times n}$, $B \in \mathbb{R}^{m\times q}$, $\mathcal{C} \in \mathbb{R}^{p\times n}$. The control function $u$ belonging to $L_2(0, \tau; \mathbb{R}^l)$.

For (1) with $m = n$, if $p(s) = \det(s\mathbb{E} - \mathcal{F})$ satisfies that $p(s) \neq 0$, it is said that the pair $(\mathbb{E}, \mathcal{F})$ is *regular*. Otherwise, it is called *singular*.

**Definition 1** [9] Consider the system (1) $(m = n)$, with $\kappa = \deg(\det(s\mathbb{E} - \mathcal{F}))$. If $\kappa = r$ is said that the DS is of *impulse-free* [9].

Thus, the DS (1) has $\kappa$ finite dynamic modes, $r - \kappa$ impulsive modes, and $n - r$ non-dynamic modes.

**Definition 2** [9] Consider the system (1). It is said that the DS is *admissible* if it is regular, impulse-free and stable.

Definition 2 allows to establish conditions for the control of DS in the sense of its stabilization [4].

On the other hand, if the system (1) is regular and impulse-free, through the algebraic-differential manipulation of the non-dynamic modes (Gauss elimination), it is possible to obtain a system descriptor of the form

$$\mathbf{E}\dot{z}(t) = \mathbb{F}z(t) + \mathbb{B}u(t), \qquad h(t) = \mathcal{C}z(t) \tag{2}$$

where $\mathbf{E} \in \Re^{r\times n}$, $\mathbb{F} \in \Re^{r\times n}$ and $\mathbb{B} \in \Re^{r\times m}$, representing the dynamic modes of the original system. Thus, the admissibility problem of the system (1) is equivalent to the admissibility of the system (2).

## 2.2 LPV Systems

**Definition 3** [3] An LPV is a dynamic system in which matrices contain functions that depend on a vector of known variant parameters.

According to Definition 3, a representative LPV model is of the form:

$$\dot{x}(t) = A(\rho(t))x(t) + B(\rho(t))u(t); \quad y(t) = C(\rho(t))x(t) \tag{3}$$

where $x(t) \in \Re^n$ are the states, $u(t) \in \Re^p$ are the controls and $y(t) \in \Re^q$ are the measured output. $\rho(t) : \mathbb{R}_+ \rightarrow \mathbb{R}^l$. If $\rho(t) = t, \;\; l = 1$, the LPV model describes an LTV system.

The typical constraints on exogenous parameters are limits on magnitudes and their indexes of variations, that is, $\forall t \geq 0$

$$\underline{\rho} \leq \rho(t) \leq \bar{\rho}, \qquad \underline{\mu} \leq \dot{\rho}(t) \leq \bar{\mu} \tag{4}$$

The system (3) can be characterized as a polytope, i.e.

$$\dot{x}(t) = \left(\sum_{i=1}^{l} A_i \rho_i\right) x(t) + \left(\sum_{i=1}^{l} B_i \rho_i\right) u(t); \;\; y(t) = \left(\sum_{i=1}^{l} C_i \rho_i\right) x(t) \tag{5}$$

where $\rho_i \in \Re, \;\; \rho_i \geq 0, \;\; i = 1, \ldots, l, \;\; \sum_{i=1}^{l} \rho_i = 1$. There $A_i, B_i, C_i, i = 1, \ldots, l$, are given matrices, representing the polytope vertices. $\rho_i \in \Re, \quad \rho_i \geq 0, \quad i = 1, \ldots, l, \;\; \sum_{i=1}^{l} \rho_i = 1$.

# 3 Robust Admissibility by PID Control of ULDS

In order to obtain the admissibility of the ULDS, a PID control is designed, without account the actuator saturation, [11, 16, 20]. Thus, consider an DS with parametric uncertainty and perturbations:

$$\begin{aligned} \mathbb{E}(\rho)\dot{z}(t) &= \mathcal{F}(\rho)z(t) + \mathcal{B}_1(\rho)\omega(t) + Bu(t) \\ h(t) &= \mathcal{C}(\rho)z(t) + \mathcal{D}_1(\rho)\omega(t), \quad y(t) = C_2(\rho)z(t) \end{aligned} \tag{6}$$

The parametric variation $\rho$ is assumed to meet the constraints defined in (4). $\mathbb{E} \in \mathbb{R}^{n \times n}$, and for all $\rho$, rank$(\mathbb{E}) = r < n$. From transformation or Gauss elimination, it can be assumed that $m = r$, so that the system is of the form given by (2), with parametric uncertainties. $\omega(t) \in L_2$ and the matrices $\mathcal{F}$, $\mathcal{B}_1$, $B$, $\mathcal{C}$, $\mathcal{D}$ and $C_2$ are of appropriate dimensions.

Consider the system (6) with, for all $\rho$, $(\mathbb{E}(\rho), \mathcal{F}(\rho), B)$ defining an stabilizable and impulse controllable finite dynamics.

*Problem 1*: Design a control $u(t)$ for the system (6) such that the corresponding closed loop system will be **admissible**.

*Problem 2*: Design a control $u(t)$ for the system (6) such that the corresponding closed loop system will be **admissible** and the effect of the perturbation $\omega(t)$ on the controlled output $h(t)$ will be minimum in the sense of the $\mathscr{H}_2$-$\mathscr{H}_\infty$ norms.

These problems demand, in addition to robust stabilization, to satisfy a robust performance index characterized under the $\mathscr{H}_2$-$\mathscr{H}_\infty$ norms. Against this background, consider a PID control for system MIMO (6),

$$u(t) = \mathbf{K_P} y(t) + \mathbf{K_I} \int y(\tau) d\tau + \mathbf{K_D} \dot{y}(t), \tag{7}$$

where $\mathbf{K_P}$, $\mathbf{K_I}$ and $\mathbf{K_D}$ are the gains of the PID MIMO. If for (6), $\mathbb{E}(\rho) = \mathbb{I}$ and $\mathbb{M} = \mathbb{I} - \mathbf{K_D} C_2(\rho) B$, that PID is of the form

$$u = \begin{pmatrix} \mathbb{M}^{-1}\mathbf{K_P} & \mathbb{M}^{-1}\mathbf{K_I} & \mathbb{M}^{-1}\mathbf{K_D} \end{pmatrix} \begin{pmatrix} y_1 \\ y_2 \\ \bar{y}_3 \end{pmatrix} \tag{8}$$

where

$$y_1(t) = y(t), \quad y_2(t) = \int_0^t y(\tau) d\tau,$$
$$y_3(t) = \frac{dy(t)}{dt}, \quad \bar{y}_3(t) = \left[ \frac{\partial C_2(\rho)}{\partial \rho} \dot{\rho} + C_2(\rho) A(\rho) \right] x(t)$$

In effect,

$$\dot{y}(t) = C_2(\rho) \dot{x}(t) = \left[ \frac{\partial C_2(\rho)}{\partial \rho} \dot{\rho} + C_2(\rho) \mathcal{F}(\rho) \right] x(t) + C_2(\rho) B u(t)$$

Since $\dot{\rho}$ appears, which is considered as a parametric variation, it must satisfy the condition (4). This is the idea for the design of the PID control, that is obtained by SOF, [10]. In order to obtain the PID controller, the results showed in [16] have been used.

## 4 Robust Control by PID and AW Compensation

In order to introduce the robustness problem in the design of the AW compensation, it is necessary to consider the saturation of the actuator, that is, for the system (6), $\mathbb{E}(\rho)\dot{z}(t) = \mathcal{F}(\rho) z(t) + \mathcal{B}_1(\rho) \omega(t) + B \sigma(u(t))$.

The non-linear function $\sigma(\circ)$ denotes the actuator saturation, which is defined by

$$\sigma(u_i) = \begin{cases} u_{i_{min}} & \text{if } u_i(t) < u_{i_{min}} \\ u_i(t) & \text{if } u_{i_{min}} \le u_i(t) \le u_{i_{max}} \\ u_{i_{max}} & \text{if } u_i(t) > u_{i_{max}}, \quad i = 1, 2, \ldots, q \end{cases} \tag{9}$$

This non-linear function describes the constraints and/or substitutions due to the operation of the actuators. The effective control applied to the process $\sigma(u)$ is a non-linear function of the output controller $u(t)$. Another type of nonlinearities as hysteresis can also be considered, [19].

As it can be seen, the linear application $\Gamma(\rho)$ that it transforms the ULDS into a system LPV does not affect the action of the control, then

$$\begin{aligned} \dot{x}(t) &= A(\rho)x(t) + \mathcal{B}_1(\rho)\omega(t) + B\sigma(u(t)) \\ h(t) &= C_1(\rho)x(t) + \mathcal{D}_1(\rho)\omega(t), \quad y(t) = C_2(\rho)x(t) \end{aligned} \tag{10}$$

In order to satisfy the control requirements, a controller PID with compensation is considered. The PID control is designed from Sect. 3. Thus, from (8), the control with AW compensation is given by

$$u(t) = \mathbf{K}_0 \begin{pmatrix} y_1 \\ y_2 \\ \bar{y}_3 \end{pmatrix} + \mathbf{E_c}[\sigma(u(t)) - u(t)] \tag{11}$$

where $\mathbf{K}_0$ is the gain of SOF. $\mathbf{E_c}$ represents the *compensation gain*, which is also a parameter design.

When saturation occurs, the signal $\vartheta = \sigma(u) - u$ is different from zero and the adverse effects are present. Taking into account the minimization of these effects by using compensation, the gain $\mathbf{K}_0$ of SOF is designed without considering the saturation, while the compensation gain $\mathbf{E_c}$ is synthesized for improving the closed loop performance.

Consider the transfer function of the disturbance $\vartheta$ to the controlled output $z_u$:

$$T_{z_u\vartheta}(s) = \left[\begin{array}{c|c} \mathbb{A} & \mathbb{B}_2 \\ \hline \mathbb{C}_2 & \mathbb{D}_2 \end{array}\right] = \mathbb{C}_2(s\mathbb{I} - \mathbb{A})^{-1}\mathbb{B}_2 + \mathbb{D}_2 \tag{12}$$

where $\mathbb{A} = \mathcal{A}_c + \mathcal{B}_c(\mathbf{K}_0)\mathcal{C}_c$, $\mathbb{B}_2 = \mathcal{B}_c + \mathcal{B}_c\mathbf{E_c}$, $\mathbb{C}_2 = \mathbf{K}_0\mathcal{C}_c$, and $\mathbb{D}_2 = 0$.

*Problem 3*: Given the dynamic system (10), design the *compensation gain* $\mathbf{E_c}$ for the controller (11), such that:

1. The closed loop system (12) be asymptotically stable.
2. The effect of $\vartheta$ on $z_u$ be minimal in some sense.

Under this formulation, the synthesis objectives can be a combination of $\mathcal{H}_2$ performance, $\mathcal{H}_\infty$ performance, and disturbance rejection, which can be described by LMIs.

## 4.1 Design of the AW Compensation

The first stage of design consists of determining the gain of feedback $\mathbf{K}_0$ for the PID, without taking into account the saturation from the actuators. The second stage of design consists of determining the AW compensation gain, for which it is necessary to take into account the saturation. Thus, consider the uncertain system (10) and the control (11), then the closed loop matrices for $T_{z_u\vartheta}(s)$ are

$$\mathbb{A} = \mathcal{A}_c(\rho) + \mathcal{B}_c(\mathbf{K}_0)\mathcal{C}_c(\rho), \quad \mathbb{B}_2 = \mathcal{B}_c + \mathcal{B}_c\mathbf{E_c}, \quad \mathbb{C}_2 = \mathbf{K}_0\mathcal{C}_c(\rho), \quad \mathbb{D}_2 = 0. \quad (13)$$

In the first stage of design, $\mathbf{K}_0$ have been obtained. Then the compensation gain $\mathbf{E_c}$ will be designed in the context of the $\mathcal{H}_2 - \mathcal{H}_\infty$ norms.

### 4.1.1 AW Compensation in $\mathcal{H}_\infty$:

The following result is obtained from the characterization of the $\mathcal{H}_\infty$-norm as LMI, [15, 22].

**Theorem 1** *Let the system* (10) *with the control* (11)*, which defines the polytope* $\mathcal{A}_{c_i}$, $\mathcal{C}_{c_i}$. *There exists a AW compensation gain, that guarantees a sub-optimal performance in* $\mathcal{H}_\infty$ *for* $T_{z_u\vartheta}(s)$ *for the system in closed loop* (12)*, if there exists* $G$, $Z$, $R$, $P_i = P_i^T \succ 0$, $\tau \gg 1$ *and the LMI following is feasible*

$$\begin{bmatrix} -G - G^T & \Phi & 0 & \mathcal{B}_c Z + \mathcal{B}_c R \\ \star & -2\tau P_i & (\mathbf{K}_0\mathcal{C}_{c_i})^T & 0 \\ \star & \star & -\mathbb{I} & 0 \\ \star & \star & \star & -\gamma^2\mathbb{I} \end{bmatrix} < 0, \quad (14)$$

*where* $\Phi = G^T\mathcal{A}_{c_i} + \mathcal{B}_c Z\mathbf{K}_0\mathcal{C}_{c_i} + P_i + \tau G^T$.

$$\textit{AW Gain:} \quad \mathbf{E_c} = Z^{-1}R, \quad (15)$$

*with* $G^T B = BZ$.

***Proof*** From the characterization of the $\mathcal{H}_\infty$-norm as LMI, [15, 22], a change of variables leads to the proof. Assuming that there is a feasible solution to the problem of extended SOF with AW compensation, according to the control (11). Then, using the dynamic closed loop matrices given in (13), the result shown is obtained by the corresponding change of variables, $R = Z\mathbf{E_c}$. □

The gain of AW compensation obtained for the transformed system is applied to the original uncertain descriptor system, because the robust stability with AW compensation for the transformed LPV system is equivalent to the robust admissibility with AW compensation for this ULDS.

An AW Compensation based-on $\mathscr{H}_2$-norm can be obtain following a similar procedure.

# 5 Numerical Evaluation

Let consider the ULDS of the Example 6.2 in [8], for which the dynamic modes, that are not affected by the algebraic modes, are given by

$$\begin{pmatrix} 0 & 0 & 2 & 0 \\ 0 & 1 & 0 & 0 \\ -1 & 0 & \alpha_1 & 0 \end{pmatrix} \dot{z} = \begin{pmatrix} \alpha_2 & 0 & 0 & 0 \\ 0 & 4 & 0 & 0 \\ -1 & 0 & 3 & 0 \end{pmatrix} z + \begin{pmatrix} 1 & 0 \\ 0 & 1 \\ 1 & 1 \end{pmatrix} u; \quad y = \begin{pmatrix} 1 & 0 & 0 & 0 \\ 0 & 0 & 1 & 0 \end{pmatrix} z \tag{16}$$

where: $\alpha_1 \in [0.8 \quad 1.2]$; $\alpha_2 \in [5.4 \quad 6.4]$. The robust admissibility for that system will be established through a PI controller, which is an appropriate controller for robust admissibility as will be shown, in addition to allowing minimum steady state error for step input. Thus,

$$\Gamma = E^T(EE^T)^{-1} = \begin{pmatrix} \frac{\alpha_1}{2} & 0 & -1 \\ 0 & 1 & 0 \\ \frac{1}{2} & 0 & 0 \\ 0 & 0 & 0 \end{pmatrix}.$$

Then,

$$\dot{x} = \begin{pmatrix} \frac{\alpha_1\alpha_2}{2} & 0 & -\alpha_2 \\ 0 & 4 & 0 \\ \frac{3}{2} - \frac{\alpha_1}{2} & 0 & 1 \end{pmatrix} x + \begin{pmatrix} 1 & 0 \\ 0 & 1 \\ 1 & 1 \end{pmatrix} u, \quad y = \begin{pmatrix} \frac{\alpha_1}{2} & 0 & -1 \\ \frac{1}{2} & 0 & 0 \end{pmatrix} x \tag{17}$$

For the robust stabilization in closed loop, the result showed in [16] is applied, then the gains of PI controller are

$$\mathbf{K_P} = \begin{pmatrix} 680.1542 & 619.5304 \\ 301.0998 & 279.0754 \end{pmatrix}, \qquad \mathbf{K_I} = (1.0e+03) \times \begin{pmatrix} 0.7683 & -1.7237 \\ 0.1906 & -0.8229 \end{pmatrix}$$

For simulation the parameters $\alpha_1 = 1$, $\alpha_2 = 5.9 + 0.5 \times \sin(5t)$ are assumed. The simulation results are shown in the Fig. 1. Figure 1a corresponds to the controlled outputs without saturation. There, the controller obtained is implemented without actuators saturation. The outputs are stabilized around the references ($\text{Ref}_1 = 2$, $\text{Ref}_2 = 5$), only affected by the temporary behavior of the parameter $\alpha_2$.

The next step is to evaluate the effect of actuator saturation, for which the limits $U_{max} = 34$; and $U_{min} = -34$ have been considered. Figure 1b presents the simulation results under conditions of actuators saturation without compensation. It is evident that, in this case, the transitory performance has been deteriorated with respect to the case without saturation. If the saturation limits are slightly smaller the system becomes unstable.

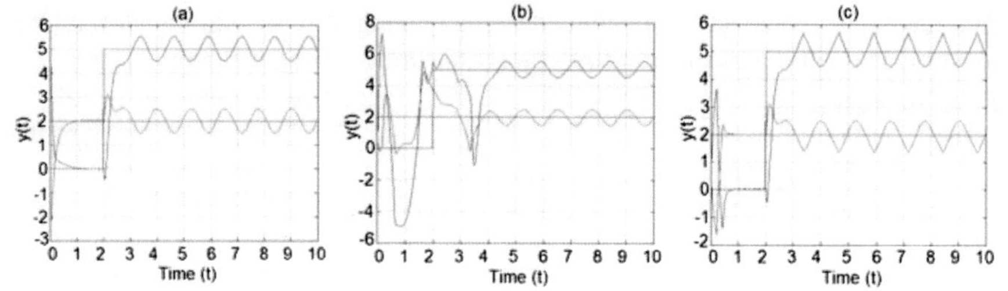


**Fig. 1** **a** Outputs without saturation. **b** Controlled Outputs without compensation. **c** Controlled outputs with saturation and compensation

Once the need to use AW compensation is demonstrated, the next step is to develop the design, which has been obtained simultaneously from the numerical evaluation of the LMIs, according to the *Problem 3* and the Theorem 1. Thus, the compensation gain obtained corresponds to:

$$\mathbf{E_c} = \begin{bmatrix} -0.3858 & -0.2173 \\ 0.2421 & -1.0192 \end{bmatrix}$$

with $\gamma = 3.1917$.

Figure 1c shows the system outputs, which has a compensation mechanism. In the presence of the saturation of the actuators, these outputs show a temporary performance that is resembled the behavior without the saturation. It is clear that the effect of the compensation is to try to bring the operation of the actuators within their performance limits.

This method of implementation allows variability, within a certain range, in the action limits of the actuators, which represents a level of robustness. Thus, the validity of the implementation is subject to this robustness that is provided by the synthesis mechanism of AW compensation. This is because although the mechanism of implementation supports changes in the behavior of the actuators, AW compensation may not. This leaves open the problem of robust synthesis of AW compensation.

## 6 Conclusions

The design of multivariable PID controllers with robust AW compensation for uncertain linear descriptor systems has been presented. The uncertainty parameters are supposed to belong to a polytope. Such as is considered in the Anti-Windup compensation techniques, the method consists of two stages: in the first one (admissibilization of DS), the PID controllers are designed without taking into account the actuator saturation for a particular class of ULDS, those that admit a linear transformation in order to transform ULDS to LPV systems. This technique allows to extend the space of descriptor systems that can be drived by PID controllers. In the second stage, a compensation gain is designed in order to minimize the adverse effects of the saturation. This compensation gain is designed taking into account the closed loop performance and the changes in the saturation limits of the actuators. The robust performance property is analyzed on the basis of the closed loop transfer of the actuator output related to the difference between the control signal effective to the process and the controller output, which is considered as a disturbance signal. Thus, the minimization of the $\mathscr{H}_\infty$ norm, characterized as LMIs, have been considered, which are described as a convex optimization problem of these transfer function, in conjunction with the specification of closed loop performance in the absence of saturation. These results can be easily extended to the case in $\mathscr{H}_2$.

## References


1. Astrom, K.J., Rundqwist, L.: Integrator windup and how to avoid it. In: 1989 American Control Conference, pp. 1693–1698. IEEE (1989)
2. Bara, G.I.: Robust analysis and control of parameter-dependent uncertain descriptor systems. Syst. & Control. Lett. **60**(5), 356–364 (2011)
3. Briat, C.: Robust control and observation of LPV time-delay systems. Ph.D. thesis, Institut National Polytechnique de Grenoble-INPG (2008)
4. Chaabane, M., Tadeo, F., Mehdi, D., Souissi, M.: Robust admissibilization of descriptor systems by static output-feedback: an LMI approach. Math. Probl. Eng. **2011**(1), 1–10 (2011)
5. Chen, J., Yu, J.: Robust control for discrete-time TS fuzzy singular systems. J. Syst. Sci. Complex. **34**(4), 1345–1363 (2021)
6. Duan, G., Yu, H.H.: LMIs in Control Systems: Analysis. Design and Applications. CRC Press, Center for Control Theory and Guidance Technology, China (2013)
7. Duan, G.R.: Analysis and Design of Descriptor Linear Systems. Springer, New York (2010)
8. Faria, F.A., Assunção, E., Teixeira, M., Cardim, R.: Robust state-derivative feedback LMI-based designs for linear descriptor systems. Math. Probl. Eng.**2010** (2010)
9. Feng, Y., Yagoubi, M.: Robust Control of Linear Descriptor Systems, vol. 194, Springer (2017)
10. Gopmandal, F., Ghosh, A.: A hybrid search based synthesis of static output feedback controllers for uncertain systems with application to multivariable pid control. Int. J. Robust Nonlinear Control. **31**(12), 6069–6090 (2021)
11. Goßmann, F., Svaricek, F.: Parameter dependent static output feedback control-an LPV approach. In: 2019 18th European Control Conference (ECC), pp. 3322–3327. IEEE (2019)
12. Kothare, M.V., Campo, P.J., Morari, M., Nett, C.N.: A unified framework for the study of anti-windup designs. Autom. **30**(12), 1869–1883 (1994)
13. Luenberger, D.G.: Dynamic equations in descriptor form. IEEE Trans. Automat. Control. **22**(3), 312–321 (1977)
14. Marir, S., Chadli, M.: Robust admissibility and stabilization of uncertain singular fractional-order linear time-invariant systems. IEEE/CAA J. Autom. Sin. **6**(3), 685–692 (2019)
15. Pipeleersa, G., Demeulenaere, B., Sweversa, J., Vandenbergheb, L.: Extended LMI characterizations for stability and performance of linear systems. Syst. & Control. Lett. **58**(7), 510–518 (2009)
16. Ríos-Bolívar, A., Narciso, F.: Admisibilización robusta de sistemas LPV descriptores vía control PID. In: Proceeding 14 Congreso Interamericano de Computación Aplicada a la Ingeniería de Procesos, CAIP 2019, vol. 1, pp. 1–10. CIT (2019)
17. Shamma, J.S.: Chap. an overview of LPV systems. Control of Linear Parameter Varying Systems with Applications, pp. 3–26. Springer (2012)
18. Tan, K.K., Wang, Q.G., Hang, C.C.: Advances in PID Control, Springer Science & Business Media (2012)

19. Tarbouriech, S., Garcia, G., da Silva Jr, J.M.G., Queinnec, I.: Stability and Stabilization of Linear Systems with Saturating Actuators, Springer Science & Business Media (2011)
20. Veselỳ, V., Ilka, A.: Generalized robust gain-scheduled PID controller design for affine LPV systems with polytopic uncertainty. Syst. & Control. Lett. **105**, 6–13 (2017)
21. Vilanova, R., Visioli, A.: PID Control in the Third Millennium, Springer (2012)
22. Wei, C.: On applications of the projection lemma to analysis and design of continuous-time systems. Master's thesis, National Sun Yat-Set University (2003)
23. Yang, C., Zhang, L., Sun, J.: Anti-windup controller design for singularly perturbed systems subject to actuator saturation. IET Control. Theory & Appl. **10**(4), 469–476 (2016)

# Observer-Based Tracking Control for PMSM Rooted on Port-Hamiltonian Systems Structural Properties

F. Ramos-García, C. Granados-Salazar, M. Rojas, and G. Espinosa-Pérez

**Abstract** The main purpose of this work is to present an observer-based control for the PMSM system. The relevance of this work is focused on the used of the PH systems structure to prove the convergence of the observer base control scheme presented by using a type of separation principle, where fist it is proved the asymptotically stability of equilibrium point of system with control, and then the convergence of the observed is proved. The numerical evaluation is exposed by MATBAL-Simlulink simulations that prove the theoretical results obtained.

## 1 Introduction

Regarding the development of output feedback control schemes using state observer, the available literature is more restricted. There are lines of research focused on determining the conditions for which the separation principle can be satisfied for systems with physical structure. For example in [3] conditions are posed to guarantee the existence of the separation principle for a class of Euler-Lagrange systems. In that same order of ideas in [1] the design of observers of nonlinear systems formed by a linear system in feedback with a static nonlinear function of the state is addressed

Authors thank the support given by PAPITT-UNAM (Programa de Apoyo a Proyectos de Investigación e Innovación Tecnológica): Proyecto DGAPA-UNAM IN109622.

F. Ramos-García (✉) · C. Granados-Salazar · M. Rojas · G. Espinosa-Pérez
Facultad de Ingeniería, Av. Universidad 3000, Ciudad Universitaria, Cd. Mx., Coyoacán CP 04510, Mexico
e-mail: ferramosgarcia03@gmail.com

C. Granados-Salazar
e-mail: christianf_gs@outlook.com

M. Rojas
e-mail: michael.rojas@ingenieria.unam.edu

G. Espinosa-Pérez
e-mail: gerardoe@unam.mx



O. Llanes-Santiago (ed.), *Proceedings of 19th Latin American Control Congress (LACC 2022)*, Studies in Systems, Decision and Control 464,
https://doi.org/10.1007/978-3-031-26361-3_4

where the linear part is a PCH system whose main result is demonstrate that with a IDA-PBC control technique it is possible to get a close-loop system whose operation point has exponential stability properties. Under the same line of research, the work of [9] designs an output feedback controller obtained using the IDA-PBC technique presented and considering that the observation scheme is designed based on the contraction theory, and in [7] An observer-based control scheme for an electrical power system is shown based on the structural properties of the IDA-PBC scheme.

These results illustrates that working in a particular class of systems to design an observer based control is an important result. Focused on the Permanent Magnet Synchronous Motor (PMSM) system, and its importance and gaining relevance on the industry, this work is based on the PMSM $dq$ model to design a tracking controller based on observer.

The outline of this paper is as follows: the Sect. 2 is focused on the mathematical model and remarks the properties of the class system that is considered. On the Sect. 3, the tracking control design is developed to be followed by the observer design in Sect. 4. The analysis of the close-loop observer-base control is explained in Sect. 5. Finally, the numeric validation is presented in Sect. 6 with MATLAB Simulink simulations and the conclusions of this work are given in Sect. 7.

## 2 Mathematical Model

In this paper, it is considered the well-known two-phase equivalent model of a three-phase PMSM expressed in $dq$ coordinates as the considered in ([8]).

The mathematical model of a PMSM represented in the $dq$ reference frame is given by

$$\begin{aligned}
\dot{x}_1 &= -\frac{R_1}{L_1}x_1 + \frac{n_p}{J_m}x_2x_3 + u_d \\
\dot{x}_2 &= -\frac{R_2}{L_2}x_2 - \frac{n_p}{J_m}(x_1+\phi)x_3 + u_q \\
\dot{x}_3 &= -\frac{n_p}{L_1}x_1x_2 + \frac{n_p}{L_2}(x_1+\phi)x_2 - \frac{r_m}{J_m}x_3 - \tau_L
\end{aligned} \tag{1}$$

where $x_1 = L_1 i_d$, $x_2 = L_2 i_q$ are the electromagnetic fluxes corresponding to the $d$ and $q$ axes, respectively, and $x_3$ is the rotor speed. The parameters $L_1$, $L_2$ and $R_1$, $R_2$ are the stator inductances and resistances, $\phi$ is the magnetic flux due to the permanent magnets, $n_p$ is the number of pole pairs, $J_m$ is the motor inertia and $r_m$ is the equivalent motor resistance. The external signals are given by the stator voltages, $u_d$, $u_q$, and the load torque $\tau_L$. It is possible to rewrite the PMSM's mathematical model in the port Hamiltonian representation as

$$\dot{x} = [J(x) - R]Px + Gu + \varepsilon \tag{2}$$

where

$$x = \begin{bmatrix} x_1 \\ x_2 \\ x_3 \end{bmatrix}; \ J(x) = \begin{bmatrix} 0 & 0 & n_p x_2 \\ 0 & 0 & -n_p(x_1 + \phi) \\ -n_p x_2 & n_p(x_1 + \phi) & 0 \end{bmatrix}$$

$$R = \begin{bmatrix} R_1 & 0 & 0 \\ 0 & R_2 & 0 \\ 0 & 0 & r_m \end{bmatrix}; \ P = \begin{bmatrix} \frac{1}{L_1} & 0 & 0 \\ 0 & \frac{1}{L_2} & 0 \\ 0 & 0 & \frac{1}{J_m} \end{bmatrix}; G = \begin{bmatrix} 1 & 0 \\ 0 & 1 \\ 0 & 0 \end{bmatrix}; \ u = \begin{bmatrix} u_d \\ u_q \end{bmatrix}; \ \varepsilon = \begin{bmatrix} 0 \\ 0 \\ -\tau_L \end{bmatrix}$$

where the PCH system in general way can be expressed as

$$\dot{x} = [J(x) - R]\nabla_x H(x) + gu + \tau_L \tag{3}$$

$$y = g^T \nabla_x H(x) = Cx \tag{4}$$

where $J(x) = -J^\top(x) \in \mathbb{R}^{n \times n}$ is the interconnection matrix, $R = R^\top \in \mathbb{R}^{n \times n}$ is the semi positive definitive dissipation matrix, $g \in \mathbb{R}^{n \times m}$ is the constant input matrix while $u \in \mathbb{R}^m$ is the vector of control input signals. The vector of measurable outputs is given by $y \in \mathbb{R}^m$, and the Hamiltonian is a scalar function given by

$$H(x) = \frac{1}{2} x^\top P x$$

where $x \in \mathbb{R}^n$ is the state vector while $P \in \mathbb{R}^{n \times n}$ is a constant symmetric positive definite matrix.

Concerning the structure for $J(x)$ it is important to point out that the formulated assumption corresponds to the case when the internal interconnection between two states is modulated by another component of the state vector. This structure can be found very often in systems with physical structure. It is important to note the next remarks

**Remark 2.1** Both the dissipation and the input matrices, $R$ and $g$ respectively, are assumed to be constant. In addition, the input matrix is assumed to be column full rank. □

**Remark 2.2** The measurable output signals $y$ are assumed to be linear combinations, determined by $C$, of the state vector $x$ which is the same as the passive output of the system. □

**Remark 2.3** The nonlinear nature of the class of systems comes from the term $J(x)Px$. In this sense, the structure of the interconnection matrix takes a fundamental role for the results presented in this paper. This structure is defined below. □

The nonlinearities that exhibit the class of systems under study are determined by the structure of the interconnection matrix $J(x)$. In the next assumption, this kind of nonlinear behavior is further specified.

**Assumption 2.1** Given the system (3) assume that matrix $J(x)$ can be decomposed as

$$J(x) = J_0 + \sum_{i=1}^{n} J_i x_i \tag{5}$$

where $J_i$, $i = 0, ..., n$, are skew-symmetric constant matrices while $x_i$ is the $i - th$ component of the state vector $x$. □

**Remark 2.4** Note that in the general conception of (5), it is assumed a dependence on all the components of the state vector. If this is not the case, as will be clear below, the presented contribution is still valid since for each $x_i$ that does not appear in $J(x)$ the number of summands is reduced but the structure of the matrix and its properties do not change. □

According to the Assumption 2.1 for PSMN the interconnection matrix has the next structure

$$J(x) = \begin{bmatrix} 0 & 0 & 0 \\ 0 & 0 & -n_p\phi \\ 0 & n_p\phi & 0 \end{bmatrix} + \begin{bmatrix} 0 & 0 & 0 \\ 0 & 0 & -n_p \\ 0 & n_p & 0 \end{bmatrix} x_1 + \begin{bmatrix} 0 & 0 & n_p \\ 0 & 0 & 0 \\ -n_p & 0 & 0 \end{bmatrix} x_2 \tag{6}$$

## 3 Control Design

The control design used in this work is the one reported in [4], in this section, the development of the proportional control design for trajectory tracking implemented in the PMSN is presented for the error coordinate dynamics system.

$$\dot{\tilde{x}} = [J(x) - R]\, P\tilde{x} + B(x_\star)P\tilde{x} + g\tilde{u} + \tau_L \tag{7}$$

where, it is define a change of coordinate $\tilde{x} = x - x_\star$ that is the control error, and $\tilde{u} = u - u_\star$.

The admissible trajectories $x_\star$ are given by the solution of a system copy

$$\dot{x}_\star = [J(x_\star) - R]\, Px_\star + gu_\star \tag{8}$$

The therm $B(x_\star)P\tilde{x}$ will be analysed as a vanishing perturbation, invoking standard perturbation theorems of [2], if the nominal system has an equilibrium point uniformly and asymptotically stable and the vanishing perturbation is bounded, the perturbed system has an asymptotically stable equilibrium point.

**Proposition 3.1** *Consider the nominal error dynamic of (7). It is define the proportional control law*

$$\tilde{u} = -Ky = \begin{bmatrix} -k_1 \frac{\tilde{x}_1}{L_1} \\ -k_2 \frac{\tilde{x}_2}{L_2} \end{bmatrix} \tag{9}$$

*If* $\left[R - gKg^T\right]_{sym} > 0$*, the equilibrium point of the error system is asymptotically stable with internal stability.* □

The system (7) with the control law (9) in closed loop is

$$\dot{\tilde{x}} = \underbrace{\begin{bmatrix} -R_1 - k_1 & 0 & n_p x_2 \\ 0 & -R_2 - k_2 & -n_p(x_1 + \phi) \\ -n_p x_2 & n_p(x_1 + \phi) & -r_m \end{bmatrix}}_{A(x)} \begin{bmatrix} \frac{\tilde{x}_1}{L_1} \\ \frac{\tilde{x}_2}{L_2} \\ \frac{\tilde{x}_3}{J_m} \end{bmatrix} \tag{10}$$

It is important to highlight that the control law applied to the PMSM system is

$$u = \begin{bmatrix} -k_1 \frac{\tilde{x}_1}{L_1} + u_{1\star} \\ -k_2 \frac{\tilde{x}_2}{L_2} + u_{2\star} \end{bmatrix} \quad \text{where} \quad u_\star = \begin{bmatrix} \dot{x}_{1\star} + \frac{R_1}{L_1} x_{1\star} - \frac{n_p}{J_m} x_{2\star} x_{3\star} \\ \dot{x}_{2\star} + \frac{R_2}{L_2} x_{2\star} + \frac{n_p}{J_m} x_{1\star} x_{3\star} + \frac{n_p \phi}{J_m} x_{3\star} \end{bmatrix}. \tag{11}$$

Given the energy function (Hamiltonian function), it is used as Lyapunov function candidate:

$$V = H(\tilde{x}) = \tilde{x}^T P \tilde{x} \tag{12}$$

Its derivative along the systems trajectories

$$\dot{V} = \tilde{x}^T P A_{sym}(x) P \tilde{x} < 0 \tag{13}$$

***Proof*** Assuring that $\dot{V}(\tilde{x}) < 0 \Rightarrow A_{sym}(x) < 0$

$$A_{sym}(x) = \begin{bmatrix} -R_1 - k_1 & 0 & 0 \\ 0 & -R_2 - k_2 & 0 \\ 0 & 0 & -r_m \end{bmatrix} < 0 \tag{14}$$

To guarantee that the matrix $A(x)$ is negative definite, all the eigenvalues of the $-A_{sym}(x)$ matrix have to be positive. Control gains must be restricted according to the following conditions $k_1 > -R_1$, and $k_2 > -2$. Considering the $K$ gain restrictions, the equilibrium point $\tilde{x} = 0$ is an asymptotically stable equilibrium point. □

# 4 Observer Design

The observation scheme used to establish observed-based control in the PMSM arises from the observation scheme reported in [5] and currently submitted in [6] where is shown a general observation scheme for a class of nonlinear port-Hamiltonian systems. The structure of the observer corresponds to the class of observers structure-preserving where the observer is represented by the next expression

$$\dot{\hat{x}} = [J(\hat{x}) - R]P\hat{x} + gu + L(y - \hat{y}) \tag{15}$$

where $\hat{y} = C\hat{x}$ while $J(\hat{x}) = -J^{\top}(\hat{x})$, $R = R^{\top} \geq 0$, with the same dimensions as in the system represented by (3), and $L$ is the gain vector of the observer.

In the case of the PMSM, the following remarks are taken into account

**Remark 4.1** Under the assumption of known parameters, measurement of $x_1, x_2$ is equivalent to measurement of electrical currents. □

**Remark 4.2** The structure of the proposed observer in this case is characterized by (15) while the output matrix takes the form and the matrix of observer gains are given by

$$C = \begin{bmatrix} \frac{1}{L_1} & 0 & 0 \\ 0 & \frac{1}{L_2} & 0 \end{bmatrix}; \quad L = \begin{bmatrix} K_1 & 0 \\ 0 & K_2 \\ 0 & 0 \end{bmatrix} \tag{16}$$

with $K_1$ y $K_2 \in \mathbb{R}$. □

Consider the system given by (3), the state observer defined by (15), and the property exhibited in the Assumption 2.1, also define the observation error as $\tilde{x}_o = x - \hat{x}$. Under these conditions the dynamics of the observation error is given by

$$\dot{\tilde{x}}_o = [J(x) - (R + LCP^{-1})]P\tilde{x}_o + B(\hat{x})\tilde{x}_o \tag{17}$$

with $B(\hat{x}) = [J_1Q\hat{x} \quad J_2Q\hat{x} \quad \cdots \; J_nQ\hat{x}]$, also where $\tilde{x}_o$ is the observer error.

**Proposition 4.1** *The characterization of the convergence properties is contained in the well-known vanishing perturbation theory, if the error nominal system given by* (17) *has an equilibrium point uniformly and asymptotically stable and the vanishing perturbation is bounded, the perturbed system has an asymptotically stable equilibrium point, ergo it is possible to guarantee that the following observation objective is met*

$$\lim_{t\to\infty} (x - \hat{x}) = 0 \tag{18}$$

□

***Proof*** The perturbation $B(\hat{x})\tilde{x}$ is considered equal to zero. Under this condition and to establish the stability properties of the equilibrium point of the nominal system, $\tilde{x} = 0$, the following candidate Lyapunov function is proposed

$$V(\tilde{x}_o) = \frac{1}{2}\tilde{x}_o^\top P \tilde{x}_o \tag{19}$$

whose derivative along of the system trajectories is

$$\dot{V}(\tilde{x}_o) = -\tilde{x}_o^\top P \left[R + LCP^{-1}\right]_{sym} P\tilde{x}_o \tag{20}$$

where sym represents the symmetric part of a matrix. If the matrix $\left[R + LCP^{-1}\right]_{sym}$ is positive definite then $\tilde{x}_o = 0$, which is the equilibrium point of the nominal error dynamics, is exponentially stable. So if $\hat{x}$ is bounded, then the equilibrium point of the error dynamics represented by (17) is asymptotically stable.

To obtain the observer gains, it is necessary to take into account the convergence analysis carried out in the last section. In this sense, consider the Lyapunov candidate function given by (19) whose derivative along the paths of the system is given by

$$\dot{V} = -\left(\frac{R_1}{L_1^2} + \frac{K_1}{L_1^2}\right)\tilde{x}_1^2 - \left(\frac{R_2}{L_2^2} + \frac{K_2}{L_2^2}\right)\tilde{x}_2^2 - \frac{r_m}{J_m^2}\tilde{x}_3^2$$

Therefore, to guarantee that the equilibrium point $\tilde{x} = 0$ has properties of asymptotic stability, the observer's gains must comply with $K_1 \geq 0$ and $K_2 \geq 0$. □

## 5 Close-Loop Based on Observer

Implementing the control law shown previously in (11) it is necessary to measure the complete state but there are some variables which are not possible to measure, for that reason it is useful to use some strategies that allow to reconstruct the unmeasured states, in this case for the control law which is defined by (11) it is necessary to write it in function of the observed states, so that the control law is given by

$$u = \begin{bmatrix} -\frac{K_1}{L_1}(\tilde{x}_{1c}) + u_{1\star} \\ -\frac{K_2}{L_2}(\tilde{x}_{2c}) + u_{2\star} \end{bmatrix} \tag{21}$$

where $\tilde{x}_c = x - x_\star$. Replacing $x$ by $\hat{x}$ therefore the control law based on the observer states is given by the next form

$$u_o = \begin{bmatrix} -\frac{k_1}{L_1}(x_1 - x_{1\star}) + u_{1\star} + \frac{K_1}{L_1}\tilde{x}_{1o} \\ -\frac{k_2}{L_2}(x_2 - x_{2\star}) + u_{2\star} + \frac{K_2}{L_2}\tilde{x}_{2o} \end{bmatrix} = u + \underbrace{\begin{bmatrix} \frac{K_1}{L_1} & 0 \\ 0 & \frac{K_2}{L_2} \\ 0 & 0 \end{bmatrix}}_{\Phi} \tilde{x}_o \tag{22}$$

If it considers the control law based on observer states it is possible to obtain the next PMSM representation

$$\dot{x} = [J(x) - R]Px + gu + \Phi\tilde{x}_o \tag{23}$$

where $u$ is a control which achieves that the closed-loop system has asymptotically stability properties.

Taking into account the observer-based closed-loop system error dynamics, is

$$\dot{\tilde{x}}_c = [J(x) - R]P\tilde{x}_c + B(x_\star) + \Phi\tilde{x}_o \tag{24}$$

while the dynamics of the observer error is defined by (17). Consider the extended system where the following auxiliary state is defined

$$z = \begin{bmatrix} \tilde{x}_c \\ \tilde{x}_o \end{bmatrix} \tag{25}$$

**Proposition** *Consider the extended system formed by the error dynamics given by* (24) *and the observation error dynamics given by* (17)*, then it is possible to guarantee that the extended system characterized by the vector* (25) *satisfies*

$$\lim_{t\to\infty} x - x_\star = 0 \quad and \quad \lim_{t\to\infty} x - \hat{x} = 0 \tag{26}$$

***Proof*** In this order it will have to prove that the closed-loop based on the observed states has stability proprieties. It proposes the next candidate Lyapunov function

$$V_{CL}(\tilde{x}_c, \tilde{x}) = \gamma_1 V_c + \gamma_2 V_o \tag{27}$$

where $\tilde{x}_c$ is the tracking error and $\tilde{x}_o$ is the observer error, and its derivative along to the system trajectories is given by the next equation

$$\dot{V}_{CL}(\tilde{x}_c, \tilde{x}_o) = \gamma_1 \dot{V}_c + \gamma_2 \dot{V}_o \tag{28}$$

then the derivative along the paths of the observer-based closed-loop system is

$$\dot{V}(\tilde{x}_c, \tilde{x}_o) = -z^\top \underbrace{\begin{bmatrix} R_o + gKg^\top & -\frac{1}{2}P\Phi \\ -\frac{1}{2}\Phi^\top P & R_o + LC \end{bmatrix}}_{\beta} z \tag{29}$$

where if $\beta>0$ then the equilibrium point $z = 0$ is asymptotically stable □

## 6 Numeric Validation

With the aim to numerically evaluate the presented contribution, the observer designed for the PMSM was implemented in a MATLAB-Simulink environment. The motor parameters were taken from ([8]) and are the following: $R_1 = 0.225\,[\Omega]$,

$R_2 = 0.225\,[\Omega]$, $r_m = 0.00003\,[\frac{\text{kg}}{\text{s}}]$, $n_p = 3$, $L_1 = 3.8\,[\text{mH}]$, $L_2 = 3.8\,[\text{mH}]$, $J_m = 0.012\,[\frac{\text{kg}}{\text{m}^2}]$, $\phi = 0.17\,[\text{Wb}]$, $\tau_L = 0$, $k_1 = 20$, $k_2 = 0.1$, $K_1 = 100$, $K_5 = 200$.

The experiment consists of establishing a desired speed profile and showing the result obtained with the control scheme based on measured states and with the control scheme based on observed states. This section shows the performance of the currents and the performance of the motor speed.

**Remark** Due to the linear constitution relationship, it is possible to analyze the behavior of the performance of the control schemes through the currents and not the magnetic fluxes. □

In Fig. 1 shows the behavior of the motor currents on the d axis, in Fig. 3 the behavior of the current on the $q$ axis is shown, and in Fig. 5 the behavior of the synchronous speed of the motor is shown, three behaviors are observed, the desired currents and desired speed, the tracking of the desired currents with the control

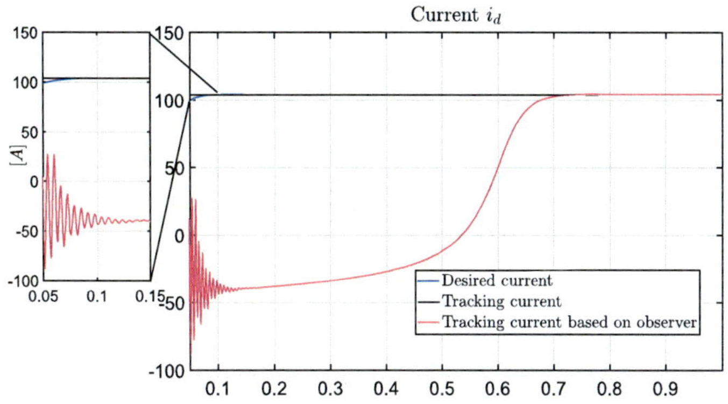


**Fig. 1** Current on d-axis behavior

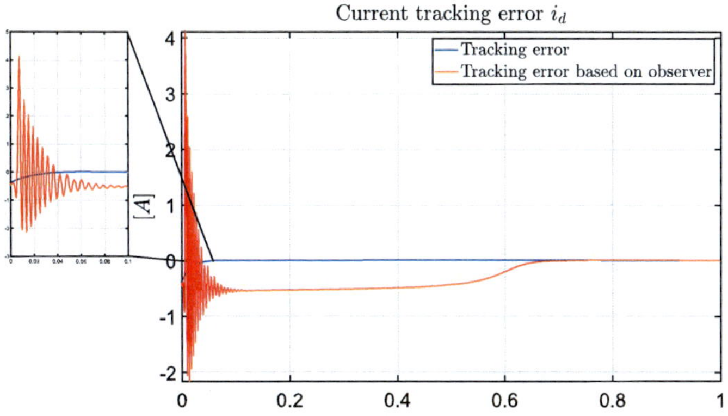


**Fig. 2** Tracking error $\tilde{i}_d$

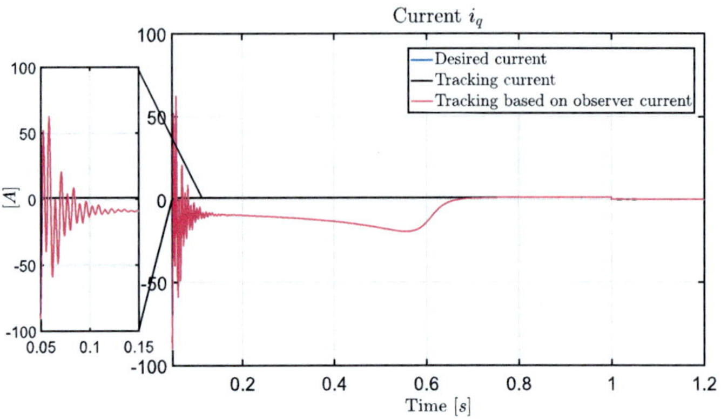


**Fig. 3** Current on q-axis behavior

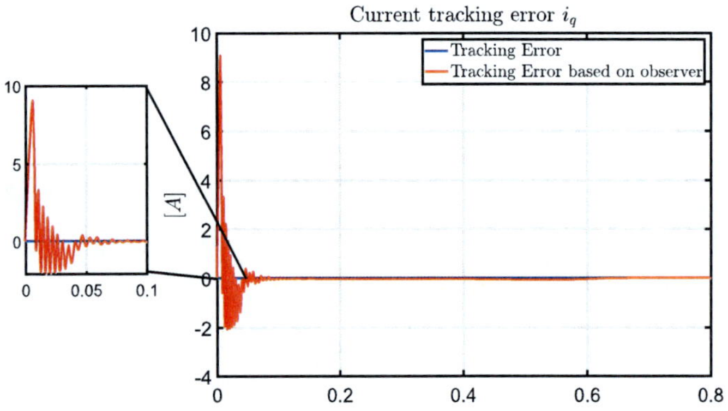


**Fig. 4** Tracking error $\tilde{i}_q$

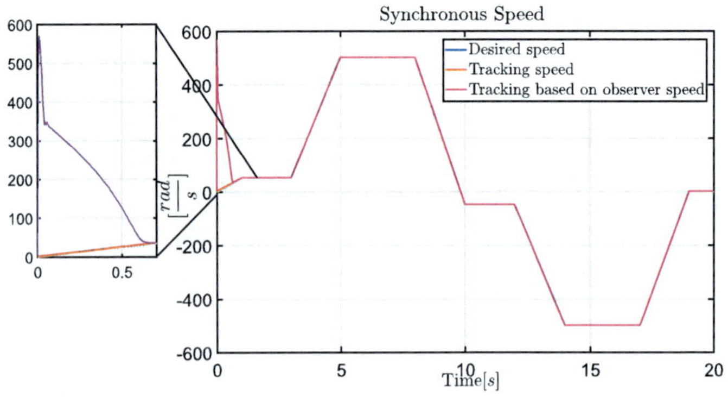


**Fig. 5** Synchronous speed behaviour

**Fig. 6** Tracking error speed

based on measured states, and tracking the currents and speed with observer-based proportional control.

Complementing the analysis of the performance of the system, Figs. 2, 4 and 6 presents the behavior of the tracking error, the tracking error of the proportional control and the tracking error of the observer-based proportional control. In both cases it can be seen that the control objective for both schemes is achieve. In the case of tracking error based on observer control, it has a different transient due to the action of the observer and its gains.

## 7 Concluding Remarks

In this work, the design of an observer-based control scheme for the PMSM is shown, as well as the proof of a type of separation principle for this system. The main points to note are the following:

- Since the structure of both the control and the observer is Hamiltonian, conditions based on the structure of the system can be established to guarantee convergence of both schemes separately.
- Using the structural features of Hamiltonian systems it is possible to establish a kind of separation principle to ensure that the observer-based control scheme converges.
- The numerical evaluation of the system is displayed establishing a desired speed profile. From this exercise the following can be concluded:
  - The control scheme based on the measured states has better performance in the transient, in this case it does not present overshoots beyond $10[A]$. and $5\,[\frac{\text{rad}}{\text{s}}]$.
  - The control scheme based on observed states presents a worse performance in the transient due to the peak phenomenon caused by the gains of the observer,

in this case the current transient has an overshoot of around 70[$A$] while in the case of velocity the transient is close to 500 [$\frac{\text{rad}}{\text{s}}$].

- After the transient, both schemes meet the control objective. In the case of currents, the control scheme based on measured states reaches the desired states in about 0.1 [s], while in the case of the observer-based control scheme it reaches the desired signal in about 0.15 [s]. In the case of speed, the measured state-based control scheme reaches the desired speed in 0.2 [s] while the observer-based control scheme reaches the desired speed in about 2 [s].

## References

1. Biedermann Bastian y Meurer, T.: Observer design for a class of nonlinear systems combining dissipativity with interconnection and damping assignment. Int. J. Robust Nonlinear Control (2021)
2. Khalil, H.: Nonlinear Systems, 3rd edn. Prentice-Hall, New Jersey (2002)
3. Loría Antonio y Panteley, E.: A separation principle for a class of Euler-Lagrange systems. In: New Directions in nonlinear observer design, pp. 229–247. Springer (1999)
4. Ramos-García, F., Espinosa-Pérez, G., Avila-Becerril, S.: On the trajectory tracking control of hamiltonian systems (2021)
5. Rojas, M., Granados-Salazar, C., Espinosa-Pérez, G.: Observer design for a class of nonlinear hamiltonian systems. IFAC-PapersOnLine **54**(19), 125–130 (2021)
6. Rojas, M., Granados-Salazar, C., Espinosa-Pérez, G.: Full-order observer design for quadratic port-controlled hamiltonian systems (2022)
7. Rojas, M., Rueda-Escobedo, J.G., Espinosa-Pérez, G., Schiffer, J.: Observer-based excitation control for transient stabilization of the single machine infinite bus system. In: 2020 59th IEEE Conference on Decision and Control (CDC), pp. 3377–3382. IEEE (2020)
8. Shah, D., Espinosa-Perez, G., Ortega, R., Hilairet, M.: An asymptotically stable sensorless speed controller for non-salient permanent magnet synchronous motors. Int. J. Robust Nonlinear Control. **24**(4), 644–668 (2014)
9. Yaghmaei, A., Yazdanpanah, M.J.: Output control design and separation principle for a class of port-hamiltonian systems. Int. J. Robust Nonlinear Control **29**(4), 867–881 (2019)

# Sliding Mode Control Based on a Generalized Reduced Fractional Order Model: Theoretical Approach

**Antonio Di Teodoro, Diego Ochoa-Tocachi, Hanna Aboukheir, and Oscar Camacho**

**Abstract** This work proposes a fractional sliding mode controller designed from a fractional FOPDT model of the real process. The new approach can be applied to nonlinear chemical processes. The controller design uses the power of Fractional Order Calculus to describe and represent the real chemical systems as a reduced order model and from it apply the sliding mode control procedure to develop the controller.

## 1 Introduction

The design of present industrial process control schemes are complicated due to the nonlinear effects and complexity of the system, and the desire for excellent performance and robustness to uncertainties in the manufacturing process, among other factors [12]. Temperature, level, pressure, and pH are the most often variables that need to be controlled in industrial applications, each with its characteristics. The majority of control systems are constructed using a process model, but modeling errors resulting from unmodeled dynamics and unknown disturbances degrade the process performance.

A. Di Teodoro · H. Aboukheir
Colegio de Ciencias e ingenierías, Departamento de Matemáticas, Universidad San Francisco de Quito USFQ, Quito, Ecuador
e-mail: nditeodoro@usfq.edu.ec

H. Aboukheir
e-mail: Hanna2k2@gmail.com

D. Ochoa-Tocachi
Departamento de Tecnología. Cuenca, ATUK Consultoría Estratégica, Ecuador, Mexico
e-mail: diego@atuk.com.ec

O. Camacho (✉)
Colegio de Ciencias e ingenierías, Universidad San Francisco de Quito USFQ, Campus Cumbayá, Casilla Postal 17-1200-841, Quito 170157, Ecuador
e-mail: ocamacho@usfq.edu.ec

 
O. Llanes-Santiago (ed.), *Proceedings of 19th Latin American Control Congress (LACC 2022)*, Studies in Systems, Decision and Control 464,
https://doi.org/10.1007/978-3-031-26361-3_5

A typical control system, or rather a suitable controller, is expected to make the process efficient, suppress the impact of external disturbances on the process, make the process stable, and optimize the entire process [20]. The uncertainties produce degradation of the control system. Conventional controllers do not have the necessary versatility to adequately compensate for the variety of dynamical complexities present in these systems. Due to the nature of these uncertainties, a more generic approach to dealing with nonlinear processes is required. Sliding Mode Control (SMC) is an ideal choice for achieving such a goal because of its robustness [2, 23].

Chemical processes tend to be of a higher order; it is common to find fourth or higher-order processes. Developing a complete model for chemical processes is rather difficult due mainly to the nonlinearity characteristics and the lack of understanding or knowledge of some process parameters [2]. The design of the Sliding Mode Controller (SMCr) depends on the order of the process model [23]. The presence of dead time is another important characteristic in the chemical process; up to now, there is not a general and practical methodology to handle it [3].

FOPDT models provide an accurate approximation of several industrial processes [14, 21]. Empirical models employing low-order linear models with deadtime are an efficient alternative strategy for process control [20]. However, these reduced-order models contain uncertainties resulting from incomplete model knowledge, and the nonlinear process effects contribute to a decline in controller performance. Even if available, it is often a complicated model, which may be along with uncertainties and time-varying behavior. Complex models lead to complicated high-order controllers which are challenging to design, implement, maintain, and diagnose.

Throughout the past few decades, fractional derivatives and integrals theory has provided an accurate description and model of various real-world phenomena observed in practical systems and situations. Moreover, mathematical phenomena of the fractional-order allow us to describe and represent real systems with greater precision than the traditional "integer" methods do [14].

As a result, differential equations based on fractional-order calculus (FOC) are more general mathematical tools than conventional integer-order differentials. This is because fractional calculus-based differential equations use a real-valued derivative or integration operator rather than relying solely on an integer-valued differentiator or integrator [1].

This work presents the design of an SMC obtained from a fractional FOPDT model of the actual nonlinear process. The design utilizes the power of FOC for describing and representing mathematical modeling phenomena systems with better precision than typical integer techniques.

# 2 Fundamentals

## 2.1 Sliding Mode Control

The Sliding Mode Control (SMC) is a nonlinear control method commonly referred to as variable structure control (VSC) [22, 23, 25].

The design of Sliding Mode Control consists of two steps. In the initial stage, a customized surface must be designed. In the second phase, a feedback control rule must be synthesized to ensure reaching the sliding surface, which must be achieved in a finite amount of time and then sliding on it; The motion of the system on a sliding surface is known as the sliding mode [23].

The controller has two components, high-frequency and low-frequency components. The high-frequency part takes the system from the initial state to the sliding surface, and the low-frequency part takes the system to the desired state. While the controlled variable is on the sliding surface, the dynamics of the plant are constrained to the equation of the sliding surface. They are robust to internal and external perturbations. The most important advantage is its insensitive to control parameters variation [8].

Summarizing, SMC follows three main steps: choice of the sliding surface, determination of the control law, and finally, establishing a convergence or reaching condition.

## 2.2 Fractional Calculus

Let $\left({}^{*}D^{\eta}_{a^+}h\right)(x)$ denote the fractional Riemann–Liouville derivative of order $\eta > 0$ (see [9, 10, 13, 17])

$$\left({}^{*}D^{\eta}_{a^+}h\right)(x) = \left(\frac{d}{dx}\right)^m \frac{1}{\Gamma(m-\eta)} \int_a^x \frac{h(t)}{(x-t)^{\eta-m+1}}\,dt, \qquad m = [\eta]+1, \qquad x > a, \quad (1)$$

where $[\eta]$ denotes the integer part of $\eta$. When $0 < \eta < 1$ and $\Gamma$ is the gamma function, then (1) takes the form

$$\left({}^{*}D^{\eta}_{a^+}h\right)(x) = \frac{d}{dx}\frac{1}{\Gamma(1-\eta)} \int_a^x \frac{h(t)}{(x-t)^{\eta}}\,dt. \quad (2)$$

The Riemann–Liouville fractional integral of order $\eta > 0$ is given by (see [9, 10, 13, 17])

$$\left(I^{\eta}_{a^+}h\right)(x) = \frac{1}{\Gamma(\eta)} \int_a^x \frac{h(t)}{(x-t)^{1-\eta}}\,dt, \quad x > a. \quad (3)$$

We denote by $I^{\eta}_{a^+}(L_1)$ the class of functions $h$, represented by the fractional integral (3) of a summable function, that is $h = I^{\eta}_{a^+}\varphi$, where $\varphi \in L_1(a, b)$. A description of this class of functions is given in [9, 10].

**Example 1** Let $\eta \in (0, 1)$, $a > 0$, $\beta > -1$ and $\eta - \beta \notin \mathbb{N}$. Then ${}^*D^{\eta}_{a^+}(x-a)^{\beta} = \dfrac{\Gamma(\beta+1)}{\Gamma(\beta+1-\eta)}(x-a)^{\beta-\eta}$.

**Theorem 1** *A function* $h \in I^{\eta}_{a^+}(L_1)$, $\eta > 0$, *if and only if*, $I^{m-\eta}_{a^+}h \in AC^m([a, b])$, $m = [\eta] + 1$ *and* $(I^{m-\eta}_{a^+}h)^{(k)}(a) = 0$, $k = 0, \ldots, s-1$.

In Theorem 1, $AC^m([a, b])$ denotes the class of functions $h$, which are continuously differentiable on the segment $[a, b]$, up to order $m-1$ and $h^{(m-1)}$ is absolutely continuous on $[a, b]$. Removing the last condition in Theorem 1, we get a class of functions that admits a summable fractional derivative.

**Definition 1** (see [9]) A function $h \in L_1(a, b)$ has a summable fractional derivative $\left({}^*D^{\eta}_{a^+}h\right)(x)$ if

$$\left(I^{m-\eta}_{a^+}h\right)(x) \in AC^m([a, b]),$$

where $s = [\eta] + 1$.

If a function $h$ admits a summable fractional derivative, then the composition of (1) and (3) can be written in the form (see [9, Theorem 2.4])

$$\left(I^{\eta}_{a^+}\,{}^*D^{\eta}_{a^+}h\right)(x) = h(x) - \sum_{k=0}^{m-1} \frac{(x-a)^{\eta-k-1}}{\Gamma(\eta-k)} \left(I^{m-\eta}_{a^+}h\right)^{(m-k-1)}(a), \qquad m = [\eta] + 1. \tag{4}$$

If $h \in I^{\eta}_{a^+}(L_1)$, then (4) can be reduced to $\left(I^{\eta}_{a^+}\,{}^*D^{\eta}_{a^+}h\right)(x) = f(x)$. However, ${}^*D^{\eta}_{a^+}\ I^{\eta}_{a^+}h(x) = h(x)$ for both cases. This is a particular case of a more general property (see [17, (2.114)])

$${}^*D^{\eta}_{a^+}\left(I^{\gamma}_{a^+}h\right) = {}^*D^{\eta-\gamma}_{a^+}h, \qquad \eta \ge \gamma > 0. \tag{5}$$

The semigroup property for the composition of fractional derivatives does not hold in general (see [17, Sect. 2.3.6]). In fact, the property:

$${}^*D^{\eta}_{a^+}\left({}^*D^{\gamma}_{a^+}h\right) = {}^*D^{\eta+\gamma}_{a^+}h, \tag{6}$$

holds whenever

$$h^{(j)}(a^+) = 0, \qquad j = 0, 1, \ldots, m-1, \tag{7}$$

and $h \in AC^{m-1}([a, b])$, $h^{(s)} \in L_1(a, b)$ and $m = [\gamma] + 1$. Thus, we can write this result in the following lemma:

**Lemma 1** *Consider* $h \in AC^{m-1}([a,b])$ *and* $h^{(m)} \in L_1(a,b)$ *then,*

$$ {}^*D^{\eta}_{a+}\left({}^*D^{\gamma}_{a+}\right)h = {}^*D^{\gamma}_{a+}\left({}^*D^{\eta}_{a+}\right)h, \tag{8}$$

*holds whenever*

$$h^{(j)}(a^+) = 0, \quad j = 0, 1, \ldots, m-1, \tag{9}$$

*where* $m = [\gamma] + 1$.

***Proof*** The proof can be found in [17, Sect. 2.3.6]. □

**Remark 1** It turns out that the Riemann-Liouville approach have certain disadvantages when trying to model, basically the problem is that the derivative of a constant is not zero. it has certain advantages over the hypotheses of the function. See [4, 5]

### 2.2.1 Caputo Derivative and Connection with Riemann–Liouville

**Definition 2** Let $\eta \geq 0$ and $m = [\eta]$. Then, we define the operator $\mathcal{D}^{\eta}_{a^+}$ by $\mathcal{D}^{\eta}_{a^+} f := I^{m-\eta}_{a^+}\left(\frac{d}{dx}\right)^m f$, whenever $\left(\frac{d}{dx}\right)^m f \in L_1[a,b]$.

**Example 2** Let $\eta \in (0,1)$, $a > 0$ and $\beta > -1$. Then $\mathcal{D}^{\eta}_{a^+}(x-a)^{\beta} = 0$ if $\beta \in \{0, 1, 2, \ldots, m-1\}$. □

## 2.3 *Connection with Riemann–Liouville*

**Lemma 2** *Let* $\eta \geq 0$ *and* $m = [\eta]$. *Assume that* $f$ *is such that both* $\mathcal{D}^{\eta}_{a^+}$ *and* ${}^*D^{\eta}_{a^+}$ *exist. Then* $\mathcal{D}^{\eta}_{a^+} f = {}^*D^{\eta}_{a^+} f - \sum_{k=0}^{m-1} \frac{(x-a)^{k-\eta}}{\Gamma(k-\eta+1)}\left(\frac{d}{dx}\right)^k f(a)$.

in consequence, we have the following lemma:

**Lemma 3** *Let* $\eta \geq 0$ *and* $m = [\eta]$. *Assume that* $f$ *is such that both* $\mathcal{D}^{\eta}_{a^+}$ *and* ${}^*D^{\eta}_{a^+}$ *exist. Then* $\mathcal{D}^{\eta}_{a^+} f = {}^*D^{\eta}_{a^+} f$ *if and only if* $\left(\frac{d}{dx}\right)^k f(a) = 0$ *for* $k = 0, \ldots, m-1$.

Some important results that allow the development of the fractional theory are related to the composition of operators (which is not always true), we must be very careful with the hypotheses that the function needs.

**Lemma 4** • *If* $f$ *is continuous and* $\eta \geq 0$, *then* $\mathcal{D}^{\eta}_{a^+} I^{\eta}_{a^+} f = 0$.
• *if* $f \in AC^m([a,b])$ *and* $m = [\eta]$, *then* $I^{\eta}_{a^+}\mathcal{D}^{\eta}_{a^+} f = \sum_{k=0}^{m-1} \frac{(x-a)^k}{k!}\left(\frac{d}{dx}\right)^k f(a)$.

- *Let* $f \in AC^k([a,b])$ *for some* $a < b$ *and some* $k \in \mathbb{N}$. *Moreover let* $\eta, e > 0$ *be such that there exists some* $l \in \mathbb{N}$ *with* $l \leq k$ *and* $\eta, \eta + \epsilon \in [l-1, l]$. *Then* $\mathcal{D}^{\epsilon}_{a^+} \mathcal{D}^{\eta}_{a^+} f = \mathcal{D}^{\epsilon+\eta}_{a^+} f$.
- *Let* $f \in AC^k([a,b])$ *for some* $k \in \mathbb{N}$. *Moreover* $\eta \in [0,k]$. *Then* ${}^*D^{k-\eta}_{a^+} \mathcal{D}^{\eta}_{a^+} f = \left(\frac{d}{dx}\right)^k f$.

All these results (Lemmas 2, 3 and 4 ) and their proofs can be found in [6]

**Remark 2** The discussion of the numerical expansion and its connection with the Grundwald-Letnikov derivative remains as future work.

## *2.4 Discussion of Fractional Parameters*

**Remark 3** To simplify our calculations, we restrict our scope to the case when the fractional parameter belongs to the interval $(0, 1)$. However, similar reasoning can be done in an arbitrary interval $(a, b)$ where $a, b \in \mathbb{R}$. In this sense, we decide to use the fractional operator as defined in (1) to simplify all possible calculations but all results can be extend to an arbitrary parameter.

Another important discussion is related with the behaviour at the boundaries $0^+$ and $1^-$ of the fractional parameters. If $\eta$ approaches 1 from the left, the fractional integral operator $I^{k-\eta}_a$ becomes the identity operator when $k = 1$ and the classical integral operator when $k = 2$ ; furthermore, the fractional differential operator $D^{\eta}_a$ becomes the classical first order one.

For example the fractional polynomials of the examples convergence to the classical ones in a pointwise sense because it is well-known that real fractional operators converge to the classical ones as $\eta$ goes to 1 (see [6, Theorem 2.20]).

## *2.5 The Laplace Transform of Fractional Derivatives*

Let $F(s) := (\mathcal{L} f)(s)$, i.e., be the Laplace transform of the function $f$ (See [9]), and $\eta > 0$.

$$\left(\mathcal{L}\, \mathcal{D}^{\eta}_{0^+} f\right)(s) = s^{\eta} F(s) - \sum_{j=0}^{l-1} f^{(j)}\left(0^+\right) s^{\eta-j-1},$$

where $l = [\eta] + 1$ and $f^{(j)}\left(0^+\right) = \frac{d^j}{dt^j} f(t)|_{t\to 0^+}$, $j = 0, ..., l-1$. Clearly, if $\eta \in (0, 1)$, then

$$\left(\mathcal{L}\, \mathcal{D}^{\eta}_{0^+} f\right)(s) = s^{\eta} F(s) - f\left(0^+\right) s^{\eta-1} \tag{10}$$

and, if $\eta \in (1, 2)$, then

$$\left(\mathcal{L}\,\mathcal{D}_{0^+}^{\eta} f\right)(s) = s^{\eta} F(s) - f\left(0^+\right) s^{\eta-1} - f'\left(0^+\right) s^{\eta-2}. \tag{11}$$

# 3 Fractional Order Sliding Mode Control Design in the Caputo Sense

In this section the mathematical development of the sliding mode control based on a FOPID surface is presented.

Since SMC is a model based controller, let us use an empirical model of the actual process. Using the reaction curve procedure, the nonlinear process can be approximated for a first-order plus dead time (FOPDT) model [21].

$$\frac{X(s)}{U(s)} = \frac{K e^{-t_0 s}}{\tau s + 1}. \tag{12}$$

## *3.1 Time Delay Approximation*

A first problem encounter in the realization of this factional controller is the time delay part [2]. In the complex settings, the natural exponential function $e^z$ can be represented as the well–known Taylor series centered at 0

$$e^z = \sum_{k=0}^{\infty} \frac{z^k}{k!},$$

which is absolutely convergent for $z \in \mathbb{C}$. A simple linear approximation is $1 + z$ where the absolute error $|e^z - 1 - z|$ equals the Taylor remainder $R_1$ given by

$$R_n(z) = \frac{1}{n-1} \int_0^z e^t (z-t)\, dt,$$

where the path of integration is a straight segment connecting 0 and $z$. Thus, this remainder is bounded by $\frac{|z|^n}{n!} \max\{1, e^z\}$. On the other hand, we have that $e^{-z} = \frac{1}{e^z}$ thus we can approximate $e^{-z}$ by $\frac{1}{1+z}$. Indeed, to see this let the approximation error, $\left|e^{-z} - \frac{1}{1+z}\right|$, be bounded by $\epsilon_{max} > 0$. Then,

$$\begin{aligned}\left|e^{-z}-\frac{1}{1+z}\right| &= \left|\frac{e^{z}-1-z}{e^{z}\left(1+z\right)}\right| \\ &= \frac{R_1(z)}{e^{Re\{z\}}\left|1+z\right|} \\ &\leq \frac{|z|}{|1+z|}\max\left\{1, e^{-Re\{z\}}\right\} \\ &= \epsilon_{max}.\end{aligned}$$

For instance, the time delay $e^{-t_0 s}$ can be approximated to $\frac{1}{t_0 s+1}$ and, in turn, the transfer function (12)

$$\frac{Ke^{-t_0 s}}{\tau s+1} \approx \frac{K}{\left(\tau s+1\right)\left(t_0 s+1\right)}, \tag{13}$$

that can be changed to a second order differential equation form [2].

**Remark 4** A discussion that seems interesting to us about the role of the fractional parameter in the exponential function is a future work. A further conjecture is that we believe that the parameter $\beta$ is not necessary. Which makes the calculation of the Laplace inverse much simpler.

### *3.2 Fractional Transfer Functions*

Let $\alpha, \beta \in \mathbb{R}$. Then, we define the fractional transfer function plus dead time as

$$\frac{X(s)}{U(s)} = \frac{Ke^{-t_0 s^{\beta}}}{\tau s^{\alpha}+1}. \tag{14}$$

where $\alpha$ and $\beta$ can be interpreted as orders of fractional derivatives and integrals. Note that transfer function (12) is obtained when $\alpha$ and $\beta$ approach to $1^{-}$.

Transfer function (14) can be approximated by (see Sect. 3.1 above)

$$\frac{Ke^{-t_0 s^{\beta}}}{\tau s^{\alpha}+1} \approx \frac{K}{\left(\tau s^{\alpha}+1\right)\left(t_0 s^{\beta}+1\right)}, \tag{15}$$

and can be changed to a fractional differential equation form. Using this approximation in equation (14) we obtain

$$\left(s^{\alpha+\beta}+\frac{1}{t_0}s^{\alpha}+\frac{1}{\tau}s^{\beta}+\frac{1}{\tau t_0}\right)X\left(s\right) = \frac{K}{\tau t_0}U\left(s\right). \tag{16}$$

For instance, take $\alpha+\beta \leq 2$. Two cases are possible in view of the value of $\alpha+\beta$. On the one hand, if $\alpha+\beta \in (0, 1]$, then it is possible to use formula (10) for all inversion in the LHS. On the other hand, if $\alpha+\beta \in (1, 2]$, the inversion of the first term can be obtained with formula (11) and (10) for the following terms. However, if function $X$ satisfies the conditions of Lemmas 2 and 3, then

$$\left(\mathcal{D}_{0+}^{\alpha+\beta}X\right)(t)+\frac{1}{t_0}\left(\mathcal{D}_{0+}^{\alpha}X\right)(t)+\frac{1}{\tau}\left(\mathcal{D}_{0+}^{\beta}X\right)(t)+\frac{1}{\tau t_0}X(t)=\frac{K}{\tau t_0}U(t). \quad (17)$$

### 3.2.1 Fractional Sliding Surface

Let define the fractional sliding surface according to (15) and (17). In the following we assume the conditions of Lemmas 2 and 3.

**Definition 3** Let $\alpha, \beta \in \mathbb{R}$, $m = [\alpha+\beta]+1$, $e(t) \in AC^{m-1}([a, b])$ and $e^{(m)} \in L_1(a, b)$. Then, the fractional slinding surface, $S_{\alpha,\beta}$, is defined as

$$S_{\alpha,\beta}[e(t)] := \left(\mathcal{D}_{0+}^{\alpha+\beta}+\lambda_1\mathcal{D}_{0+}^{\alpha}+\lambda_2\mathcal{D}_{0+}^{\beta}+\lambda_0\right)I^1[e(t)]. \quad (18)$$

**Remark 5** In the classical sense, our operator converge to similar result as presented in [2], when $n = 2$ as $\alpha, \beta$ approach to $1^-$. Binomial theorem can be used to generalize $n$. This is a fascinating topic for further research. Just emphasis fractional computation for outside-interval parameters (0, 1) is much more difficult to calculate.

**Remark 6** Note that conditions of Lemmas 2 and 3 allow us to commute the classical integral operator $I^1$ in the fractional sliding surface operator in Definition 3, i.e.,

$$\left(\mathcal{D}_{0+}^{\alpha+\beta}+\lambda_1\mathcal{D}_{0+}^{\alpha}+\lambda_2\mathcal{D}_{0+}^{\beta}+\lambda_0\right)I^1 = I^1\left(\mathcal{D}_{0+}^{\alpha+\beta}+\lambda_1\mathcal{D}_{0+}^{\alpha}+\lambda_2\mathcal{D}_{0+}^{\beta}+\lambda_0\right). \quad (19)$$

Classical derivation of operator Definition 3 will be needed in the following:

$$\frac{d}{dt}S_{\alpha,\beta}[e(t)] = \left(\mathcal{D}_{0+}^{\alpha+\beta}+\lambda_1\mathcal{D}_{0+}^{\alpha}+\lambda_2\mathcal{D}_{0+}^{\beta}+\lambda_0\right)[e(t)] \quad (20)$$

where we have used the result in Remark 6 followed by the application of the Fundamental Theorem of Calculus.

The next theorem show how we can construct the equivalent fractional control law $U_e^{\alpha,\beta}q(t)$.

**Theorem 2** *Let $\alpha, \beta \in \mathbb{R}$, $m = [\alpha+\beta]+1$, $e(t) \in AC^{m-1}([a, b])$ and $R^{(m)}(t)$, $X^{(m)}(t) \in L_1(a, b)$. Then, the fractional control law $U_e^{\alpha,\beta}q(t)$ based in the fractional sliding surface Definition 3 is*

$$U_{eq}^{\alpha,\beta}(t) = \frac{\tau t_0}{K}\begin{pmatrix} \mathcal{D}_{0^+}^{\alpha+\beta}R(t) + \lambda_1\mathcal{D}_{0^+}^{\alpha}R(t) + \lambda_2\mathcal{D}_{0^+}^{\beta}R(t) + \lambda_0 R(t) + \\ \left(\frac{1}{t_0} - \lambda_1\right)\mathcal{D}_{0^+}^{\alpha}X(t) + \left(\frac{1}{\tau} - \lambda_2\right)\mathcal{D}_{0^+}^{\beta}X(t) + \left(\frac{1}{\tau t_0} - \lambda_0\right)X(t) \end{pmatrix}. \tag{21}$$

***Proof*** Let $e(t) = R(t) - X(t)$. First, we apply operator (20) to function $e(t)$ to obtain

$$\frac{d}{dt}S_{\alpha,\beta}e(t) = \frac{d}{dt}S_{\alpha,\beta}R(t) - \frac{d}{dt}S_{\alpha,\beta}X(t) \tag{22}$$

by linearity of this fractional operator. Next, $\frac{d}{dt}S_{\alpha,\beta}e(t) = 0$ guides us to

$$\begin{aligned} &\mathcal{D}_{0^+}^{\alpha+\beta}R(t) + \lambda_1\mathcal{D}_{0^+}^{\alpha}R(t) + \lambda_2\mathcal{D}_{0^+}^{\beta}R(t) + \lambda_0 R(t) = \\ &\mathcal{D}_{0^+}^{\alpha+\beta}X(t) + \lambda_1\mathcal{D}_{0^+}^{\alpha}X(t) + \lambda_2\mathcal{D}_{0^+}^{\beta}X(t) + \lambda_0 X(t). \end{aligned} \tag{23}$$

Second, we combine Eq. (17) and (23) and we get

$$\begin{aligned} &\mathcal{D}_{0^+}^{\alpha+\beta}R(t) + \lambda_1\mathcal{D}_{0^+}^{\alpha}R(t) + \lambda_2\mathcal{D}_{0^+}^{\beta}R(t) + \lambda_0 R(t) + \\ &\mathcal{D}_{0^+}^{\alpha+\beta}X(t) + \frac{1}{t_0}\mathcal{D}_{0^+}^{\alpha}X(t) + \frac{1}{\tau}\mathcal{D}_{0^+}^{\beta}X(t) + \frac{1}{\tau t_0}X(t) = \\ &\mathcal{D}_{0^+}^{\alpha+\beta}X(t) + \lambda_1\mathcal{D}_{0^+}^{\alpha}X(t) + \lambda_2\mathcal{D}_{0^+}^{\beta}X(t) + \lambda_0 X(t) + \\ &\qquad\qquad\qquad\qquad\qquad\qquad\qquad\qquad \frac{K}{\tau t_0}U(t). \end{aligned}$$

The result follows directly. □

### 3.2.2 Stability Considerations

Concerning stability in fractional order systems and sliding mode control, there are an important number of scientific publications associated with this topic; we recommend to interested readers the following Refs. [7, 11, 15, 16, 18, 19, 24].

## 4 Conclusions

This paper presented a theoretical approach design of a Sliding Mode Control based on fractional order calculus for nonlinear processes. The nonlinear process was linearized using empirical modeling procedures, which means using the reaction curve procedure, which produced a reduced-order model. The controller was designed using the reduced order model, which can make a possible implementation easier if we consider a complete model. The future works are directed to get a tuning equation set using bioinspired optimization methods, and the implementation and comparison with other SMC approaches.

**Acknowledgements** This research was supported by Colegio de Ciencias e Ingenierías, Universidad San Francisco de Quito USFQ, through the Poli-Grants Program under Grant POLI002OC.

## References

1. Arunshankar, J., et al.: Control of nonlinear two-tank hybrid system using sliding mode controller with fractional-order pi-d sliding surface. Comput. Electr. Eng. **71**, 953–965 (2018)
2. Camacho, O., Smith, C.A.: Sliding mode control: an approach to regulate nonlinear chemical processes. ISA Trans. **39**(2), 205–218 (2000)
3. Báez, E., Bravo, Y., Leica, P., Chávez, D., Camacho, O.: Dynamical sliding mode control for nonlinear systems with variable delay. In: 2017 IEEE 3rd Colombian conference on automatic control (CCAC), pp. 1–6. IEEE (2017)
4. Ceballos, J., Coloma, N., Di Teodoro, A., Ochoa-Tocachi, D.: Generalized fractional Cauchy-Riemann operator associated with the fractional Cauchy-Riemann operator. Adv. Appl. Clifford Algebras **30**(5), 1–22 (2020)
5. Ceballos, J., Coloma, N., Di Teodoro, A., Ochoa-Tocachi, D., Ponce, F.: Fractional multicomplex polynomials. Complex Anal. Oper. Theory **16**(4), 1–30 (2022)
6. Diethelm, K.: The analysis of fractional differential equations: An application-oriented exposition using operators of caputo type (2004)
7. Eray, O., Tokat, S.: The design of a fractional-order sliding mode controller with a time-varying sliding surface. Trans. Inst. Meas. Control. **42**(16), 3196–3215 (2020)
8. Gambhire, S., Kishore, D.R., Londhe, P., Pawar, S.: Review of sliding mode based control techniques for control system applications. Int. J. Dyn. Control **9**(1), 363–378 (2021)
9. Kilbas, A., Marichev, O., Samko, S.: Fractional integrals and derivatives (Theory and Applications) (1993)
10. Kilbas, A.A., Srivastava, H.M., Trujillo, J.J.: Theory and Applications of Fractional Differential Equations. North-Holland Mathematics Studies, vol. 204. Elsevier, Amsterdam (2006)
11. Labbadi, M., Boubaker, S., Djemai, M., Mekni, S.K., Bekrar, A.: Fixed-time fractional-order global sliding mode control for nonholonomic mobile robot systems under external disturbances. Fractal Fraction **6**(4), 177 (2022)
12. Liptak, B.G.: Instrument Engineers' Handbook, Volume Two: Process Control and Optimization. CRC press (2018)
13. Miller, K.S., Ross, B.: An Introduction to the Fractional Calculus and Fractional Differential Equations. Wiley (1993)
14. Muresan, C.I., Ionescu, C.M.: Generalization of the fopdt model for identification and control purposes. Processes **8**(6), 682 (2020)
15. Petras, I.: Stability of fractional-order systems with rational orders (2008). arXiv:0811.4102
16. Petráš, I.: Fractional-Order Nonlinear Systems: Modeling, Analysis and Simulation. Springer Science & Business Media (2011)
17. Podlubny, I.: Fractional-order systems and fractional-order controllers. Inst. Exp. Phys. Slovak Acad. Sci. Kosice **12**(3), 1–18 (1994)
18. Qin, Z., Wu, R., Lu, Y.: Stability analysis of fractional-order systems with the Riemann-Liouville derivative. Syst. Sci. Control Eng.: An Open Access J. **2**(1), 727–731 (2014)
19. Rivero, M., Rogosin, S.V., Tenreiro Machado, J.A., Trujillo, J.J.: Stability of fractional order systems. Math. Prob. Eng. **2013** (2013)
20. Seborg, D.E., Edgar, T.F., Mellichamp, D.A., Doyle III, F.J.: Process Dynamics and Control. Wiley (2016)
21. Smith, C.A., Corripio, A.B.: Principles and Practices of Automatic Process Control. Wiley (2005)
22. Utkin, V., Guldner, J., Shi, J.: Sliding Mode Control in Electro-Mechanical Systems. CRC Press (2017)

23. Utkin, V., Poznyak, A., Orlov, Y.V., Polyakov, A.: Road Map for Sliding Mode Control Design. Springer (2020)
24. Vinagre, B.M., Monje, C.A.: Introducción al control fraccionario. Revista Iberoamericana de Automática e Informática Industrial **3**(3), 5–23 (2010)
25. Zinober, A.S.: Variable Structure and Lyapunov Control, vol. 193. Springer (1994)

# Tuning and Optimization Software for Non-linear PID Controllers with Guaranteed Robustness

**Mercedes Ramírez Mendoza, Liset Mayo Martí, Guillermo González Yero, and Pedro Albertos Pérez**

**Abstract** In current industries there is a marked need to satisfactorily tune process controllers, mainly caused by the lack of knowledge and time of the technical personnel. Therefore, it is important that any advanced method for control tuning should be simple and friendly, otherwise there is little chance to be used. This article presents a software tool in MATLAB® that facilitates the tuning optimization of nonlinear Proportional Integral Derivative (NPID) controllers with guaranteed robustness. It comprises different evolutionary algorithms to solve a constrained multiobjective optimization problem. It uses the Cultural Algorithm by default, achieving a suboptimal and efficient performance of the NPID for robustness region (NPID-RR) controllers in few iterations and with less computation time with respect to other implemented algorithms. It is a fast, interactive and easy to modify software tool for the user, being of great value for the understanding of the control method, allowing the experimentation in the subject and being very efficient in a large set of common systems in industrial processes, including an actual process of the Cuban steel industry.

**Keywords** PID controllers · Tuning methods · Robustness · Evolutionary algorithms · Optimization

M. R. Mendoza (✉)
Department of Automation Engineering, Faculty of Electrical Engineering, Universidad de Oriente, Ave. Las Américas, Santiago de Cuba, Cuba
e-mail: mramirez@uo.edu.cu

L. M. Martí · G. G. Yero
Development and Innovation Group, Industrial Zone, Circunvalante Norte Km 3½, Acinox Las Tunas, Las Tunas, Cuba
e-mail: guillermo@acinoxtunas.co.cu

P. A. Pérez
Institute of Automation and Industrial Computing, Polytechnic University of Valencia, C/Vera, Valencia, Spain
e-mail: pedro@aii.upv.es

 
O. Llanes-Santiago (ed.), *Proceedings of 19th Latin American Control Congress (LACC 2022)*, Studies in Systems, Decision and Control 464,
https://doi.org/10.1007/978-3-031-26361-3_6

# 1 Introduction

A large process industry may have between five hundred and five thousand control loops, and PID controllers are used in 90–97% of cases. This dominance is largely a result of the properties of the PID controller, such as: easy to understand, good performance for most processes, easy to tune; accompanied by its tradition and its ability to be combined with other more advanced control schemes [1]. In previous studies it has been reported that only 20–30% of controllers in the process industry are satisfactorily tuned; two of the main reasons for that are the lack of knowledge as well as the available time of the technical staff [1, 2]. As a consequence, many parameters are set to default values. Therefore, it is important that any advanced method for controller tuning should be simple and friendly, otherwise there is little chance to be used.

Current research aims to improve the performance of robust PID controllers by taking into account trade-offs between criteria that are generally in conflict, such as load disturbance attenuation, robustness, and noise sensitivity, which are recognized in the literature as reasonable control objectives for the industry. Since the beginning of the twenty-first century, Aström and Hägglund, who have been the fathers of advanced PID control, have proposed methods based on optimization of load disturbance rejection under robustness constraints [3], and they have pointed out the great value of the availability of software providing the possibility to experiment with them and at the same time be able to use the program to obtain controllers in a real plant, incorporating it into an auto-tuning procedure [4] for people with a moderate knowledge of PID controllers.

There are several authors who have developed software tools that simplify the tuning of PID controllers, making the design methods more widely used and attractive, many of these are collected in [5, 6]. Several of them have chosen the Matlab® platform due to its powerful graphical environment, complex calculation capacity and especially for its important academic contribution in the study of control systems. Among them, SWORD (Software-based optimal robust design) [1, 7], Optimal Robust Tuning (hSIRO) [8], PIDag [9] and DIGITROLM [10], all with excellent performance, can be mentioned.

Appropriating the theory of control with guaranteed robustness and taking advantage of the benefits of NPID control, which achieves performance unattainable by linear PIDs to manage compromise relationships, a new concept of control arises, the NPID with guaranteed robustness, which satisfies all the PID control design specifications and manage online their trade-offs online.

The nonlinear PID for robustness region (NPID-RR) design method was derived from this concept, the robustness-disturbance rejection relationship [11], but unlike the widely used PIDs, the tuning is not a simple task, therefore, although the method offers tempting results, they risk to be poorly used in industrial or academic applications. So it would be helpful to design a simple and fast software tool for their tuning. With this goal in mind, the main objective of this paper is to design a software tool in MatLab® that facilitates the optimal tuning of the NPID-RR controllers, using a multi-objective optimization method that provides efficient performance. In order to illustrate the procedure, an NPI-RR applic ation is developed.

## 2 NPI-RR Controller Tuning

Recent research aims to improve the performance of robust PID controllers [12] by meeting multiple requirements, specifications and design objectives, in most cases in conflict with each other [13]. A suitable mechanism in this scenario is the definition of an optimization problem, where the cost function reflects the balance between the objectives desired by the designer [7, 13].

In the NPI-RR [11] (and similarly in the NPID-RR) a multi-objective optimization method is used to measure the performance of the controller, whose control law is expressed as follows:

$$C(s) = \left(1 + \frac{k_{i-rate}}{s}\right)\Delta K(e) + K^{\circ}{}_{min} + \frac{k^{\circ}{}_{i-min} + \Delta k_{i-min}}{s} \tag{1}$$

where $K^{\circ}{}_{min}, k^{\circ}{}_{i-min}$, are the minimum values of the proportional and integral gains of $C(s)$. In addition, $\Delta k_{i-min}$ is the residual of a function evaluated in the nominal design corresponding to $K^{\circ}{}_{min}, k^{\circ}{}_{i-min}$.

The MOO function is defined through the combination of three convex cost functions: the Integral Absolute Error (IAE), the Integral Quadratic Error (ISE) as an indirect measure of the overshoot, and the Total Variation (TV) to take into account the changes in the control signal,

$$min[f_{ISE,IAE,TV}] = min\left\{\sum\nolimits_{k=1}^{3} w_k \cdot f_k(i,q)\right\} \quad i = \varepsilon, \ldots, 1, q = 0, \ldots, 1 \tag{2}$$

where $f_k$ represents the different functions to calculate IAE, ISE and TV respectively and where $w_k$ are the corresponding weights.

Multi-objective optimization design involves three stages: problem definition, which involves identifying the decision variables, the design objectives and the constraints (if any), the approximation of the Pareto front and the selection of the solution to implement. This selection is made based on the set of calculated Pareto optimal solutions, where it is possible to analyze and compare the benefits between solutions and, from them, select an appropriate compromise solution.

Among the multi-objective optimization methods, the Evolutionary Algorithms and metaheuristics, such as the Tabu search and the Simulated Annealing [14], stand out. The fact that EA's work with a set of potential solutions at a time, allows them to generate a set of elements in the efficient frontier in a single iteration, in addition, they are less susceptible to the shape or the continuity of the efficient frontier. Like other heuristic methods, they do not guarantee to find the optimal efficient solutions, but they do guarantee a good approximation of the frontier in a computationally reasonable time [14]. This is the main reason that motivates the study and analysis of these algorithms in the present work for their application in the solution of the NPI-RR optimization problem.

The *multiobjective evolutionary algorithms* (MOEAs) allow handling large search spaces, generating multiple Pareto solutions in a single run and avoiding, through proper implementation, sensitivity towards Pareto convex regions. The two main questions to be taken into account when applying a MOEA in solving this type of problem are: how to carry out the assignment of the aptitude or adaptation function of each individual (fitness) and consequently the final selection, with the purpose of directing the search towards the Pareto-optimal set whereas maintaining some diversity in the population in the Pareto-optimal front, in order to obtain a set of non-dominated solutions with a uniform and extended distribution. The way to solve the previously mentioned difficulties has given rise to a considerable number of different methods during the last decade. An extensive review of these methodologies can be found in [15].

## 3 Performance of the NPI-RR Using Different Evolutionary Algorithms

In order to tune the non-linear function of the NPI-RR for the application in the Acinox Las Tunas mold level control (MLC), several optimization methods such as heavy sum, GA and PSO have been previously used. Even small computation times have been achieved for obtaining the optimal parameters, it is still a considerable time and a tool showing the results as quickly as possible is required. For this reason, other evolutionary algorithms were implemented to solve the NPI-RR multi-objective optimization problem: Genetic Algorithm [13], Differential Evolution [16], NSGA-II [17], and Cultural Algorithm [18].

To compare the results obtained with the last mentioned optimization algorithms, an experimentation process was carried out with the case study of the billet MLC of ACINOX Las Tunas [19] with the purpose of making a fair comparison. They were programmed in the MatLab® mathematical assistant, (R2015a), using a DELL laptop with Intel® runs™ i3-3227U processor at 1.90 GHz with 8 GB of RAM, Windows 10 Enterprise, 64 bits.

The main quantitative measures of the quality of the solutions obtained were defined as the cost function and the computation time required to reach the optimal parameters of the controller.

The time required by the algorithms for 30 iterations is illustrated in Fig. 1. This test is very valuable in our application, since we intend to develop a software tool that facilitates the tuning of the NPI-RR controller being as fast as possible. Therefore, the priority is not to find the optimal value of the multiobjective function, but to find minimum values that do not require many generations.

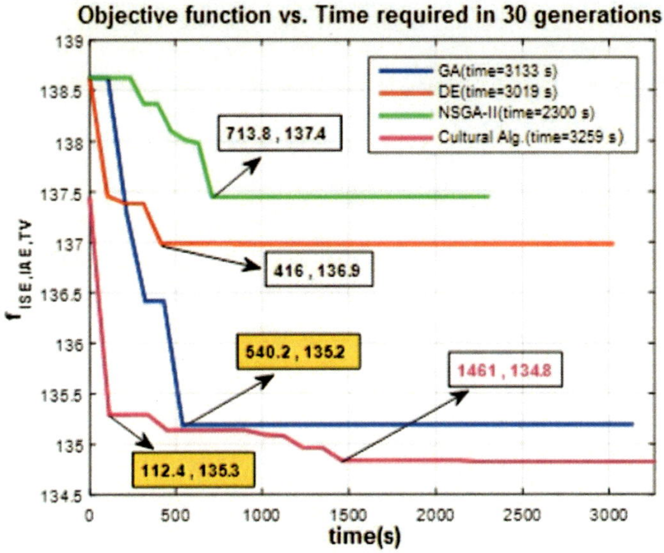


**Fig. 1** Behavior of the Cost Function during the time required to perform 30 iterations by using the different implemented optimization algorithms

In Fig. 1, the moment where the algorithms find the minimum values of $f_{ISE,IAE,TV}$ is shown. As it can be seen, the Cultural Algorithm takes the longest time to perform the 30 iterations, despite achieving the optimal value of the cost function in generation 14 with a time of approximately 25 min, which continues to be a significant time for our application. The GA stands out because it achieves a value $f_{ISE,IAE,TV}$ close to the optimum and in less time, approximately nine minutes. However, the result of the Cultural Algorithm at close to two minutes after starting its execution is tempting since it achieves a value $f_{ISE,IAE,TV}$ that it is not optimal but it is very close to the optimum.

All these results are quantified in Table 1, adding the optimum fit point $(e_h, K_h)$ of the non-linear function of the NPI-RR in each case. Unlike the algorithms used in previous research, they not only reduce the computation time, but also achieve superior results for $f_{ISE,IAE,TV}$ compared to the weighted sum algorithm. The NPI-RR controller that uses the Cultural Algorithm decreases the ISE, IAE and TV performance indices with the optimal value of the cost function, which it achieves in generation 14, despite consuming most of the computing time.

Figure 2 shows the response of the test case MLC system, using the NPI-RR controllers from Table 2.4, together with the responses of the system controlled by the linear PI of minimum gains $(\overset{\circ}{K}_{min}, \overset{\circ}{k}_{i-min})$, corresponding to a value of $M_{s-min} = 1.5$ and the one of maximum gains $(\overset{\circ}{K}_{max}, \overset{\circ}{k}_{i-max})$, which satisfies the robustness requirement $M_{s-max} = 1.9$.

It can be observed that the optimal NPI-RR controllers obtained by the different algorithms provide responses similar to that of the maximum gains PI controller,

**Table 1** Comparison of the performance of the NPI-RR with various optimization algorithms (In bold the optimal one)

| Method | Function (1)$f_{ISE,IAE,TV}$ | Computation Time (s) | gene | ISE | IAE | TV | Opt. Pt ($e_h$, $K_h$) |
|---|---|---|---|---|---|---|---|
| GA | 135.19 | 540.2 | 6 | 1345.6 | 172.07 | 74.05 | (2.02,0.049) |
| DE | 136.97 | 416 | 5 | 1407.1 | 179.15 | 71.35 | (4.63,0.049) |
| NSGA-II | 137.44 | 713.8 | 10 | 1409.3 | 179.46 | 71.83 | (4.82,0.049) |
| **Cultural Algorithm** | **134.82** | **1461** | **14** | **1340.4** | **171.68** | **73.87** | **(1.35,0.049)** |

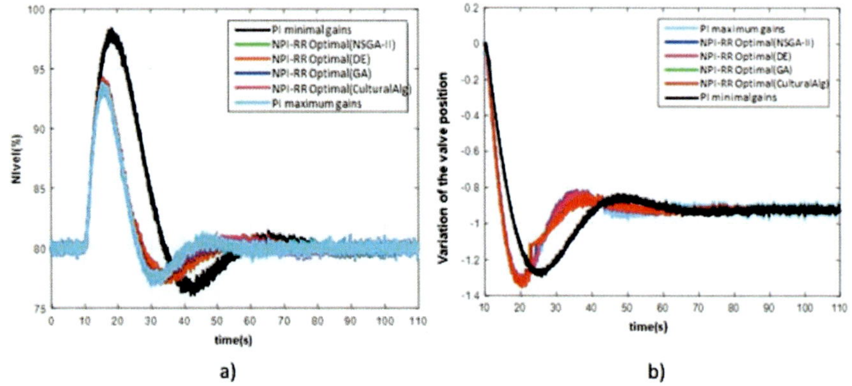


**Fig. 2** Temporal responses of the MLC against the critical disturbance $d_c = -8$ mm/s using different controllers; **a** Level in mold; **b** Control output

in terms of rejection of the disturbance, since the overshoots practically coincide. However, the Cultural Algorithm achieves the optimal value of $f_{ISE,IAE,TV}$ among all the proposed algorithms, which allows an efficient performance of the NPI-RR controller with 98.64% of the maximum value of rejection of disturbances, the shortest settling time and a decrease in 23.94% of the control effort with respect to the controller with the highest rejection of disturbances. For these reasons, the Cultural Algorithm is proposed to be used in the software tool for tuning the NPI-RR controller presented below.

## 4 Description of the Proposed Software Tool

The proposed software tool aims to facilitate the tuning of the NPI-RR controller for use in both industry and academia, and also to be part of a self-tuning approach in the future. It has been called SWONT-RR (SoftWare-based Optimal Non linear Tuning in Robust Region) and it was developed in MatLab® version 2015a [20]. It was designed so that it can be used in the test set used in [21] and in other researches on PID control, made up by using 134 representative models of the process industry.

The designed tool faithfully follows the steps involved in the two procedures making up the design method of the NPI-RR controller, exposed in [11], substituting the heavy sum optimization method used in step three, item (e), of the procedure for fitting the nonlinear function, by the Cultural Algorithm evolutionary optimization method, due to the significant improvements shown in the previous section.

Among the requirements to correctly work with SWONT-RR, a MatLab® version similar to or newer than the one used to develop the software should be used. The user must provide the model of the plant in the Laplace domain. The software only operates with linear, stable open-loop systems, since it is precisely a requirement of the SWORD software [1, 7], which is used in the proposed tool to obtain the linear PI controllers with maximum and minimum gains satisfying a defined robustness region.

In addition to the main file, which is described by the flowchart in Fig. 3, the tool consists of various Matlab® functions and Simulink models. For example, the model named Simulink 1 executes the first step of the procedure for fitting the nonlinear function of the NPI-RR.

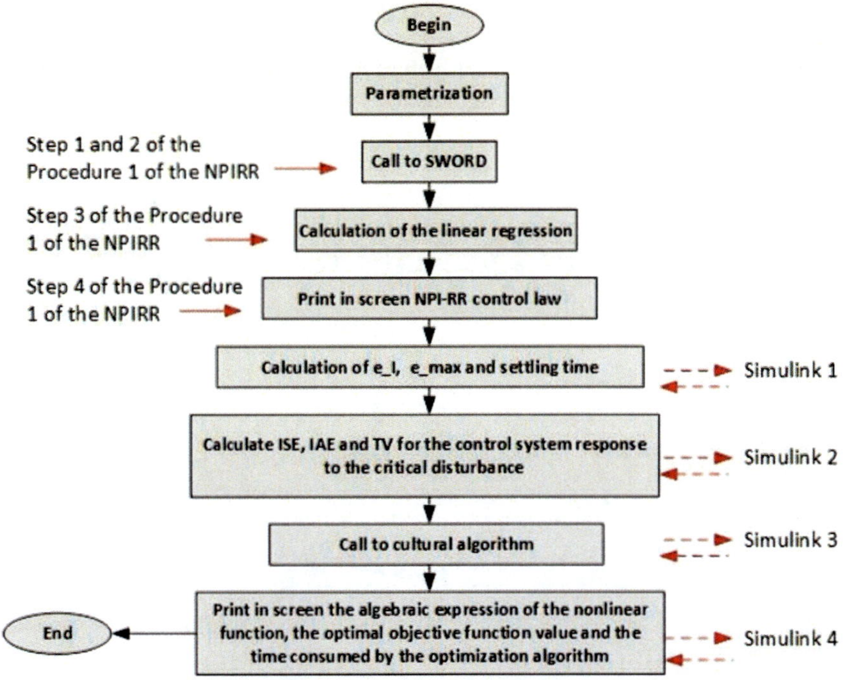


**Fig. 3** General flow diagram of the main file of the SWONT-RR software tool

## 4.1 Results

At the end of each software execution, all the relevant information about the NPI-RR controllers is stored in the structure called "NPI_RR".

During the execution of the program, different information and data on the progress of the software is printed on the screen keeping the user attentive. For instance: the starting and ending times of the adjustment methods of the linear PI controllers by means of SWORD and of the optimization algorithm, the control law of the NPI-RR for the process to be controlled, the values of the parameters of the non-linear function $e_h$ and those $K_h$ corresponding to the best controller of each generation of the optimization algorithm and finally the optimal parameters that are inserted in the algebraic expression of the nonlinear function. In addition, the optimal value of the multi-objective function and the computation time used by the tool are printed. These results are accompanied by seven graphs that facilitate the understanding of the NPI-RR control method and make visible the progress of the optimization algorithm. Below are examples of some of these graphs (Figs. 4 and 5), for a typical process model (case P4 of the test set [21]) with transfer function:

$$G(s) = \frac{1}{(1+s)^4} \tag{3}$$

In general, to validate the practical feature of the SWONT-RR tool, an analysis of its behavior was carried out through the test set made up of 134 processes, which

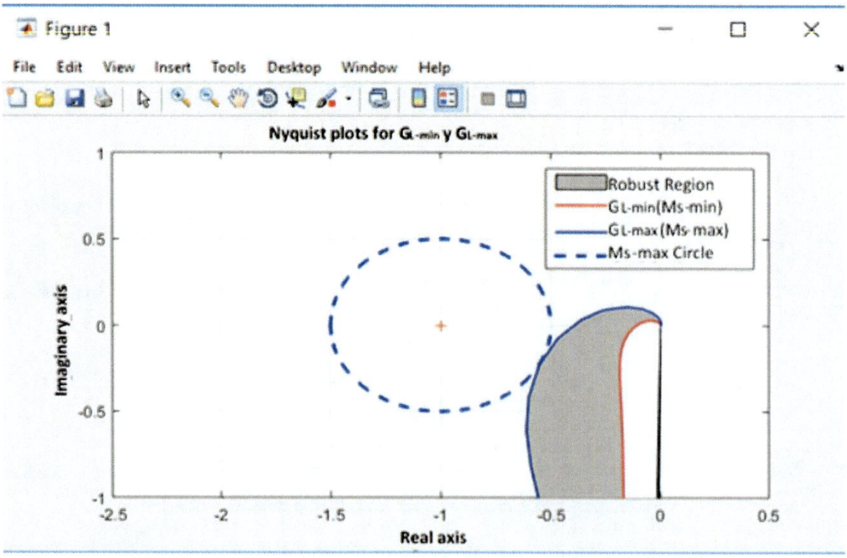


**Fig. 4** Region of robustness defined by $1.2 \leq M_s \leq 2.0$ for the process $P4$ in the frequency domain using the Nyquist trajectories of the closed—loop system

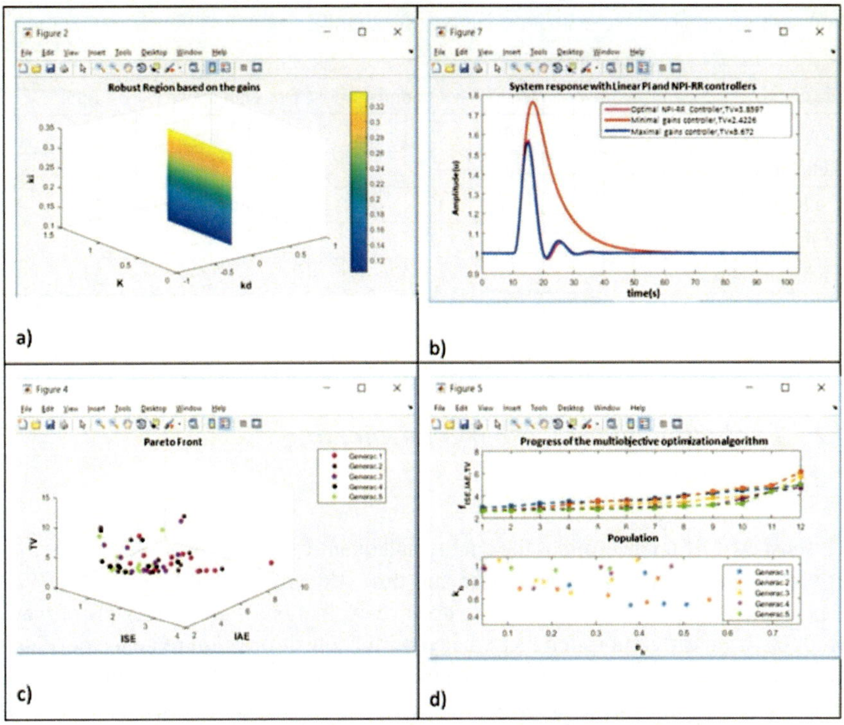


**Fig. 5** Progress of the Culture Algorithm optimization for a population of 12 NPI-RR controllers and five generations. **a** Robustness region defined by $1.2 \leq M_s \leq 2.0$ for the process $P4$ based on proportional, integral and derivative gains; **b** Temporary responses of the process $P4$ against the critical unit step disturbance using different linear PI controllers with minimum and maximum gains and optimal NPI-RR; **c** Pareto front; **d** Cost function and parameters $e_h$ and $K_h$ of each controller of the populations of the five generations

has been previously discussed. The comparison of the results obtained by the NPI-RR by using the Cultural Algorithm, tuned by means of the SWONT-RR, with the NPI-RR that uses the previously tuned weighted sum method [19], shows in all cases a similarity in the values of $f_{IAE,ISE,TV}$, although the tuning through SWONT-RR tends to be better in most cases. This supports the efficient performance of the optimization method used to solve the multi-objective problem of the controller object of study in the programmed software tool. The detailed description of all the tests carried out, as well as the corresponding codes, can be found in [22].

**Table 2** Evaluation of the cost according to ISO/IEC 9126–1

| – | Sub-Features | | | |
|---|---|---|---|---|
| Features | comply | does not comply | Does not apply | Total |
| Functionality | 4 | 1 | 0 | 5 |
| Reliability | 1 | 1 | 1 | 3 |
| Usability | 0 | 0 | 5 | 5 |
| Efficiency | 3 | 0 | 0 | 3 |
| Maintainability | 4 | 0 | 1 | 5 |
| Portability | 1 | 1 | 2 | 4 |
| Total | 13 | 3 | 9 | 25 |

### 4.2 *Evaluation of the Cost of Implementing the Computational Tool*

The cost evaluation determines the quality and quantity of resources needed in terms of money, effort, capacity, knowledge and time. Regarding quality, ISO/IEC 9126 was created for the specification and extensive evaluation of the quality of software products from different criteria associated with acquisition, requirements, development, use, evaluation, support, maintenance, quality assurance and software audit. We are going to base ourselves on the ISO/IEC 9126–1: Quality Model, which classifies software quality in a structured set of internal and external quality characteristics in terms of functionality, reliability, usability, efficiency, maintainability and portability. First, we take from a population of 135 processes, 8 processes of the test set randomly from each type of model and the real process of the MLC of Acinox Las Tunas, which due to its characteristics is part of the P6 model of the test set and were individually tuned using the SWONT-RR tool, the results according to ISO/IEC 9126–1 are shown in the Table 2.

As conclusions of a total of 25 sub-characteristics, 52% are met, 8% are not met and 36% do not apply due to the short period of use or the time the software was created, therefore different users are required to apply the tool in other industrial processes of our company or others.

Material costs are very low, since a computer and the Matlab software license are required. What it does require is time for analysis, study and knowledge.

## 5 Conclusion

The tuning of the NPID-RR controller can lead to consider its application cumbersome and, therefore, it risks to be poorly used in industrial or academic applications, even though the method is an alternative to improve the PID control. This work deals and illustrates the design of a software tool in MatLab® that facilitates the optimal tuning of the NPI-RR controllers, using a multi-objective optimization method providing efficient performance.

To achieve this, the performance of the NPI-RR was evaluated using different optimization algorithms: Genetic Algorithm, Differential Evolution Algorithm, NSGA-II and Cultural Algorithm, and, as a test case the Acinox Las Tunas billet MLC. The results showed that the Cultural Algorithm allows an efficient behavior of the NPI-RR in few generations and requiring a minimum computation time, since it finds suboptimal controllers before convergence, which exceed those obtained by the rest of the implemented algorithms. It was determined to use, by default, the Cultural Algorithm in the proposed software tool, as well as a value of maximum iterations equal to five, as a stop condition.

A simple, interactive, fast and easily modifiable software tool was developed for the tuning of the optimal NPI-RR, in an industry representative test set, using the tools provided by MatLab®. The software is accompanied by a user guide that explains how to use it, based on different process examples. These facilities will increase the applicability and practical utility of this novel control method, based on the requirements of rejection of disturbances, robustness, control effort and their commitment relationships, where tuning optimization is performed based on global performance indicators both in academia as well as industry.

## References


1. Garpinger, O.: Analysis and Design of Software-Based Optimal PID Controllers, Ph.D. Thesis, Department of Automatic Control, Lund University, Sweden (2015)
2. Garpinger, O., Hägglund, T.: A Software Tool for Robust PID Design, presented at the 17th IFAC World Congress, Seoul, South Korea (2008)
3. Åström, K.J., Hägglund, T.: Control PID Avanzado. PEARSON EDUCACIÓN, S.A, Madrid (2009)
4. Garpinger, O.: Design of robust PID controllers with constrained control signal activity. Department of Automatic Control, Lund University, Sweden, Licentiate Licentiate (2009)
5. Li, Y., Ang, K.; Chong, G.: PID control system analysis and design. Control Syst. IEEE **26**(1), 32–41 (2006)
6. Oviedo, J.J.E., Boelen, T., van Overschee, P.: Robust advanced PID control (RaPID)—PID tuning based on engineering specifications. IEEE Control Syst. Mag. 26:15–19 (2006)
7. Garpinger, O., Hägglund, T.: Software-based optimal PID design with robustness and noise sensitivity constraints. J. Process Control **33**, 90–101 (2015)
8. Ramírez, Y.B.: Herramienta para la sintonización de controladores PI/PID de uno y dos grados de libertad. Facultad de Ingeniería, Universidad de Costa Rica, Tesis de Grado (2012)
9. Acevedo, B.R.: Desarrollo de una herramienta en Matlab para Sintonización de Controladores PID, utilizando algoritmos genéticos basado en técnicas de optimización multiobjetivo. Revista Colombiana de Tecnologías Avanzadas, vol. 2 (2014). ISSN: 1692–7257
10. Ramírez, J.A.: Aplicación de diseño, simulación, identificación de sistemas e implementación de controladores PID—DIGITROL, Revista Politécnica, vol. 23, pp. 27–37 (2016). ISSN 1900–2351
11. González, G., Ramírez, M., Albertos, P.: Robust nonlinear adaptive mould level control for steel continuous casting. IFAC-PapersOnLine 51(25), 164–170 (2018)
12. Vilanova, R., Alfaro, V.M.: Control PID robusto: Una visión panorámica. Revista Iberoamericana de Automática e Informática Industrial, vol. 8, pp. 141–158 (2011)

13. Reynoso, G., Sanchis, J., Blasco, X., Martínez, M.: Algoritmos Evolutivos y su empleo en el ajuste de controladores del tipo PID: Estado Actual y Perspectivas. Revista Iberoamericana de Automática e Informática industrial **10**, 251–68 (2013)
14. Coello, C.A., Lamont, G.B., Van Veldhuizen, D.A.: Evolutionary Algorithms for Solving Multi-Objective Problems, 2nd edn. Springer, Berlin, Heidelberg, New York, Hong Kong London, Milan, Paris, Tokyo (2007). ISBN: 978–0–387–33254–3
15. Coello, C.A., Van Veldhuizen, D.A., Lamont, G.B.: Evolutionary Algorithms for Solving Multiobjective Problems. Kluwer Academic Publishers, New York (2002)
16. Zapata, M., Mezura, E., Portilla, E.A.: Evolución diferencial con memoria de parámetros para la optimización de mecanismos de cuatro barras. Res. Comput. Sci. **134**, 9–22; rec. 2017–03–02; acc. 2017–05–06 (2017)
17. Peñuela, C.A., Granada, M.: Optimización multiobjetivo usando un algoritmo genético y un operador elitista basado en un ordenamiento no dominado (NSGA-II). Scientia et Technica Año XII, vol. 35, Universidad Tecnológica de Pereira (2007)
18. Reynolds, R.G.: An introduction to cultural algorithms. In: Presented at the Third Annual Conference on Evolutionary Programming (1994)
19. González, G.: Modelado y Control de nivel en un molde de vaciado continuo. Departamento de Control Automático, Universidad de Oriente, Santiago de Cuba, Cuba, Tesis de Doctorado (2017)
20. Benítez, J., Hueso, J.L.: Introducción a MATLAB, ed. Departamento de Matemática Aplicada Universidad Politécnica de Valencia (2015)
21. Åström, K.J., Hägglund, T.: Advanced PID Control, ed. ISA—The Instrumentation, Systems, and Automation Society (2005)
22. Mayo, L.: Sintonía y Optimización de los controladores PI no lineales con robustez garantizada mediante software. Ingeniería Automática, Universidad de Oriente, Santiago de Cuba, Cuba, Master (2019)

# Sliding Mode Control Based on a Nonlinear PID Surface for Chemical Processes

**Mateo Vásquez, Joseline Yanascual, Marco Herrera, Francisco Rossomando, and Oscar Camacho**

**Abstract** In this work a nonlinear PID sliding surface is used to design a Sliding Mode Controller. A linear sliding surfaces have certain disadvantages; for example, the magnitude of the control input required to keep the system states on the sliding surface usually increases in direct proportion to the magnitude of the tracking error. Thus, the nonlinear sliding surface may provide better performance if the nonlinearity is appropriately selected. The proposed controllers are applied to a nonlinear mixing tank to test its performance. It have been compared against a PID controller and a Sliding mode control through the simulations and performance indices, ISE, TVu, maximum overshoot, and settling time.

## 1 Introduction

Sliding Mode Control (SMC) is a nonlinear control method that modifies the nonlinear system's dynamics by applying a discontinuous control signal that forces the system to slide along manifolds with desired motions in the system state [19]. This is because the state feedback control law is not a continuous function of time. Therefore, such a system can switch from one continuous structure to another based on the current position in the state space [18].

M. Vásquez · J. Yanascual · M. Herrera
Departamento de Automatización y Control Industrial, Escuela Politécnica Nacional, Quito, Ecuador
e-mail: marco.herrera@epn.edu.ec

F. Rossomando
Instituto de Automática, Universidad Nacional de San Juan, San Juan, Argentina
e-mail: frosoma@inaut.unsj.edu.ar

O. Camacho (✉)
Colegio de Ciencias e Ingenierias, Universidad San Francisco de Quito USFQ, Campus Cumbayá, Casilla Postal 17-1200-841, Quito 170157, Ecuador
e-mail: ocamacho@usfq.edu.ec

 
O. Llanes-Santiago (ed.), *Proceedings of 19th Latin American Control Congress (LACC 2022)*, Studies in Systems, Decision and Control 464,
https://doi.org/10.1007/978-3-031-26361-3_7

During the sliding modes theory progress, the range of control problems and design methods to solve them has expanded significantly [19]. Despite their great diversity, the basic principles of the SMC theory have continued unaffected since the creation of the necessary mathematical tools. The main control methods in sliding modes allow the decomposition of the original system into independent subsystems of lower dimensions that are insensitive to various types of uncertainty in the description of the control object. These properties predetermine the sliding modes theory's success for a broad class of dynamic systems. SMC has advanced to deal with uncertainties and has become well-known within the control community, to the point that SMC is being used in several engineering fields. For example, one can find applications in electrical and mechanical engineering [5, 6, 17], in robotics [3, 11, 14, 16, 21] and in controlling chemical processes [1, 2, 4, 7, 10, 12, 13, 15, 20].

The synthesis control procedure in the SMC is divided into two phases [18]. The first point is necessary to choose the manifold in the system's state space with the desired trajectories. In the second step, the control is selected to ensure the existence of a sliding mode on this preselected manifold. Sliding mode control provides insensitivity to parameter variations and complete rejection of disturbances when is on the sliding phase.

The control is selected as a function of the sliding surface ($\sigma$) and its time derivatives, thus selecting ($\sigma$) such that the sliding mode has desired properties [2]. The selection of ($\sigma$) can usually be reduced to choosing the parameters of the sliding surface that determine the closed loop response and performance of the system.

The PD sliding surface is used as a sliding surface for the conventional SMC. Integral action is added to obtain a PID control structure because the integral term can eliminate the steady-state error resulting from the boundary layer [18].

Despite their simplicity, linear sliding surfaces have certain disadvantages. For instance the magnitude of the control input required to keep the system states on the sliding surface usually increases in proportion to the magnitude of the tracking error. Another problem is the replacement of the nonlinear system characteristics with linear dynamics arising from the control law obtained by the linear sliding surface. The linear dynamics of the linear sliding surface may not fit the global dynamics of the controlled system. Therefore, a nonlinear sliding surface may provide better performance if the nonlinearity is appropriately selected.

In this work, a PID sliding surface with nonlinear tuning terms, is used for designing a Sliding Mode Control. The PID sliding surface is called NPID [8]. The NPID is a function defined by parts. One function is nonlinear, and the other is linear. The proposed approach looks for a faster reaching time to improve the performance of the SMC controller. Therefore, the contributions of this study can be summarized as follows: i. The NPID sliding surface algorithm is proposed, in our knowledge, for the first time to improve the reaching time in the Sliding Mode Controller. ii. The performance evaluation on a nonlinear chemical process with variable parameters is used to demonstrate the proposed algorithm's success in achieving the performance indexes.

This work is divided as follows: section two describes fundamentals and SMC controllers, in section three, the proposed design is shown, section four presents the application by simulation in a nonlinear chemical process, and finally, some conclusions are presented.

## 2 Background

This section presents some fundamental concepts for the mathematical development of the new control approach. The proposed nonlinear PID by Han and a brief about sliding mode control theory.

### 2.1 *Nonlinear PID Proposed by Han*

The NPID proposed by Han [8] consists of a function defined by parts. One function is nonlinear, and the other is linear, where the error and $\delta$ parameter represent working intervals. This is to obtain better results for reference changes and disturbances [22].

According to Han [8], the mathematical expression of the NPID controller is given by Eq. (1).

$$NPID = K_p\psi\left(e_p, \alpha_p, \delta_p\right) + K_i\psi\left(e_i, \alpha_i, \delta_i\right) + K_d\psi\left(e_d, \alpha_d, \delta_d\right) \tag{1}$$

$\psi(e, \alpha, \delta)$ is a nonlinear function described as follows:

$$\psi(e, \alpha, \delta) = \begin{cases} |e|^{\alpha} sgn(e) & when\ |e| > \delta \\ \delta^{\alpha-1} \cdot e & when\ |e| \leq \delta \end{cases} \tag{2}$$

The $\delta$ parameter defines the linear area in the nonlinear function where the controller acts as a conventional PID. The $e$ is the error between the reference and the controlled variable, and $\alpha$ parameter must be chosen to take in mind the desired response of the proportional, integral, and derivative parts [22].

### 2.2 *Sliding Mode Control (SMC)*

The Sliding Mode Control (SMC) methodology, commonly referred to as variable structure control (VSC), is a nonlinear control method [18]. The design of Sliding Mode Control consists of two steps. In the initial stage, a customized surface must be designed.

While on the sliding surface, the plant's dynamics are constrained to the surface's equations and are robust to internal and external perturbations.

In the second phase, a feedback control rule must be devised to ensure convergence of a system's trajectory to the sliding surface; hence, the sliding surface must be achieved in a finite amount of time. The motion of the system on a sliding surface is known as the sliding mode [19]. The most important advantage is its insensitive to control parameters variation.

The controller has two components, high-frequency and low-frequency components. The high-frequency part takes the system from the initial state to the sliding surface, and the low-frequency part is to take the system to the desired state.

Therefore, in the following part are explained the three main steps: choice of the sliding surface, determination of the control law, and finally, establishing a convergence or reaching condition.

### 2.2.1 Sliding Surface Choice

Different sliding surfaces have been suggested in the literature, and each works better for a certain application. The most common way to get the sliding regime that guarantees the state will move toward its reference is to use the surface [2, 18].

$$\sigma(t) = \left(\frac{d}{dt} + \lambda\right)^n \int e(t)\, dt \tag{3}$$

Where $e(t)$ is the difference between the reference $R(t)$ and the controlled variable $X(t)$ ; $e(t) = R(t) - X(t)$. The objective of the control is to keep the sliding surface at zero.

$\sigma(t)$ is the sliding surface; the designer chooses this. In this paper, we are using a PID surface, where $n$ depends on the order of the system and $\lambda$ is an adjustable parameter [2, 18].

### 2.2.2 Control Law Determination

The structure of a sliding mode controller consists of two parts; one called the continuous part $(U_e q(t))$ and the second one named discontinuous or nonlinear part $U_n(t)$

$$U_{SMC}(t) = U_e q(t) + U_n(t) \tag{4}$$

$U_e q(t)$: It is obtained from the equivalent control method [18]. It is used to keep the variable to be controlled on the sliding surface $\sigma(t) = 0$. The equivalent control is deduced, considering that the derivative of the surface is zero.

The sliding condition is given by:

$$\frac{d\sigma(t)}{dt} = 0 \tag{5}$$

And combining previous equation and the model of the process system, $U_{eq}(t)$ is obtained.

$U_n(t)$: It is the discontinuous control part, it allows the system to reach the sliding surface. It incorporates a nonlinear element that includes the switching element of the control law; therefore $U_n(t)$ contains the switching element, and it is given by:

$$U_n(t) = K_D sign(\sigma(t)) \tag{6}$$

$K_D$ is a tuning parameter responsible for the reaching mode.

The chattering phenomenon is the main concern in SMC by jumps of the sliding surface [19]. In practice, chattering provokes inadmissible effects such as excitation of system dynamics neglected in modeling, actuator vibration with wear, and performance degradation. With the aim of reducing such effects, it has been proposed several approaches,one of those possibilities lies on smoothing the non-linear switching function using soft logistic activation functions, as the sigmoid function preferentially [2]. Therefore, the discontinuous part of the proposed controller uses here the sigmoid function as follows:

$$U_D(t) = K_D \frac{\sigma(t)}{|\sigma(t)| + \delta_{sig}} \tag{7}$$

### 2.2.3 Convergence Condition

The convergence condition, often known as the attractiveness condition, ensures that the system's dynamics will always converge on the sliding surface. Formulating a Lyapunov function $V(t) > 0$ with finite energy is required. The candidate Lyapunov function is defined as follows.

$$V(t) = 1/2\sigma^2(t) \tag{8}$$

It is sufficient to make sure that the derivative of the function $V(t)$ is negative for it to be possible for the function $V(t)$ to diminish. Therefore, the condition of convergence can be written as follows:

$$\frac{dV(t)}{dt} = \frac{d\sigma(t)}{dt}\sigma(t) < 0 \tag{9}$$

## 3 Sliding Mode Control Design Approach

In this section the mathematical development of the sliding mode control based on a nonlinear PID surface is presented.

Since SMC is a model based controller, let us use an empirical model of the actual process using the reaction curve procedure [2], therefore the resulting FOPDT model is as follows:

$$\frac{X(s)}{U(s)} = \frac{Ke^{-t_0 s}}{\tau s + 1} \tag{10}$$

Following the procedure as is described in [2], the order of the system, $n$, is equal to 2; therefore, the following surface is proposed:

$$\sigma(t) = \frac{de(t)}{dt} + \lambda_0 \psi\left(e_p, \alpha_p, \delta\right) + \lambda_1 \psi\left(e_i, \alpha_i, \delta\right) \tag{11}$$

Where: $e_i$ is the integral error, $e_p$ is the proportional error. $\psi(e, \alpha, \delta)$ is a function that consists of two parts, a nonlinear part and a linear part, the intervals of these two are defined by the parameter; the nonlinear function is shown in Eq. (2).

In order to design of SMC, it should be considered that the controller output has two parts,one continuous and one discontinuous.

## 3.1 Continuous Part Design

Applying the equivalent control procedure [18]:

$$\frac{d^2 e(t)}{dt^2} + \lambda_0 \psi\left(e_p, \alpha_p, \delta_p\right) + \lambda_1 \psi\left(e_i, \alpha_i, \delta_i\right) = 0 \tag{12}$$

It is necessary to know what function takes $\psi$ because each consists of two functions. Therefore, it is needed to consider every possible case, which is four.

$$\frac{KU(t)}{\tau t_0} = \frac{d^2 X(t)}{dt^2} + \frac{\tau + t_0}{\tau t_0}\frac{dX(t)}{dt} + \frac{1}{\tau t_0}X(t) \tag{13}$$

The continuous part $U_c$ is designed, and several cases are analyzed.

### 3.1.1 If $|e| \leq \delta$ and $|e_i| \leq \delta$

The surface has the following form:

$$\sigma(t) = \frac{de(t)}{dt} + \lambda_0 \delta^{\alpha_p - 1} e(t) + \lambda_1 \delta^{\alpha_i - 1} \int e(t)dt \tag{14}$$

Following the equivalent control procedure, as the derivative of the sliding surface equals zero, and considering that the error is the difference between the reference and the transmitter output, it is obtained:

$$\frac{d^2R(t)}{dt^2} - \frac{d^2X(t)}{dt^2} + \lambda_0\delta^{\alpha_p-1}\left(\frac{dR(t)}{dt} - \frac{dX(t)}{dt}\right) + \lambda_1\delta^{\alpha_i-1}e(t)dt = 0 \quad (15)$$

The reference derivates are eliminated, avoiding kicking to the final control element but not affecting the controller performance [2]; the continuous part of the controller is:

$$U_c(t) = \frac{dX(t)}{dt}\left(\frac{\tau + t_0}{K} - \frac{\lambda_0\tau t_0\delta^{\alpha_p-1}}{K}\right) + \frac{\lambda_1\tau t_0\delta^{\alpha_i-1}}{K}e(t) + \frac{X(t)}{K} \quad (16)$$

If $\lambda_0 = \frac{\tau+t_0}{\tau t_0}\left(\frac{1}{\delta^{\alpha_p-1}}\right)$ the equivalent control law is given by:

$$U_c(t) = \frac{\lambda_1\tau t_0\delta^{\alpha_i-1}}{K}e(t) + \frac{X(t)}{K} \quad (17)$$

The value of $\lambda_1$ is obtained from the discriminant term, matching to zero the surface derivate of the surface:

$$\frac{d^2e(t)}{dt^2} + \lambda_0\frac{d\psi\left(e_p, \alpha_p, \delta\right)}{dt} + \lambda_1\frac{d\psi\left(e_i, \alpha_i, \delta\right)}{dt} = 0 \quad (18)$$

$$s^2 + \lambda_0\delta^{\alpha_p-1}s + \lambda_1\delta^{\alpha_i-1} = 0 \quad (19)$$

$$\Delta = {\lambda_0}^2\delta^{2\alpha_p-2} - 4\lambda_1\delta^{\alpha_i-1} \geq 0 \quad (20)$$

$$\lambda_1 \leq \frac{{\lambda_0}^2}{4}\delta^{2\alpha_p-\alpha_i-1} \quad (21)$$

$$\lambda_1 = \frac{K_\lambda}{4}\left(\frac{\tau + t_0}{\tau t_0}\right)^2\frac{1}{\delta^{\alpha_i-1}} \quad (22)$$

with $0 < K_\lambda \leq 1$. Finally, the $U_c$ control law is given by:

$$U_{eq}(t) = \frac{K_\lambda}{4K}\frac{(\tau + t_0)^2}{\tau t_0}e(t) + \frac{X(t)}{K} \quad (23)$$

The next cases takes the same procedure as the first one, the table shows the parameter's values depending of $\psi$ function (Table 1).

**Table 1** Tuning parameters for continuous control action

| | $\|e_i\| \leq \delta$ | $\|e_i\| > \delta$ |
|---|---|---|
| $\|e\| \leq \delta$ | $\lambda_0 = \left(\frac{\tau+t_0}{\tau t_0}\right)\frac{1}{\alpha_p \|e\|^{\alpha_p-1}}$, $\lambda_1 = \frac{K_\lambda}{4}\left(\frac{\tau+t_0}{\tau t_0}\right)^2 \frac{1}{\delta^{\alpha_i-1}}$ | $\lambda_0 = \frac{\tau+t_0}{\tau t_0}\left(\frac{1}{\delta^{\alpha_p-1}}\right)$, $\lambda_1 = \frac{K_\lambda}{4}\left(\frac{\tau+t_0}{\tau t_0}\right)^2 \frac{1}{\alpha_i \|e_i\|^{\alpha_i}}$ |
| $\|e\| > \delta$ | $\lambda_0 = \frac{\tau+t_0}{\tau t_0}\left(\frac{1}{\alpha_p \|e\|^{\alpha_p-1}}\right)$, $\lambda_1 = \frac{K_\lambda}{4}\left(\frac{\tau+t_0}{\tau t_0}\right)^2 \frac{1}{\delta^{\alpha_i-1}}$ | $\lambda_0 = \frac{\tau+t_0}{\tau t_0}\left(\frac{1}{\alpha_p \|e\|^{\alpha_p-1}}\right)$, $\lambda_1 = \frac{K_\lambda}{4}\left(\frac{\tau+t_0}{\tau t_0}\right)^2 \frac{1}{\alpha_i \|e_i\|^{\alpha_i-1}}$ |
| $U_{eq}(t) = \frac{K_\lambda}{4K}\frac{(\tau+t_0)^2}{\tau t_0}e(t) + \frac{X(t)}{K}$ | | |

## 3.2 Convergence or Reaching Condition

The candidate Lyapunov condition [18] as is presented in Sect. 2.2.3 is applied to satisfy the reaching mode. After some calculations is obtained the following result that $K_D > 0$.

For the discontinuous part, it is necessary $\delta_{sig}$. It is obtained in a heuristic way, and follows the next equation:

$$\delta_{sig} = \delta_{sig0} + m\,|\Delta R| \tag{24}$$

# 4 Simulation Results

In this part a nonlinear mixing tank with variable parameters [9] is used for simulation. The proposed controller is compared against an SMC as proposed by [2] and a PID controller. All the nonlinear equations model and operation points appears in [9].

## 4.1 Model Identification Process

Following the procedure as presented by [2], the nonlinear process can be approximated for a reduced model. The identified FOPDT is as follows:

$$G(s) = \frac{-0.822e^{-3.79s}}{2.07s + 1} \tag{25}$$

The tuning parameters for the proposed approach are given in Table 2. The values were heuristically obtained.

**Table 2** Tuning parameters for the first proposed controller of the mixing tank

| Parameter | Value |
|---|---|
| $\alpha_p$ | 1.1 |
| $\alpha_i$ | 1.2 |
| $\alpha_d$ | – |
| $\delta$ | 0.01 |
| $\delta_{sig}$ | $-2.7\,\lvert\Delta R\rvert + 1.4$ |
| $K_D$ | 1.1 |
| $K_\lambda$ | 1 |

## 4.2 Simulation Tests

Two tests were carried out to compare the controllers, a reference change to the maximum allowed value by the physical conditions of the plant, for various reference changes, that is, ascending and descending reference changes in a range 150–190 °F, and test to rejection of disturbances.

Figure 1 shows the temperature response of the variable height mixing tank for the proposed controllers and two conventional controllers, an SMC with a linear PID surface and a PID tuned by Dahlin's method. It is observed that the best waveforms are of the two proposed controllers since its response does not present oscillations and has low overshoot and short settling times. Moreover, according to Fig. 1, the

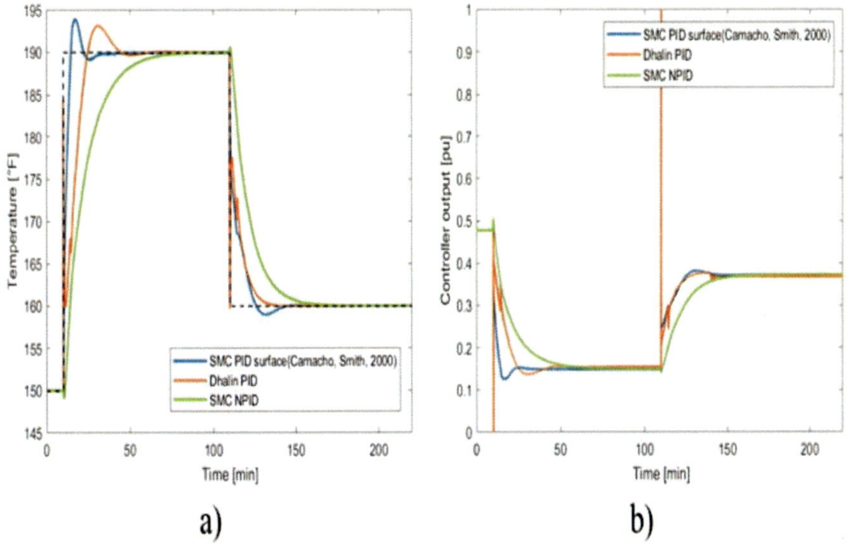


**Fig. 1** **a** Output process **b** Control signal for reference change

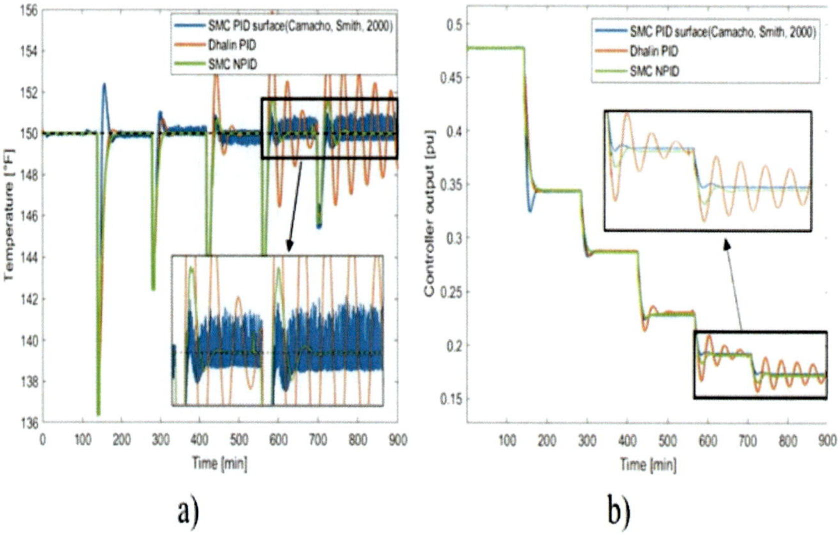


**Fig. 2** **a** Output process **b** Control signal for a disturbance

**Table 3** Performance indexes

| Controller | ISE | | | TVu | | | MP (%) | | | Ts (s) | | |
|---|---|---|---|---|---|---|---|---|---|---|---|---|
| | PID | SMC | SMC-NPID | PID | SMC | SMC-NPID | PID | SMC | SMC-NPID | PID | SMC | SMC-NPID |
| Set point changes | 2.31 | 2 | 1.84 | 0.29 | 0.04 | 0.08 | 3.98 | 3.29 | 1.29 | 46 | 50.14 | 46.6 |
| Disturbance | 0.48 | 0.3 | 0.31 | 0.21 | 0.12 | 0.31 | 3.91 | 1.63 | 1.12 | 84.82 | 38.93 | 40.09 |

proposed controllers are less aggressive than the conventional ones. Below is the plant's response with the different controllers to different reference changes.

The hot water flow is taken in this case as a disturbance of the system, and in Fig. 2 the response of the system and controllers, respectively is shown.

The simulations results show that the three SMC controllers present fewer oscillations but chattering due to the controller's shape. Finally, the PID controller has the slowest response.

The following table presents the performance indexes ISE, TVu, maximum over the peak, and settling time for a positive step input in the reference value, positive and negative reference changes (Table 3).

## 5 Conclusions

This work showed a control proposal based on a nonlinear PID surface. The nonlinear function parameters were considered part of the design in the mathematical development. Furthermore, to obtain better controller performance under different reference changes, a variable delta of the sigmoid function was added, depending on the reference change. Two conventional controllers, Dahlin-based PID and SMC, were also designed to compare the results obtained in the proposed controllers, and performance indexes, ISE, TVU, maximum peak, and settling time, were used as performance quantifiers. For reference changes, the two controllers presented a small number of oscillations. In the case of disturbances, the response of the proposed controllers is more robust and smoother than the conventional controllers. According to the values of the performance indexes, it is observed that the most suitable are the two proposed controllers.

**Acknowledgements** This work was supported by the Universidad San Francisco de Quito through the Poli-Grants Program under Grant 17461. This research was supported by Colegio de Ciencias e Ingenierías, Universidad San Francisco de Quito USFQ, through the Poli-Grants Program under Grant POLI002OC.

## References

1. Camacho, O.: A predictive approach based-sliding mode control. IFAC Proc. Vol. **35**(1), 381–385 (2002)
2. Camacho, O., Smith, C.A.: Sliding mode control: an approach to regulate nonlinear chemical processes. ISA Trans. **39**(2), 205–218 (2000)
3. Capito, L., Proaño, P., Camacho, O., Rosales, A., Scaglia, G.: Experimental comparison of control strategies for trajectory tracking for mobile robots. Int. J. Autom. Control **10**(3), 308–327 (2016)
4. Cargua-Sagbay, D., Palomo-Lema, E., Camacho, O., Alvarez, H.: Flash distillation control using a feasible operating region: a sliding mode control approach. Ind. & Eng. Chem. Res. **59**(5), 2013–2024 (2020)
5. Dhanasekar, R., Kumar, S.G., Rivera, M.: Sliding mode control of electric drives/review. In: 2016 IEEE International Conference on Automatica (ICA-ACCA), pp. 1–7. IEEE (2016)
6. Dominguez, X., Camacho, O., Leica, P., Rosales, A.: A fixed-frequency sliding-mode control in a cascade scheme for the half-bridge bidirectional dc-dc converter. In: 2016 IEEE Ecuador Technical Chapters Meeting (ETCM), pp. 1–6. IEEE (2016)
7. Espín, J., Castrillon, F., Leiva, H., Camacho, O.: A modified smith predictor based-sliding mode control approach for integrating processes with dead time. Alex. Eng. J. **61**(12), 10119–10137 (2022)
8. Han, J.: Nonlinear pid controller. Acta Automat. Sin. **20**, 487–490 (1994)
9. Iglesias, E.: Using fuzzy logic to enhance control performance of sliding mode control and dynamic matrix control (2006)
10. Mehta, U.V., Kaya, I.: Smith predictor with sliding mode control for processes with large dead times. J. Electr. Eng. **68**(6), 463–469 (2017)
11. Mera, M., Ríos, H., Martínez, E.A.: A sliding-mode based controller for trajectory tracking of perturbed unicycle mobile robots. Control Eng. Pract. **102**, 104548 (2020)

12. Mihoub, M., Nouri, A.S., Abdennour, R.B.: Real-time application of discrete second order sliding mode control to a chemical reactor. Control Eng. Pract. **17**(9), 1089–1095 (2009)
13. Musmade, B.d., Patil, Y., Munje, R., Patre, B.: Design of sliding mode control scheme for nonlinear processes for improved performance. In: 2015 International Conference on Industrial Instrumentation and Control (ICIC), pp. 992–996. IEEE (2015)
14. Proaño, P., Capito, L., Rosales, A., Camacho, O.: Sliding mode control: implementation like pid for trajectory-tracking for mobile robots. In: 2015 Asia-Pacific Conference on Computer Aided System Engineering, pp. 220–225. IEEE (2015)
15. Rasul, T., Pathak, M.: Control of nonlinear chemical process using sliding mode control. In: 2016 IEEE 1st International Conference on Power Electronics, Intelligent Control and Energy Systems (ICPEICES), pp. 1–5. IEEE (2016)
16. Salinas, L.R., Santiago, D., Slawiñski, E., Mut, V.A., Chavez, D., Leica, P., Camacho, O.: P+ d plus sliding mode control for bilateral teleoperation of a mobile robot. Int. J. Control Autom. Syst. **16**(4), 1927–1937 (2018)
17. Sira-Ramírez, H., Llanes-Santiago, O.: Sliding mode control of nonlinear mechanical vibrations. J. Dyn. Syst. Meas. Control **122**(4), 674–678 (2000)
18. Slotine, J.J.E., Li, W., et al.: Applied Nonlinear Control, vol. 199. Prentice hall Englewood Cliffs, NJ (1991)
19. Utkin, V., Poznyak, A., Orlov, Y.V., Polyakov, A.: Road Map for Sliding Mode Control Design. Springer, Berlin (2020)
20. Wang, H., Wang, Z., Zhou, Q., Liang, J., Yin, Y., Su, W., Wang, G.: Optimization and sliding mode control of dividing-wall column. Ind. & Eng. Chem. Res. **59**(45), 20102–20111 (2020)
21. Yang, H., Liu, L., Wang, Y.: Observer-based sliding mode control for bilateral teleoperation with time-varying delays. Control Eng. Pract. **91**, 104097 (2019)
22. Zaidner, G., Korotkin, S., Shteimberg, E., Ellenbogen, A., Arad, M., Cohen, Y.: Non linear pid and its application in process control. In: 2010 IEEE 26th Convention of Electrical and Electronics Engineers in Israel, pp. 000574–000577. IEEE (2010)

# Fault Diagnosis in Industrial Processes

# Exploratory Data Analysis on the Poincaré Features of the Electrical Torque Oriented to the Severity Diagnosis of a Gearbox Tooth Breakage

**Mariela Cerrada, María-Victoria Mejía, Renato Ortega, Diego Cabrera, and Rene-Vinicio Sánchez**

**Abstract** This work develops the exploratory data analysis of the Poincaré features extracted from the electrical torque of an induction motor which is coupled to a gearbox under different severity levels of the failure mode related to the tooth breakage. This research was developed because of the sensibility of the electrical torque faced to the mechanical faults in the motor load. Several features were extracted from the Poincaré plots of the electrical torque, and feature selection was applied by using different techniques such as ReliefF, Chi-2 and a cluster validity index called Composing Density Between and With clusters $CD_{bw}$. Boxplots and t-SNE plots were used to visualize the data distribution and perform the data analysis regarding nine different fault severity levels of the tooth breakage. These graphics show Poincaré features are discriminative to separate some set of fault severities, and they can be used as inputs for machine learning based classification models.

## 1 Introduction

One of the elements for the success of industrial production is the maintenance to avoid unexpected plant shutdown and improve the useful life of the devices involved in the process. Condition-based maintenance is an alternative into the maintenance activities to known early the health state of a device [1]. Particularly, gearboxes are important rotating machines, because of their role in the mechanical power transmission. Most of the complex industrial process are dependent on such device,

M. Cerrada (✉) · M.-V. Mejía · R. Ortega · D. Cabrera · R.-V. Sánchez
GIDTEC, Universidad Politécnica Salesiana, Cuenca, Ecuador
e-mail: mcerrada@ups.edu.ec

M.-V. Mejía
e-mail: mmejiac6@est.ups.edu.ec

D. Cabrera
e-mail: dcabrera@ups.edu.ec

R.-V. Sánchez
e-mail: rsanchezl@ups.edu.ec

 
O. Llanes-Santiago (ed.), *Proceedings of 19th Latin American Control Congress (LACC 2022)*, Studies in Systems, Decision and Control 464,
https://doi.org/10.1007/978-3-031-26361-3_8

then, methods and techniques for condition monitoring are reported in the literature [2, 3]. Several signals can be used for gearbox monitoring, being the vibration the most popular [4, 5]. However, the searching of new informative signals for fault diagnosis of rotating machines is still an open problem. The electrical torque provided by the induction motor coupled to a mechanical device such as a gearbox has been identified as a fault sensitive signal. The performance of that signal regarding the vibration signal was compared in [6], by analyzing the Fast Fourier Transform (FFT) and the power spectrum, showing that such electrical torque exhibits evident change when the tooth breakage appears.

On the other hand, nowadays, data-driven approaches are essential tools to propose fault classifiers for complex systems [7, 8]; this approach has been widely reported where vibration, current, or acoustic emission signals has been used with success [9]. Although deep learning approaches have facilitated the development of data-driven based fault classifiers, shallow models adjusted by machine learning techniques still attract interest because of their interpretability [10]. Besides the signal selection, feature extraction and selection are challenged tasks for developing shallow classifiers, and this is the problem addressed in this work by using Poincaré plots (PP) and its related features extracted from such electrical torque signal [11, 12]. After feature extraction, feature selection is performed by using feature rankings from techniques such as ReliefF, Chi-2 and a cluster validity index $CD_{bw}$ [9, 13]. Boxplots and t-SNE plots are used to show that Poincaré features generate proper data clusters in the feature space, useful for the further development of shadow classifiers.

## 2 Problem Statement and Preliminaries

Gearboxes in many cases are closed and difficult to inspect directly and detect possible damages. Then, to avoid the stoppage of the machinery, condition monitoring techniques are applied by extracting usually statistical indicators such as RMS, standard deviation, kurtosis, among others. In this paper, we propose the use of the electrical torque because of its sensibility to gearbox faults, specifically the level of broken tooth, and the extraction of other indicators different from the statistical ones, i.e., eleven features extracted from PP. Then, the exploratory data analysis leads to understand the behavior of each of the features faced to the failure development. The first step before starting the exploratory data analysis is the processing of the raw data and adapt it to calculate the Poincaré features. As the available data was not statically enough, data augmentation was used and, after that, the autocorrelation signal was investigated to build the proper PP. Then, over each piece of signal, the lag is calculated. The average of the first lag is considered for all the signals, no matters the level of severity. After that, the PP is obtained. Finally, eleven features are computed to obtain the completed corpus of data. In the next sections, this process is described more precisely.

## 2.1 Test Bed and Data Acquisition

The test bed for measuring the current and voltage signals is depicted in Fig. 1; it is composed by a gearbox powered by a 2 Hp induction motor, with nominal speed of 1100 rpm.

A constant load is coupled to the gearbox through a magnetic break adjusted up to 3 loads: L1, L2, L3 corresponding to no-load and loads due to the voltage set on 10 and 20 V. The one stage gearbox has a faulty pinion due to different severity levels of tooth breakage in one tooth, see Fig. 1. Several sensors are attached properly on the gearbox, such as acoustic emission (AE) sensors, current clamp (CVC) meters, voltage sensors (VLL and VLN), encoders (E), and accelerometers (A), which are connected to a DAQ (NIcDAQ), and the samples are collected by a signal processing software developed in LabVIEW. All sensors have a sample frequency of 50 KHz, excepting the acoustic emission sensors that has 1 MHz. For this work, we only use the current and voltage signals of the rotor to calculate the electrical torque, by Eq. 1:

$$T_e = n_p(\overrightarrow{\lambda}_e \times \overrightarrow{i}_e) \tag{1}$$

where $n_p$ is the number of poles of the induction motor, $\lambda_e$ is the stator flux, $i_e$ is the rotor current, and the arrows stand for the spatial vector representation. The flux is calculated according to Eq. 2:

$$\overrightarrow{\lambda}_e(t) = \int_0^t \left(\overrightarrow{v}_e(\tau) - R_e \overrightarrow{i}_e(\tau)\right) d\tau \tag{2}$$

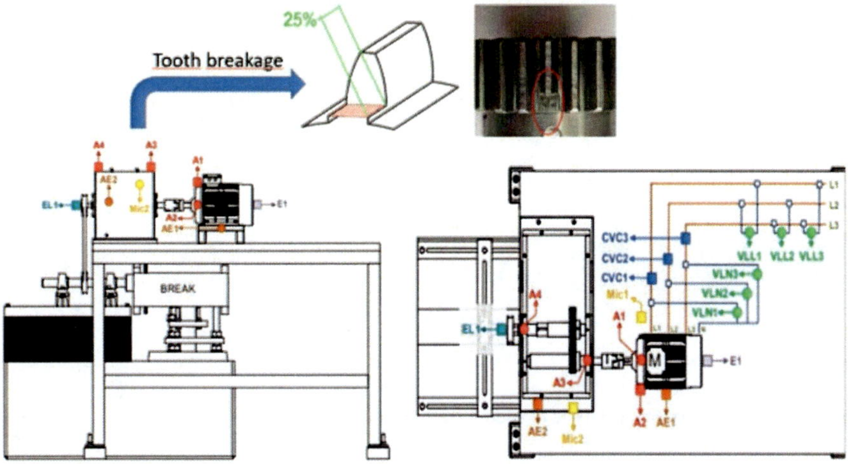


**Fig. 1** Test bed and sensors placement

where $v_w$ is the rotor voltage, and $R_e$ is the equivalent resistance of the rotor inductances. Samples of 10 s length and 10 repetitions under each load configuration were recorded for each of the nine severity levels (classes) of tooth breakage, from P1 for the healthy state to P9 for the 100% of tooth breakage, with 12.5% of tooth loosing each level, from P2 to P8; then, 30 signals for each class is stored in the database, for a total of 270 signals.

The sensitive behavior of the electrical torque faced to the fault severity of broken tooth has been identified in [6], by detecting the change of the amplitude in the frequency response at the gear meshing frequency $f_{GM}$, pinion frequency $f_p$, and gear frequency $f_g$, after applying Fast Fourier Transform on the raw time series of the electrical torque, under different loads $L_1$, $L_2$, $L_3$, as illustrated in Fig. 2.

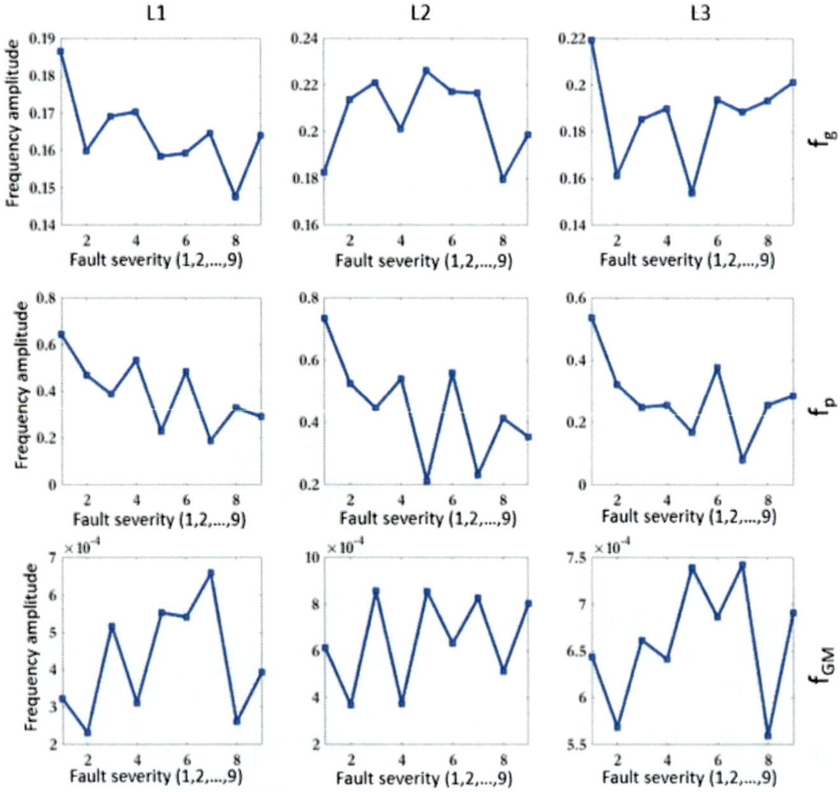


**Fig. 2** Fault severity sensitivity identified from the electrical torque

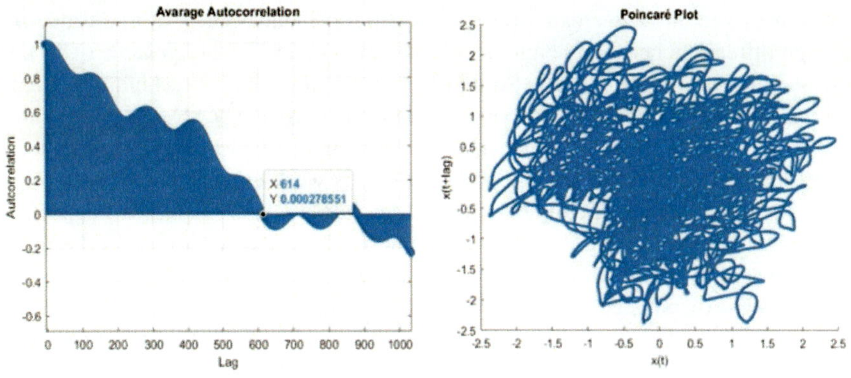


(a) (b)

**Fig. 3** Autocorrelation of $x(t+\tau)$ versus $x(t)$ for the level of fault P1

## 2.2 *Feature Extraction*

PP is a plot to quantify the self-similarity of a time series to discover a chaotic behavior; it has been widely used for analyzing biomedical signals [14]. Being the gearbox a highly nonlinear system, this signal processing tool is useful for feature extraction. Let $x(t)$ be a times series and $x(t+\tau)$ its corresponding lagged representation, by considering the lag $\tau$. The Poincaré plot can be built in simple manner by plotting $x(t+\tau)$ versus $x(t)$ [14]. The lag $\tau$ can be selected as the time for which $x(t)$ and $x(t+\tau)$ has, or is near to, zero-crossing autocorrelation. A signal can have several zero-crossing, but usually the first zero-crossing is selected. For a signal database, the lag selection can be based on the autocorrelation average.

Figure 3a shows a typical plot of autocorrelation of $x(t+\tau)$ versus $x(t)$ for different lags, showing the first zero-crossing occurs at $\tau = 614$, when a electrical torque signal is due to the healthy state of the gearbox tooth (named P1). After that, a PP of the signal is obtained, as in Fig. 3b.

Finally, eleven features, or attributes, are computed from the PP such as the standard deviation (SD1) of the points generating perpendicular lines along $y = x$ the standard deviation (SD2) of the points generating perpendicular lines along $y = -x$, the convex hull, among other features related to the shape of the Poincaré plot (i.e., SD, SDC, SDC1, SDC2, SD1up, SD1down, Cup, Cdown). For more details, see [15].

## 2.3 *Feature Selection*

ReliefF and Chi-2 are two of the supervised feature ranking techniques using in this work [9]. ReliefF algorithm estimates the quality of the attributes regarding the ability to discriminate neighbor instances, by calculating their K-nearest-neighbors.

A weighted vector W is created for each feature to identify the feature producing the high difference between neighbors of different classes. The high value of W the best qualification to the feature. Similarly, Chi-2 is a statistical technique for feature ranking that measure the independence of the occurrence between attributes and classes, if a non-dependence between an attribute and a class is not identified, then the attributed is discarded. On the other hand, this work applies also a clustering-based feature selection that implement a cluster validity assessment. This approach is performed in two phases: (i) the first phase applies ANOVA to retain the most informative individual features, and (ii) the second phase identifies the best group of those selected features generating proper cluster structures of classes in the feature space by using a cluster validity index ($CD_{bw}$) to assess the compactness and separation of clusters [13]. These feature selection techniques are supported by extensive statistical and mathematical formulations. Due to the lack of space, we suggest the reading of the works in [9, 13].

## 3 Main Results

Once the database of current and voltage signals is available, we compute the electrical torque, according to Eq. 1. A data augmentation procedure was implemented to increase the number of samples, by applying a sliding window along to the whole signal, with an overlapping size of 0.1 s and 1 s length; then, each original sample produce 90 new signals. Due to the electrical torque has a DC component, a zero mean transformation was applied to eliminate the offset, and this allowed calculating properly the autocorrelation of the signals. The average lag $\tau = 726$ was selected from the average auto-correlation of all the signals. After computing the PP for each signal, the eleven features were extracted and rescaling over the interval [0, 1]. Thus, we have a rescaled corpus matrix of 24300 samples and 11 features.

### 3.1 *Exploratory Data Analysis over all Features*

Figure 4 shows the boxplot for all the features in the case of P1 and P9. The figure shows that the data distributions in case of P1 is quite different from the case P9. More of the data are over 0.5 for all the features in P1, and bellow of 0.5 for P9. This is a significant behavior for fault diagnosis purposes. Figure 5 shows the data distribution for the features $SD$ and $C_{up}$ in case of all fault severities. We can see that $SD$ is a discriminative feature for the set of severities P1-P2-P6, P1-P4-P5 and P7-P8-P9. Instead, it looks that $C_{up}$ is not a proper discriminative feature. Figure 6 show the t-SNE plots with all the Poincaré features to visualize and compare different levels of faults. In this case, it seems that the P1-P2, P1-P7 and P1-P9 distance metrics in general give a significant separation of clusters, this means the points of each fault are

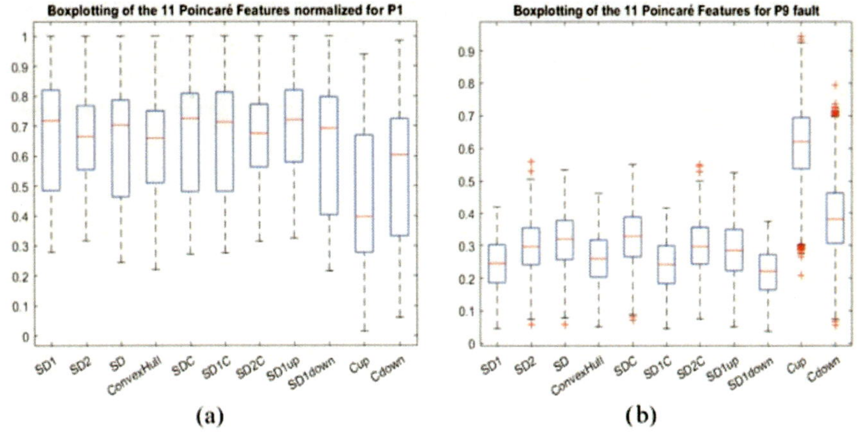


**Fig. 4** Data distribution: **a** P1, **b** P9

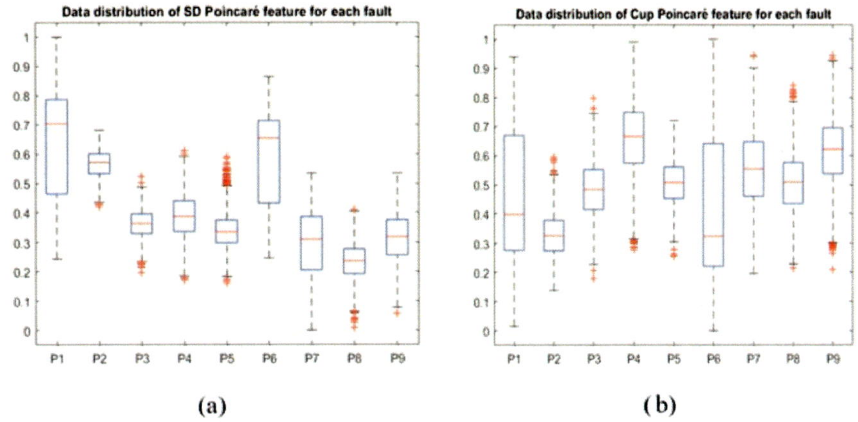


**Fig. 5** Data distribution: **a** SD, **b** Cup

separated. However, in Fig. 6b the distance metric does not give a good separation of points to their respective clusters because they are very close to each other. In general, with t-SNE, we can see how samples are spaced apart and grouped together.

### 3.2 *Exploratory Data Analysis over the Set of Features*

After applying the feature selection by ReliefF and $CD_{bw}$, the best result is obtained with $CD_{bw}$. Chi-2 was not able to rank the features. The feature ranking for ReliefF and $CD_{bw}$ are shown in Fig. 7a and b. The first figure shows a bar chart to compare

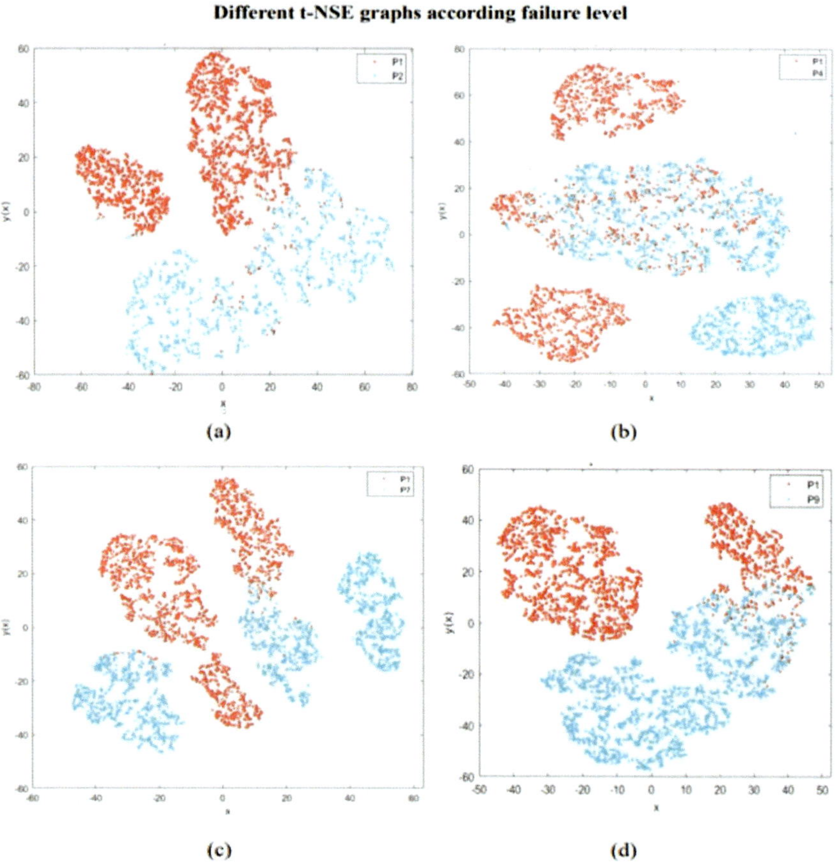


**Fig. 6** t-NSE graphic for four different fault levels with all Poincaré features: **a** t-NSE plot for P1 and P2 fault levels, **b** t-NSE plot for P1 and P4 fault levels, **c** t-NSE plot for P1 and P7 fault levels, **d** t-NSE plot for P1 and P9 fault levels

the relevance of each feature regarding the ReliefF method, where the three most important features are $SD1$, $SD1_{up}$ and $SD1C$. On the other hand, Fig. 7b presents the result for $CD_{bw}$ according to number of used features. The features names are the following with its respective number in parentheses: Dispersion of points SD1 (1), Dispersion of points SD2 (2), Ratio SD (3), Convex-hull area (4), Centered axis Poincaré ratio SDC (5), Dispersion of points SD1C (6), Dispersion of points SD2C (7), Superior asymmetry SD1up (8), Inferior asymmetry SD1down (9), Superior asymmetry ratio Cup (10) and Inferior asymmetry ratio Cdown (11). Therefore, the three most important features selected here shown in Fig. 7b are [4, 5, 11], which means these features are $SDC$, $C_{down}$ and the Convex-hull area, respectively. Lastly,

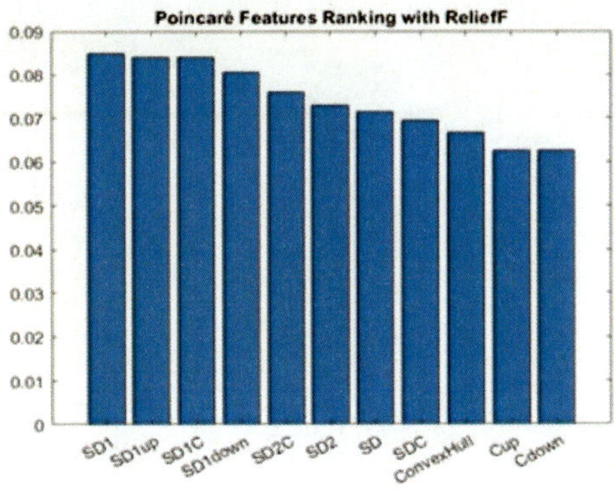


(a)

| *Number of used features* | *Features selected with CDbw* |
|---|---|
| 3 | [5,11,4] |
| 4 | [5,6,11,4] |
| 5 | [5,10,2,8,4] |
| 6 | [3,10,2,7,8,4] |
| 7 | [3,1,10,2,7,8,4] |
| 8 | [3,1,6,10,2,7,8,4] |
| 9 | [5,1,6,11,10,9,2,7,8] |
| 10 | [3,1,6,11,10,9,2,7,8,4] |
| 11 | [3,5,1,6,11,10,9,2,7,8,4] |

(b)

**Fig. 7** Results of feature ranking: **a** Results with ReliefF, **b** Results with $CD_{bw}$

Fig. 8 shows the data distribution for the set of the 3 best attributes selected by $CD_{bw}$. To visualize the distribution regarding P1 (healthy state), the figure shows different sets of faults. We can see that this set of 3 features produces separated clusters among the set of fault severity.

## Conclusions

Electrical torque and Poincaré features arise as an alternative for fault diagnosis. The data distribution of the Poincaré features shows that they can discriminate among different faults. As seen in the Boxplot and t-SNE graphs, there are significant differences between some set of features. The feature ranking methods such as ReliefF and $CD_{bw}$ show different groups of Poincaré features that were selected according to the importance of each one. However, some set of features are more discriminative than others for diagnosis purposes. This ability must be analyzed and quantified in the phase of developing a classification model. The methods of exploratory data analysis were employed because they are practical tools able to provide information about the data distribution, that reveals the discriminative features oriented to fault classification.

The results described in this paper should allow us to understand better the Poincaré features of this specific dataset of the electrical torque, and also to estimate the features that might work the best when using machine learning based classifiers. Future work aims at developing the shallow classifiers with those significant selected features, and it will be possible to determine which are the features that work the best depending on the chosen classifier. It is possible to generate another exploratory data analysis for electrical torque, however; in this work, Poincaré plots, t-SNE, Box-

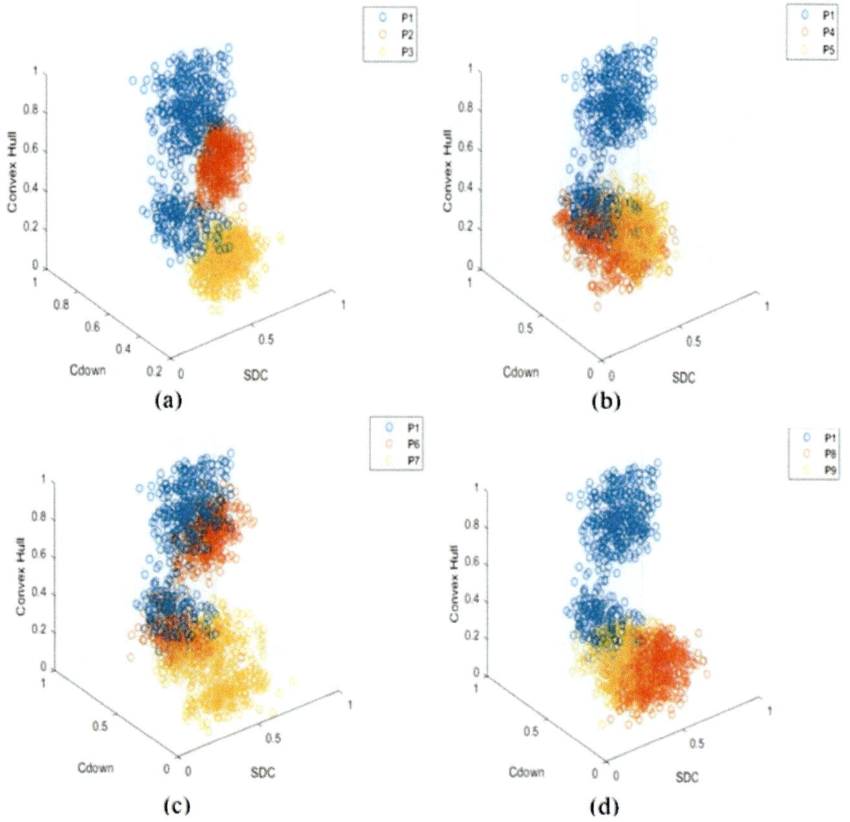


**Fig. 8** Data distribution after feature selection with CDbw: **a** Data distribution with P1-P2-P3, **b** Data distribution with P1-P4-P5, **c** Data distribution with P1-P6-P7, **d** Data distribution with P1-P8-P9

plots, and 3D graphs were applied. There are still more ways to analyze the data and define what are the features that could be selected and used in future works.

Next works will be addressed to analyze the fault data variation under time varying loads. However, an average of the expected time varying load can be assumed as constant, and any change can be considered like a deviation to other average value. Then, it is expected that Poincare Plots can detect those deviations through the change of the plot even under time varying characteristics.

## References


1. Alaswad, S., Xiang, T.: A review on condition-based maintenance optimization models for stochastically deteriorating system. Reliab. Eng. Syst. **157**, 54–63 (2017)
2. Salameh, J.P., Cauet, S., Etien, E., Sakout, A., Rambault, L.: Gearbox condition monitoring in wind turbines: a review. Mech. Syst. Signal Process. **111**, 251–264 (2018)
3. Cerrada, M., Trujillo, L., Hernández, D.E., Correa Zevallos, H.A., Macancela, J.C., Cabrera, D., Vinicio Sánchez, R.: AutoML for feature selection and model tuning applied to fault severity diagnosis in spur gearboxes. Math. Comput. Appl. **27**, 1–25 (2022)
4. Wang, T., Han, Q., Chu, F., Feng, Z.: Vibration based condition monitoring and fault diagnosis of wind turbine planetary gearbox: a review. Mech. Syst. Signal Process. **126**, 662–685 (2019)
5. Vishwakarma, M., Purohit, R., Harshlata, V., Rajput, P.: Vibration analysis & condition monitoring for rotating machines: a review. Mater. Today: Proc. **4**, 2659–2664 (2017)
6. Ortega Lucero, L.R.: Estimación del par de carga en motores de inducción basado en su modelo matemático y orientado a la detección de fallos en cajas de engranajes. Bachelor's thesis. Universidad Politécnica Salesiana. Ecuador (2021)
7. Zhao, R., Yan, R., Chen, Z., Mao, K., Wang, P., Gao, R.X.: Deep learning and its applications to machine health monitoring. Mech. Syst. Signal Process. **115**, 213–237 (2019)
8. Liu, R., Yang, B., Zio, E., Chen, X.: Artificial intelligence for fault diagnosis of rotating machinery: a review. Mech. Syst. Signal Process. **108**, 33–47 (2018)
9. Sánchez, R.V., Lucero, P., Vásquez, R.E., Cerrada, M., Cabrera, D.: A comparative feature analysis for gear pitting level classification by using acoustic emission, vibration and current signals. IFAC-PapersOnLine **52**, 346–352 (2018)
10. Cirrincione, G., Kumar, R.R., Mohammadi, A., Kia, S.H., Barbiero, P., Ferretti, J.: Shallow versus deep neural networks in gear fault diagnosis. IEEE Trans. Energy Convers. **35**, 1338–1347 (2020)
11. Medina, R., Macancela, J.C., Lucero, P., Cabrera, D., Sánchez, R.V., Cerrada, M.: Gear and bearing fault classification under different load and speed by using Poincaré plot features and SVM. J. Intell. Manuf. **35**, 1–25 (2020)
12. Medina, R., Alvarez, X., Jadán, D., Cerrada, M., Sánchez, R.V., Macancela, J.C.: Poincaré plot features from vibration signal for gearbox fault diagnosis. In: IEEE Second Ecuador Technical Chapters Meeting (ETCM), pp. 1–6 (2017)
13. Peña, M., Cerrada, M., Cabrera, D., Sánchez, R.V.: Fast feature selection based on cluster validity index applied on data-driven bearing fault detection. In: 2020 IEEE ANDESCON, pp. 1–6 (2020)
14. Brennan, M., Palaniswami, M., Kamen, P.: Do existing measures of Poincare plot geometry reflect nonlinear features of heart rate variability? IEEE Trans. Biomed. Eng. **48**, 1342–1347 (2001)
15. Peña, M., Cerrada, M., Medina, R., Cabrera, D., Sánchez, R.V.: Poincaré plot features and statistical features from current and vibration signals for fault severity classification of helical gear tooth breaks. J. Comput. Inf. Sci. Eng. **23**, 1–13 (2022)

# A Proposal of Training Data Selection for Remaining Job Cycle Time Prediction in Industrial Monitoring: A Real Case Study

**Alejandro García-Celis, Adriana Villalón-Falcón, Randy Piñero-Aguilar, Fernando Suárez-Concepción, Alberto Prieto-Moreno, and Orestes Llanes-Santiago**

**Abstract** The adoption of Industry 4.0 paradigm by manufacturing companies requires intelligent planning of all company processes, especially production processes. In terms of production efficiency, predicting the remaining job cycle time (RJCT), which is the time it takes to complete the current cycle of operations, is key for planning of operations within industrial processes. Although a wide set of methods have been developed for RJCT prediction, applying these methods to real industrial data is difficult when trying to discriminate between normal and not normal behavior of an operation (start-up, fault) from all the data, due to, for example, the lack of labeling. In this paper, a procedure is proposed to the selection of data according to their level of similarity, which allows grouping data characterizing the process dynamics, in order to use the normal data to train RJCT prediction models. The proposed procedure is applied to a real water purification process in the pharmaceutical industry, and neural networks are used as models to predict the RJCT of process phases. The proposal validation with the case study shown satisfactory results.

---

A. García-Celis · A. Villalón-Falcón (✉) · O. Llanes-Santiago
Universidad Tecnológica de La Habana José Antonio Echeverría, CUJAE, Marianao, La Habana, Cuba
e-mail: avillalon@automatica.cujae.edu.cu

A. García-Celis
e-mail: alejandro.gc@automatica.cujae.edu.cu

O. Llanes-Santiago
e-mail: orestes@tesla.cujae.edu.cu

R. Piñero-Aguilar · F. Suárez-Concepción · A. Prieto-Moreno
EMSI FARMA, La Habana, Cuba
e-mail: rpinero@emsifarma.cu

F. Suárez-Concepción
e-mail: fsuarez@emsifarma.cu

A. Prieto-Moreno
e-mail: aprieto@emsifarma.cu

 
O. Llanes-Santiago (ed.), *Proceedings of 19th Latin American Control Congress (LACC 2022)*, Studies in Systems, Decision and Control 464,
https://doi.org/10.1007/978-3-031-26361-3_9

## Acronyms

| | |
|---|---|
| CNN | Convolutional Neural Network |
| DBSCAN | Density-Based Spatial Clustering of Applications with Noise |
| DTW | Dynamic Time Warping |
| MLP | Multi Layer Perceptron |
| RJCT | Remaining Job Cycle Time |

# 1 Introduction

The adoption of Industry 4.0 paradigm by manufacturing companies enables smart manufacturing, improved supply chain performance, higher organizational efficiency, and business productivity [1]. Achieving these goals requires intelligent planning and execution of all company processes, especially, production processes.

In terms of production efficiency, predicting the remaining job cycle time of operations is key for planning within industrial processes. Remaining job cycle time (RJCT) is defined as the time required to complete an operation, and its prediction is a critical task [2]. It is common to see that the real finishing time of an operation does not match the planned time due to unexpected production disturbances. The prediction of the RJCT is the dynamic response to the changing state of production, anticipating the deviation from production planning and trigger scheduling decisions in the manufacturing system [3, 4].

A literature review shows that RJCT prediction methods generally use information about event logs, instead of measurements from the physical process variables [4–6]. Another approach is to use initial conditions to infer its duration without predicting the RJCT in real time, limiting the ability to react to possible variations in the process conditions during the course of the operation [2, 7]. Deep learning methods have taken prominence in the field of RJCT prediction due to the increasing computational capabilities and availability of historical data. These methods use the measurements of process variables as input [8]. Hybrid approaches based on machine learning and classification focus on creating clusters based on time-series similarity, improving the accuracy and precision of RJCT prediction after the classification process [7, 9, 10].

As shown, a wide set of methods have been developed for RJCT prediction. However, applying these methods to real processes triggers a series of problems, in the authors' experience, many of them related to data preprocessing. Industrial data has all kinds of quality and structures. One of the most difficult jobs when working with industrial data is the discrimination of what is the normal behavior of an operation from what is not normal (start-up, fault), due to, for example, the lack of labeling. In the area of anomaly detection, several methods are based on the principle of labeling the process data associated with the normal operation. Anomalies are detected when the process data cannot be identified as coming from a normal operation. Probabilistic methods assume that areas of higher density are more likely to belong to normal behaviors. The distance-based method assumes that data belonging to a normal operation are gather in a cluster, while the anomaly data are far away from that cluster [11].

Hybrid approaches in RJCT prediction, based on classification and individual training by category, treat the labeling problem in order to reduce the prediction error. In general, clustering techniques find different categories according to the different final products of the productive process. Based on this, a procedure is proposed to discriminate normal data from all data based on cluster analysis, in order to use the normal data to train RJCT prediction models.

The main contributions of this paper are summarized as follows: (1) A sequence selection procedure based on cluster analysis with the distance between sequences is proposed to select similar sequences that exhibit the process normal behavior to train RJCT prediction models. (2) This study applies the proposed procedure and state-of-the-art machine learning techniques to a water purification process in the pharmaceutical industry, where the proposed procedure is validated with real industrial data, offering a practical solution to support decision-making within production processes planning.

The paper is structured as follows. Section 2 presents the proposed procedure for sequence selection. Section 3 presents the methodology for remaining job cycle time prediction. Section 4 presents the case study, and discusses the results obtained from the proposed procedure validation. Section 5 corresponds to the conclusions and future work.

## 2 Proposed Procedure for Data Selection

The proposed procedure for selecting normal operation data from all data is presented in Fig. 1.

**Data normalization:** The Min-Max normalization is used to ensure an equal contribution from each variable to the computation of distances between sequences. Since the variables that have higher absolute values will have higher values of distance between their executions (independently of their level of similarity). The data is returned in a [0, 1] range. The normalization is given by the maximum and minimum values of a variable.

**Fig. 1** Proposed procedure for data selection

**Compute distances:** To obtain a measure of dissimilarity between the different executions that conform the dataset, dynamic time warping (DTW) is used. This method minimizes the effects of displacement and distortion over time in the computation of the distance that separates one time series from another [12]. For each execution, considering the analysis of each variable independently, the average distance, to the rest of the sequences is calculated. This constitutes a mapping, where to each time series, representing the evolution of the execution, is associated with the average distance to the rest of the executions. The duration of each execution is added to the dataset, since it may be representative to discriminate the executions belonging to the start-up or to some failure. Due to the difference between the values of distances and durations, Z-score normalization is used to obtain a representation of the data in a common range, given by the mean and standard deviation of the attribute.

**Clustering:** Density-based spatial clustering of applications with noise (DBSCAN) uses a minimum density level estimate to cluster data. The algorithm is based on a threshold for the number of neighbors, minPts, within the jpgilon radius for a given distance measure. Due to the ability of this algorithm to detect outliers values [13], it is suitable for determining executions that do not represent process dynamics. In addition, the algorithm does not require prior knowledge of the number of clusters [14], so it is able to detect if there are different types of behavior of the process sequences automatically.

A decrease in the jpgilon value decreases the allowed radius threshold for data clustering and therefore increases the similarity between the executions of the same cluster. This reduces the prediction error of the model. However, as the jpgilon value decreases, the number of values detected as noise increases, which can lead to a greater loss of information. For the DBSCAN algorithm, it is desirable to have at least one parameter that allows capturing different knowledge from the data [13]. In this case, jpgilon represents the degree of freedom, which defines the maximum allowed distance for cluster formation. Therefore, it is proposed to select the value of jpgilon starting from an initial value and by gradually increasing this value to retrain the model, and to compare each prediction error, with the permissible threshold specific to the process. The maximum jpgilon value that, after retraining, achieves a prediction error below the acceptable level according to the industrial process, is a good approximation for the construction of the dataset representative of the process.

## 3 Job Cycle Prognostic

The job cycle prognostic constitutes the prediction of the RJCT of an operation when the system is running in normal conditions. The data of the system in healthy operation, are much more numerous than those representing some failure [15], so it is possible to build a model that generalizes the behavior of the operation and predict its remaining cycle time using the information extracted from the process. In order to obtain a RJCT prediction model the methodology presented in Fig. 2 is used.

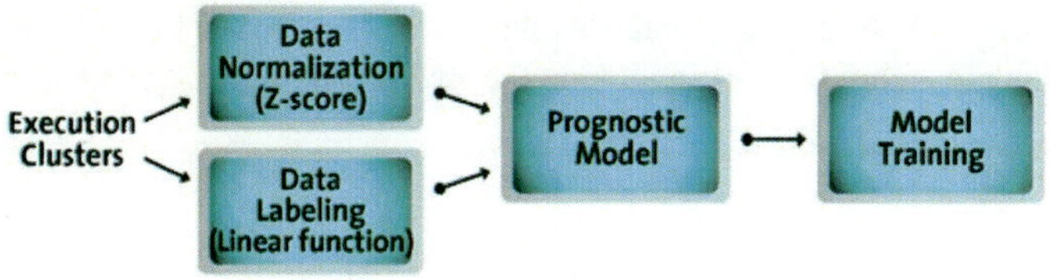


**Fig. 2** Scheme for obtaining RJCT prediction models

**Data labeling:** For RJCT predictions, it is more convenient to use labels representing a linear function, since it is a continuous process. From the beginning of the operation, the process variables present a variation in their parameters that corresponds to the course of the process.

$$L = EOL - C \tag{1}$$

Where $EOL$ is the final time instant of the operation; and $C$ is the current time instant. In label construction, cycles are used as the unit of measurement, converting the time units to discrete values that represent an extension of time.

**Data normalization:** Due to the possible different ranges of values of the measurements used, it is necessary to equate the contribution of each variable to the network, using a common range, in order to accelerate the learning process, avoid outliers and noisy results. Therefore, the data preparation stage ensures more robust results and benefits the learning capability of the algorithm [16].

The Z-score normalization is one of the most popular, it converts the input values to a common range given by the mean and standard deviation of the attribute [17, 18].

**Prognostic model:** Neural networks have numerous applications in the field of time series predictions, and are widely used for their capability to model highly nonlinear, complex, and multidimensional systems without prior knowledge of their physical behavior. Generally, a deep network model, designed for time series prediction, consists of three main parts: (1) an input layer, where sensor data is received; (2) n-hidden layers, which are selected considering the type of features to be extracted from data; and (3) an output layer, which learns the final relationship between the extracted features and the prediction variable (RJCT) [19]. For MLP, in prediction tasks, generally, three-layer architectures are used with: only one hidden layer, and an output layer with one neuron and its linear activation function [18, 20]. Feature extraction in the CNN is performed considering the spatial dependencies on the local regions of data; consists of two sub-structures: (1) multiple convolutional layers; and (2) fully-connected layer for regression. In this paper, a time window is adopted for data preparation to utilize the multivariate time information [21].

**Model training:** Once the executions and the model structure are selected, the model is trained to predict the RJCT using the sensor measurements. For the neural network models selected in this study, mean square error is used as loss function in the optimization algorithm for training, using the RMSProp optimizer with a mini-batch approach.

## 4 Case Study: Water Purification by Reverse Osmosis

In this section, a prognostic model is obtained based on the proposed procedure to predict the RJCT of water purification process phases. Water purification process using reverse osmosis generally executes two main phases: (1) water recirculation, and (2) main production of purified water.

The *recirculation phase* involves the flow of water through the reverse osmosis stages and its subsequent return to the pretreatment tank; a process that is continuously repeated until the water meets the quality requirements. In the *main production phase*, the purified water is generated and released into the distribution line. Knowing when the purified water is going to be available and when it's not, and the accuracy in the prediction of each phase duration has a great impact and contributes to the correct scheduling of production systems. It can also improve the planning of production capacities in real time and avoid wasting raw materials. The objective is to predict the remaining time for the completeness of each phase, using the values of the measurements acquired from the sensors.

The historical data collected from process variables are organized in executions. An execution represents the running of one phase. Each execution is labeled with the phase it belongs to, but not whether it was completed successfully or not. The proposed procedure is used to determine the executions that were executed successfully to train the RJCT prediction models.

After selecting the executions corresponding to the normal behavior of each phase, the models presented in Sect. 3 are trained. Hyper-parameters for each model are tuned by grid search. In experiments, a k-fold cross validation with $k = 10$ is performed to evaluate the generalization capacity of each model [20]. A Wilcoxon statistical test, with a significance level ($\alpha$) of 0.05, is performed to compare the prediction errors in each partition.

### *4.1 Recirculation Phase*

**Data selection:** The clustering analysis creates a single cluster, in addition to the isolated executions. This indicates that there is a single mode of behavior that defines the dynamics of the recirculation phase and that the executions that do not follow this evolution must be part of one of the events outside the process (start-up, failure). In order to tune the data selection, the values 2.0, 2.5 y 3.0 of jpgilon are used. A comparison is performed by training the models with the datasets formed with each value of jpgilon (see Fig. 3). As the level of dissimilarity between the executions

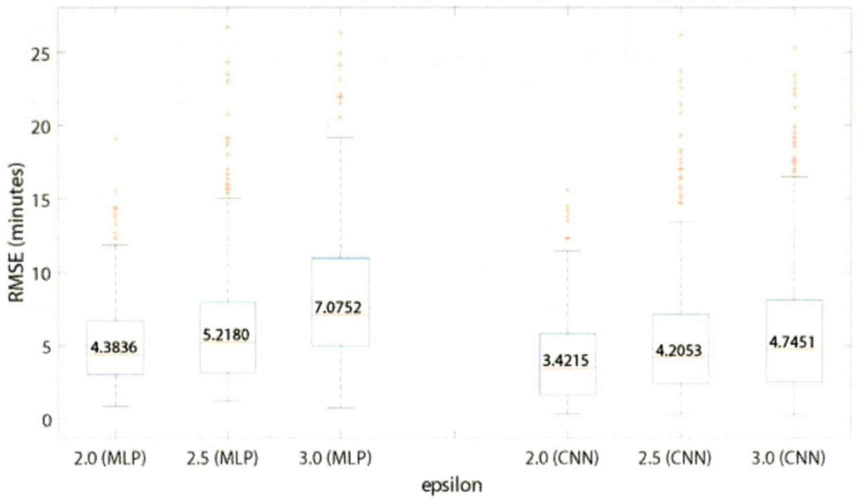


**Fig. 3** Epsilon selection

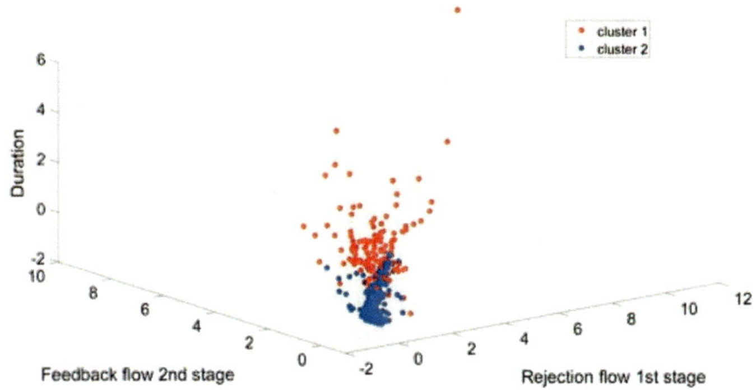


**Fig. 4** Data clustering (Recirculation phase)

increases, the prediction error increases. The results obtained for an jpgilon value of 2.0 show a lower mean and variance of the prediction error, therefore, this value is selected to build a dataset that represents the recirculation phase (see Fig. 4).

**Prognostic model:** The configuration of the MLP model in Sect. 3 is implemented by grid search, tuning the number of neurons in the hidden layer and the learning rate. The analysis shows superior results for combinations of 8 neurons in the hidden layer (sigmoid activation function), a learning rate of 0.01. For the clustered data in the execution's selection analysis, a grid search with the same ranges of values is also performed. The best results are obtained for 10 neurons in the hidden layer and a learning rate of 0.1. For both cases, a total of 50 training iterations and a batch size of 400 are used, due to the high number of samples in the executions.

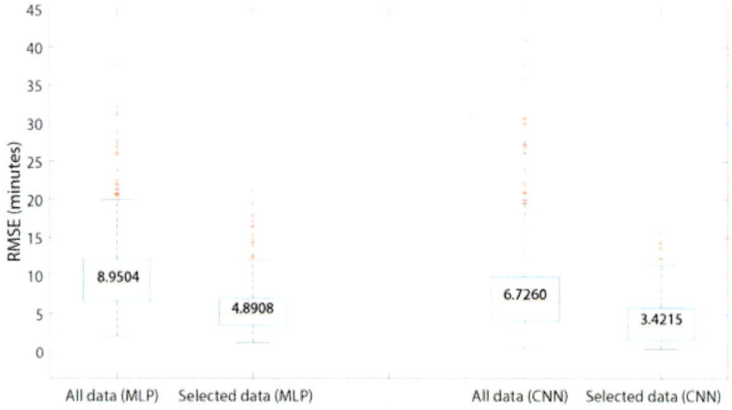


**Fig. 5** RMSE comparison (Recirculation)

For the CNN model described in Sect. 3, a data window with a size of 30 (cycles) is used, determined by the minimum number of samples of the executions. The one-dimensional convolutional layers have a filter with a shape (12 × 1), where the length (12) represents 1 min, a time interval sufficient for the algorithm to detect degradation patterns. A filter in the last convolutional layer is used with a size smaller than that of the preceding layers (6 × 1), where the length represents half a minute (6). The number of filters, epochs and the learning rate are determined through a grid search. For the fully connected layer, 10 units are selected in the hidden layer, while for the output layer a node with a linear activation function is maintained. The batch size is kept at 400. The best results are obtained for a learning rate of 0.01, 5 filters and 30 epochs. With the subset formed after the elimination of executions, the grid search is maintained with the same ranges of values, resulting in the best combination for 5 filters, the initial learning rate of 0.01, and a total of 50 training epochs.

Figure 5 shows the prediction errors for the datasets formed before and after the execution selection analysis, using the proposed neural network architectures. It is observed that the results differ when all executions and the selected executions formed after the analysis are used. As the p-value is much lower than 0.05 (see Table 1) in both cases, it can be affirmed that there are significant differences between both results, resulting superior the one achieved after the execution selection analysis,

**Table 1** Wilcoxon test (Recirculation phase)

| Experiment | MLP all data | MLP selected data | CNN all data | CNN selected data |
|---|---|---|---|---|
| Average | 8.95 | **4.89** | 6.73 | **3.42** |
| Standard deviation | 5.78 | **3.50** | 6.75 | **3.05** |
| Test p-value | 1.30e-43 | | 5.60e-29 | |

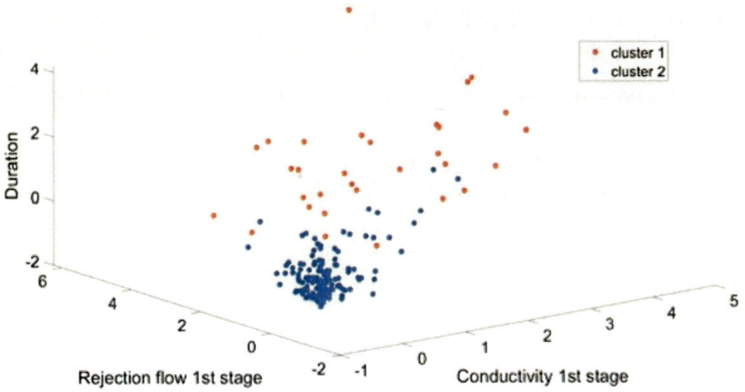


**Fig. 6** Data clustering (Main Production)

given by the decrease of the average error and variability. The difference between these two cases in terms of the average error has a significant impact on the industry, allowing the planning of activities with greater precision and avoiding the waste of prime materials.

## *4.2 Main Production Phase*

As in the previous phase, only one cluster of data and isolated executions are found (see Fig. 6). The same criteria as for the Recirculation phase analysis are used. The prediction error presents a significant difference before and after the execution selection, where a lower error is achieved after the execution selection (see Fig. 7).

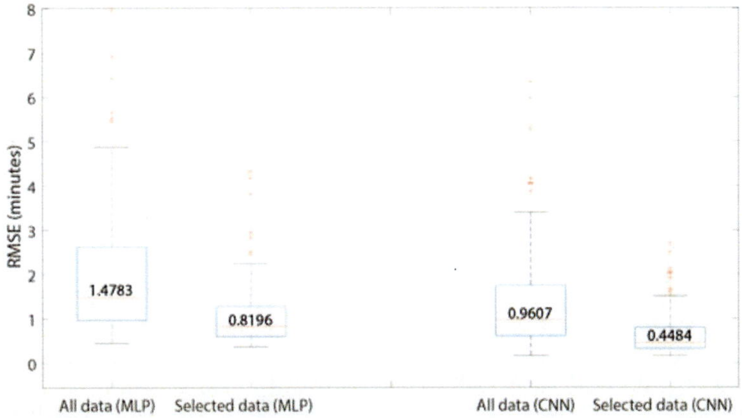


**Fig. 7** RMSE comparison (Main Production)

## 5 Conclusions and Future Work

In this paper, a procedure that allows grouping the data according to the similarity between the executions that represents the behavior of an operation is proposed. This improve the RJCT prediction model reliability and allows the application of scheduling strategies in an effective way. In addition, it helps to identify phase duration deviations in real time, which is indicative of the system health status and allows the detection of anomalies. The proposed procedure is applied to a real water purification process for the prediction of the time to complete the recirculation and main production phases. Satisfactory results are achieved by eliminating executions with a level of dissimilarity above the tolerable that decreases the prediction error in the neural network models used.

Future work will investigate the possibility of focusing execution selection on reducing the error towards the end of the job cycle, since these tasks are considered critical and late predictions are severely penalized.

**Acknowledgements** The authors Adriana Villalón-Falcón, Alberto Prieto-Moreno and Orestes Llanes-Santiago acknowledge the financial support provided by the Project PN223LH004-23, of the National Program of Research and Innovation Automation, Robotics and Artificial Intelligence (ARIA) from the Ministry of Science, Technology and Environment (CITMA) of Cuba.

## References

1. Seif, A., Tan, Z.H.E.: Implementing Industry 4.0, pp. 1–24. Springer, Berlin (2021)
2. Lee, G.M., Gao, X.: Appl. Sci. **11**(16), 7428 (2021)
3. Fang, W., Guo, Y., Liao, W., Ramani, K., Huang, S.: Int. J. Prod. Res. **58**(9), 2751 (2020)
4. Choueiri, A.C., Sato, D.M.V., Scalabrin, E.E., Santos, E.A.P.: J. Manuf. Syst. **56**, 188 (2020)
5. Van Dongen, B.F., Crooy, R.A., van der Aalst, W.M.: OTM Confederated International Conferences "On the Move to Meaningful Internet Systems", pp. 319–336. Springer, Berlin (2008)
6. Van der Aalst, W.M., Schonenberg, M.H., Song, M.: Inf. Syst. **36**(2), 450 (2011)
7. Wang, J., Zhang, J., Wang, X.: IEEE Trans. Semicond. Manuf. **31**(1), 173 (2018)
8. Gelmereanu, C., Morar, L., Bogdan, S.: Procedia Econ. Finance **15**, 1563 (2014)
9. Chen, T., Wang, Y.C.: Robot. Comput.-Integr. Manuf. **26**(1), 83 (2010)
10. Chen, T.: Appl. Soft Comput. **40**, 558 (2016)
11. Zhang, X., Mu, J., Zhang, X., Liu, H., Zong, L., Li, Y.: Patt. Recogn. **121**, 108234 (2022)
12. Senin, P.: Information and Computer Science Department University of Hawaii at Manoa Honolulu, USA, vol. 855, no. 1–23, p. 40 (2008)
13. Schubert, E., Sander, J., Ester, M., Kriegel, H.P., Xu, X.: ACM Trans. Database Syst. (TODS) **42**(3), 1 (2017)
14. Ester, M., Kriegel, H.P., Sander, J., Xu, X., et al.: KDD'96, pp. 226–231. AAAI Press, Palo Alto (1996)
15. Biggio, L., Kastanis, I.: Front. Artif. Intell. **3**, 578613 (2020)
16. Thakkar, U.R.: Remaining useful life prediction of a turbofan engine using deep layer recurrent neural networks. Ph.D. thesis, Carleton University (2021)
17. Ellefsen, A.L., Bjørlykhaug, E., Æsøy, V., Ushakov, S., Zhang, H.: Reliab. Eng. & Syst. Safety **183**, 240 (2019)

18. Alberto-Olivares, M., Gonzalez-Gutierrez, A., Tovar-Arriaga, S., Gorrostieta-Hurtado, E: 2019 16th International Conference on Electrical Engineering, Computing Science and Automatic Control (CCE), pp. 1–6. IEEE (2019)
19. Villalón-Falcón, A., Prieto-Moreno, A., Quiñones-Grueiro, M., Llanes-Santiago, O.: Comput. Appl. Math. **41**(1), 1 (2022)
20. Aggarwal, C.C., et al.: Neural Networks and Deep Learning, vol. 10, p. 978. Springer, Berlin (2018)
21. Li, X., Ding, Q., Sun, J.Q.: Reliab. Eng. & Syst. Safety **172**, 1 (2018)

# Energy-Efficient Wavelet Transform Implementation for Fault Diagnosis

**Iliover Vega-Gonzalez and Jorge Ramírez-Beltrán**

**Abstract** In recent years an increase in the use of Wavelet Transform has been observed, especially in the fields of fault diagnosis in biomedical, power quality and hydraulic applications. Many of the aforementioned systems require a certain degree of autonomy to remain performing an event detection algorithm for periods of several days or even months. Applications with these constraints in power consumption usually are implemented using microcontrollers, however, these platforms are not typically conceived to perform high-complexity processing algorithms like a multi-level Wavelet Transform. In this paper the implementation of a four-level Wavelet Transform using Low-Energy Accelerator (LEA) module is shown; LEA module is present in some of the most recent Texas Instruments' MSP430 family of microcontrollers. In practical tests performed to the system, a reduction in the average current consumption of 31.28% relative to the current consumption of the same system disabling the LEA module is obtained. In addition, a remarkable reduction in the time required to perform the Wavelet Transform is observed, which allows an increase to the sampling frequency for which the practical tests were performed.

## 1 Introduction

In recent years the use of the Wavelet Transform as a tool for fault detection in monitoring systems has increased, mostly due to the fact that it gives the possibility to obtain an optimal time–frequency resolution in all frequency ranges [7].

Some of the fields in which the Wavelet Transform has been most commonly used to detect events are: biomedical applications [2, 5, 6, 17, 20], power quality monitoring in electrical networks [3, 7, 10, 18, 19, 22] and hydraulic networks [12, 13, 15, 25, 26]. Most of these cases require the designed system to be kept operating autonomously for long periods of time without an external power supply,

I. Vega-Gonzalez (✉) · J. Ramírez-Beltrán
Universidad Tecnológica de La Habana José Antonio Echeverría, Cujae, Calle 114, #11901 / Ciclovía y Rotonda, La Habana, Cuba
e-mail: ivegag@cime.cujae.edu.cu



O. Llanes-Santiago (ed.), *Proceedings of 19th Latin American Control Congress (LACC 2022)*, Studies in Systems, Decision and Control 464,
https://doi.org/10.1007/978-3-031-26361-3_10

which imposes strong power constraints. For instance, pressure-based leak detection in water pipes often requires sensors to be located on the periphery of a water distribution system, where no external power source is available [16] and physical access is difficult. In such cases, the reduction in power consumption extends the time required to access the site and change the battery.

The most common low-power data acquisition and processing systems employ microcontrollers, since the use of a desktop or single-board computer implies a current demand above 100 mA, which would not allow an event detection system to remain monitoring for several days. The main disadvantage of using a low-power microcontroller to detect events that require the implementation of a Wavelet Transform is that they are typically not designed to perform such a high level of processing complexity. Therefore, the microcontroller takes longer to execute these algorithms, which limits the sampling rate of the system and increases current consumption by keeping the CPU (Central Processing Unit) active for a longer period. Algorithms that require complex operations such as a convolution or a transform are typically implemented in Digital Signal Processors, either synthesized in a Programmable Logic Device or as commercial hardware [8, 23, 24]. In either case, their high power consumption has a negative impact on autonomy as well as portability, by having to add some extra hardware component to perform the computations, instead of doing all tasks with the controlling unit.

Recently, several studies have been made of microcontroller-based systems that make efficient use of their smart peripherals to implement a Wavelet Transform with low power consumption. An implementation of a Haar Wavelet Transform [1] using integer optimization [4] on a Texas Instruments MSP430F169 microcontroller is shown in [11]. The implementation shown in that work is optimized to minimize the number of multiplications so that the CPU executes the Haar Wavelet Transform faster. However, the fact of using all the time the CPU to implement the Haar Wavelet Transform is inconvenient because, in microcontrollers, the CPU is the main source of current consumption and the energy saving strategies are aimed at reducing the CPU's time in active mode. In [17] a MSP430FR5969 microcontroller is used to implement a four-level Wavelet Transform decomposition. This work takes advantage of the hardware resources that the MSP430FR59xx series microcontrollers have to perform digital signal processing operations, such as hardware multipliers and three Direct Memory Access (DMA) channels. It also benefits from the reduction in current consumption that implies the use of a microcontroller with FRAM (Ferroelectric Random Access Memory) as non-volatile storage memory. However, the time used to implement the Wavelet Transform, the signal's compression and its storage (0.920 s in total) represents a major limitation in applications where higher sampling frequencies are required, as it is the case for burst detection in hydraulic networks (200–400 Hz [21, 26, 28]).

Currently some families of Texas Instruments microcontrollers include a Low-Energy Accelerator (LEA) module [14], which is a low-power co-processor that is capable of performing operations without any CPU intervention and triggering an interrupt when the operation is completed. This paper shows the implementation of a four-level Wavelet Transform and compares its performance and the reduction in power consumption with the implementation using only hardware multipliers and DMA controller to perform the operations. The aim of the work is to verify the feasibility of implementing Wavelet-based fault diagnosis algorithms that must fit energy and speed constraints.

## 2 Wavelet Transform in Microcontrollers

This section describes the main aspects regarding the implementation of the Wavelet Transform in the selected microcontroller. First, the chosen Wavelet Transform algorithm is shown, followed by an explanation of the LEA module incorporated in some of the Texas Instruments' MSP430 microcontrollers to execute high-complexity digital signal processing.

### *2.1 Discrete Wavelet Transform*

Wavelet Transform analysis uses functions called Wavelets that allow transforming the signal of interest into a representation that provides the signal information in a more useful way [1]. By means of the Discrete Wavelet Transform (DWT), a signal can be represented by approximation coefficients and detail coefficients, with knowledge of the Wavelet function from which these coefficients were obtained (e.g. Haar, Daubechie, Coifflet, etc.). The approximation and detail coefficients can be obtained by Eqs. (1) and (2) respectively [1]:

$$S_{m,n} = \int_{-\infty}^{\infty} x(t)\phi_{m,n}(t)dt \tag{1}$$

$$T_{m,n} = \int_{-\infty}^{\infty} x(t)\psi_{m,n}(t)dt \tag{2}$$

where $x(t)$ is the input signal, $S_{m,n}$ and $T_{m,n}$ represent the approximation and detail coefficients respectively; m and n are integers controlling the Wavelet function's dilation and translation respectively; $\psi_{m,n}(t)$ represents the Wavelet function for an m, n pair and $\phi_{m,n}(t)$ represents the scaling function associated with a Wavelet for an m, n pair.

From Eqs. (1) and (2) it follows that the approximation and detail coefficients can be determined by Eqs. (3) and (4) respectively [1].

$$S_{m+1,n} = \frac{1}{\sqrt{2}} \sum_{k} c_k S_{m,2n+k} \tag{3}$$

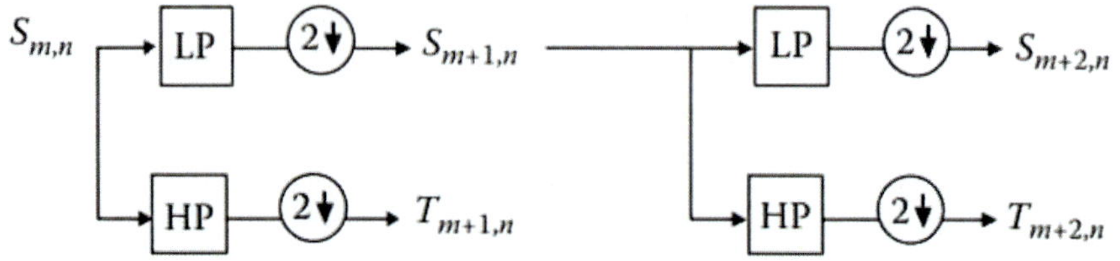


**Fig. 1** Filtering scheme to obtain the approximation and detail coefficients of the Wavelet Transform through successive stages [1]

$$T_{m+1,n} = \frac{1}{\sqrt{2}} \sum_k b_k S_{m,2n+k} \tag{4}$$

where $c_k$ and $b_k$ are scale coefficients.

In Eqs. (3) and (4) it is observed that the approximation and detail coefficients of each level of decomposition can be obtained from the coefficients of the preceding level, this is the reason why this method is known as the Fast Wavelet Transform. It is observed that Eqs. (3) and (4) can be interpreted as filters with impulse response $c_k/\sqrt{2}$ and $b_k/\sqrt{2}$. In this case $c_k/\sqrt{2}$ represents a low-pass filter and $b_k/\sqrt{2}$ represents a high-pass filter. Following this analysis, the Wavelet decomposition for each level can be performed following the scheme shown in Fig. 1 [1], in that figure the circle with the 2 $\downarrow$ inside indicates the operation of taking every two values of the filtered signal, LP and HP represent the low-pass and high-pass filters respectively.

## *2.2 Low-Energy Accelerator Module*

The LEA module is a 32-bit hardware accelerator specific to the latest MSP430 microcontroller families with FRAM. This accelerator can implement signal processing, matrix multiplication and other operations such as FIR, Infinite Impulse Response and Fast Fourier Transform filters, which usually takes a large amount of time and energy to execute in the main program. The LEA module in the MSP430FR599X family of microcontrollers has 4 KB of Random Access Memory shared with the CPU so that the results of operations performed in the LEA can be used in the main process when the LEA indicates that it has finished an operation. This allows the CPU to remain in a low-power mode during the time the LEA module is processing the input signal.

Figure 2 shows the block diagram of the LEA system. Inside it has a core that is independent of the microcontroller's CPU. This core executes the functions configured in its internal registers and exchanges information with the CPU through its shared RAM. It can also interact directly with the peripherals of the microcontroller,

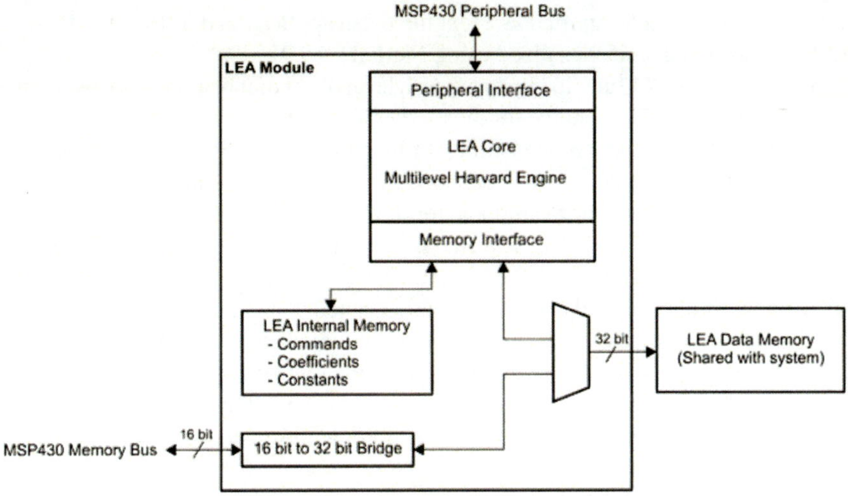


**Fig. 2** Block diagram of LEA system [9]

such as the DMA controller. Once an operation is finished, the DMA controller can move a data buffer without waking up the CPU from a low power consumption mode.

The decomposition scheme in Fig. 1 shows that to obtain each level of decomposition from the original signal, two Finite Impulse Response (FIR) filtering operations and two downsampling operations are required. Both operations are supported by the Digital Signal Processing Library for MSP430 Microcontrollers[1], which is a library optimized by the manufacturers to work with the LEA module.

## 3 Comparison of Power Consumption Between Algorithm Without LEA and with LEA

To evaluate the performance of the designed system, the case study that was chosen is the detection of leaks in metallic pipes using accelerometers. This application was chosen since it is a very useful tool for fault diagnosis in hydraulic networks and it requires a sampling frequency around 3000 Hz, which is 1500 times higher than the one implemented in [17], thus allowing to make a comparison in more extreme conditions.

The Wavelet Transform is executed on an MSP430FR5994 microcontroller and current is measured with an UT81B scopemeter[2], which is able to exchange data with a PC through an optically isolated USB transfer cable. Incoming data from the

[1] http://www.ti.com/tool/msp-dsplib.

[2] https://www.epanorama.net/newepa/2011/05/04/ut-81b-multimeteroscilloscope/.

scopemeter is read and exported as a text file by an application developed in Python. All the measured data is visualized using MatLab.

A Wavelet Coiflet 1 function up to a fourth level of decomposition was used for the Wavelet Transform. This choice for the mother Wavelet follows the results obtained in [25, 26], where authors demonstrate, through the use of Lipschitz coefficient, its superiority over other Wavelets to enhance the singularities generated by bursts in water pipes. Choosing four decomposition levels allows a direct comparison with the work presented in [17]. The Wavelet Transform was implemented in the microcontroller in two ways: (i) using the LEA module and (ii) using the microcontroller's hardware multiplier with the LEA module disabled. The algorithm of the second case was also designed so that the hardware multiplier interacts with a channel of the microcontroller's DMA controller and with the low power modes.

The microcontroller's firmware was designed so that in the main loop the system is acquiring 256 samples while remaining in a low power mode. Once a buffer is filled with 256 samples, the system exits the low power mode and proceeds to obtain the approximation and detail coefficients for up to a fourth level of decomposition. Figures 3 and 4 show the results of the measurements using the LEA module and disabling it. In both cases it is observed how between time instants $t_1$ and $t_2$, the consumption regime varies. This increase in consumption is due to the fact that it is when the microcontroller exits the low current consumption mode and the Wavelet Transform is executed.

To determine the system's average current it should be taken into account that the current waveform is periodic with a period of 256/3000 s (the Wavelet Transform is executed when 256 samples have been acquired at a sampling frequency of 3000 Hz), therefore the average current should be determined in an interval of 85.33 ms. In both cases, with LEA module and without LEA module, it is concluded that the average current consumption values when the system is in the low power consumption mode are very similar, therefore, the main difference lies in the determination of the

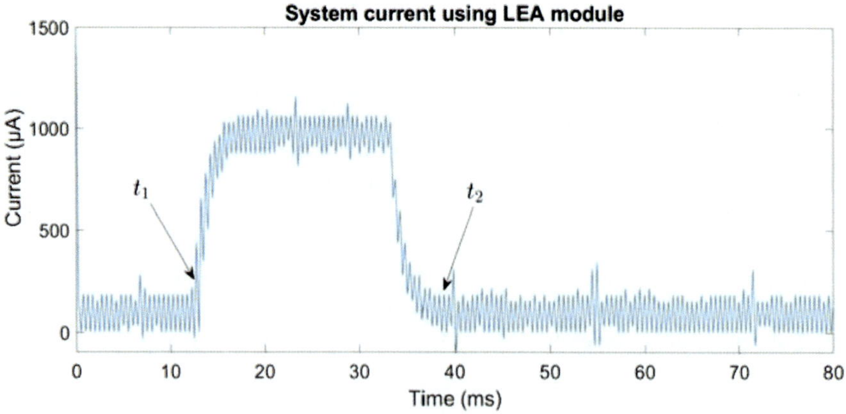


**Fig. 3** System current implementing a Wavelet Transform using LEA module

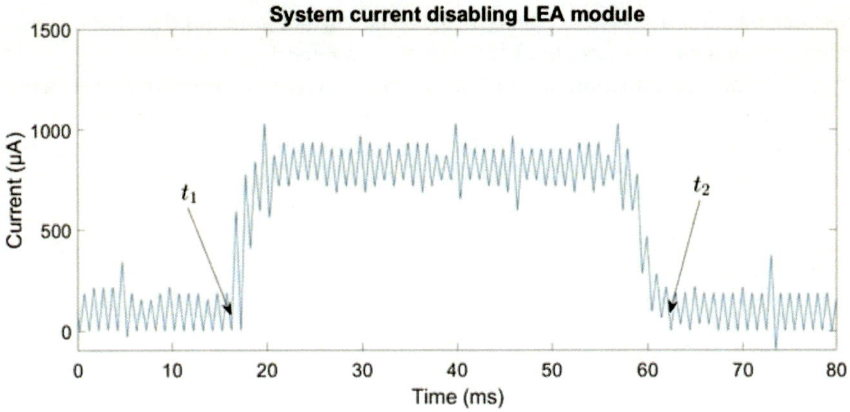


**Fig. 4** System current implementing a Wavelet Transform with the LEA module disabled

approximation and detail coefficients. To compare the power consumption of such operation, in both cases the current–time product must be determined, since the system consumption is not only influenced by the average current but also by the time spent in executing the operation [27]. Taking this into account, the average current in a period of acquisition and processing of the input signal is determined by (5).

$$I_{avg} = \frac{\Gamma_W + \Gamma_{LP}}{\frac{256}{f_s}} \tag{5}$$

where $I_{avg}$ is the average system current for one acquisition cycle, $\Gamma_W$ is the current–time product of the Wavelet Transform implementation, $\Gamma_{LP}$ is the current–time product of the system in the low-power mode, and $f_s$ is the system sampling frequency.

Substituting the measurement results into (5), Table 1 can be filled. It is observed in Table 1 that by implementing the Wavelet Transform with the LEA module, the execution time was significantly reduced, which would allow the system to reach even higher sampling frequencies. In this case the average current during the time of the Wavelet Transform calculation was higher when the LEA module was used,

**Table 1** Average system current with LEA module and without LEA module

| Condition | $t_2 - t_1$ (ms) | $\Gamma_W$ (μAs) | $\Gamma_{LP}$ (μAs) | $I_{avg}$ (μA) | %[a] |
|---|---|---|---|---|---|
| With LEA | 27.0846 | 18.0227 | 8 | 304.9535 | 68.18 |
| Without LEA | 46.1442 | 29.8098 | 8.3571 | 447.2689 | 100 |

[a] The '%' column refers to the percentage that the corresponding $I_{avg}$ value represents with respect to the maximum value of both cases

however the time $t_2 - t_1$ was shorter, thus, the current–time product had a lower value than in the case where the LEA module is not used.

In addition, the reduction in the time required to run the Wavelet Transform using LEA module resulted in a 31.82% decrease in average current consumption, which allows the system to remain autonomous for a considerably longer period of time than would be the case if no LEA module was used to implement the Wavelet Transform.

For the performed tests, measurements were made by analyzing the functional units, as suggested in [27], since it allows a direct comparison between both methods for implementing the Wavelet Transform. However, the main benefits of implementing a Wavelet Transform using a LEA module can only be obtained when the rest of the system's units are properly optimized for low power consumption, otherwise, a reduction in current consumption in the order of 140 μA, might not be significant. The same effect occurs with sampling frequency, for low sampling frequencies, the reduction is not as significant as for values in the order of 3 kHz, as it was shown in the case study.

## 4 Conclusions

The use of LEA modules incorporated in the latest Texas Instruments microcontrollers is highly recommended by manufacturers to perform digital signal processing functions. The implementation of a Wavelet Transform using the LEA module embedded in the MSP430FR5994 microcontroller resulted in a 31.82% reduction in the average current consumption compared to the implementation of the Wavelet Transform using a hardware multiplier. The use of the LEA module also allows to increase the system's sampling rate by reducing the time it takes for the system to find the approximation and detail coefficients for the specified decomposition level.

The main benefits of the presented work are obtained in systems entirely designed for low power operation and sampling frequency in the order of 3 kHz.

The obtained results show the feasibility of implementing Wavelet-based fault diagnosis algorithms for systems where a large autonomy and high sampling rate values are desired, as it is the case for burst detection in water pipes and leak detection in metallic pipes using accelerometers.

## References

1. Addison, P.S.: The Illustrated Wavelet Transform Handbook: Introductory Theory and Applications in Science, Engineering. Medicine and Finance. CRC Press, Boca Raton (2017). https://doi.org/10.1201/9781315372556
2. Bali, J.S., Nandi, A.V.: ECG signal based power aware system for obstructive sleep apnea detection. In: 2017 International Conference on Recent Trends in Electrical, Electronics and Computing Technologies (ICRTEECT), pp. 59–63 (2017). https://doi.org/10.1109/ICRTEECT.2017.43

3. Beniwal, R.K., Aggarwal, A., Saini, R., Saini, S.: Detection of anomalies in the quality of electricity supply. Int. J. Future Revolut. Comput. Sci. & Commun. Eng. **4**(2), 6–10 (2018)
4. Calderbank, A., Daubechies, I., Sweldens, W., Yeo, B.L.: Wavelet transforms that map integers to integers. Appl. Comput. Harm. Anal. **5**(3), 332–369 (1998). https://doi.org/10.1006/acha.1997.0238
5. Cervigón, R.: Biomedical applications of the discrete wavelet transform. In: Discrete Wavelet Transforms-Biomedical Applications. InTech (2011)
6. Deepu, C.J., Heng, C.H., Lian, Y.: A hybrid data compression scheme for power reduction in wireless sensors for IoT. IEEE Trans. Biomed. Circuits Syst. **11**, 245–254 (2017). https://doi.org/10.1109/tbcas.2016.2591923
7. Gursoy, M.I., Yilmaz, A.S., Ustun, S.V.: A practical real-time Power Quality event monitoring applications using discrete wavelet transform and artificial neural network. J. Eng. Sci. Technol. **13**(6), 1764–1781 (2018)
8. Gutiérrez-Gnecchi, J.A., Morfin-Magana, R., Lorias-Espinoza, D., del Carmen Tellez-Anguiano, A., Reyes-Archundia, E., Méndez-Patiño, A., Castañeda-Miranda, R.: DSP-based arrhythmia classification using wavelet transform and probabilistic neural network. Biomed. Signal Proc. Control **32**, 44–56 (2017). https://doi.org/10.1016/j.bspc.2016.10.005
9. Instruments, T.: MSP430FR58xx, MSP430FR59xx, and MSP430FR6xx Family User's Guide (2017)
10. Kapisch, E.B., Silva, L.R.M., Martins, C.H.N., Barbosa, A.S., Filho, L.M.A., Duque, C.A., Tavil, A.E., de Souza, L.A.R.: An implementation of a Power System Smart Waveform Recorder using FPGA and ARM cores. Measurement **90**(Supplement C), 372–381 (2016). https://doi.org/10.1016/j.measurement.2016.04.062
11. Knezevic, S., Stojanovic, R.: MSP430 implementation of wavelet transform for purposes of physiological signals processing. In: Education and Research Conference (EDERC), 2012 5th European DSP, pp. 119–123. IEEE (2012)
12. Kumar, A., Kumar, R.: Time-frequency analysis and support vector machine in automatic detection of defect from vibration signal of centrifugal pump. Measurement **108**, 119–133 (2017). https://doi.org/10.1016/j.measurement.2017.04.041
13. Kumar, D., Tu, D., Zhu, N., Hou, D., Zhang, H.: In-line acoustic device inspection of leakage in water distribution pipes based on wavelet and neural network. J. Sens. **2017**, 5789510 (2017). https://doi.org/10.1155/2017/5789510
14. Lingam, S., Lee, S.j., Zipperer, J., Goel, M.: Low energy accelerator processor architecture (2018)
15. Ng, K.S., Chen, P., Tseng, Y.: A design of automatic water leak detection device. In: 2017 IEEE 2nd International Conference on Opto-Electronic Information Processing (ICOIP), pp. 70–73. IEEE, Singapore (2017). https://doi.org/10.1109/OPTIP.2017.8030701
16. Pointl, M., Fuchs-Hanusch, D.: Assessing the potential of LPWAN communication technologies for near real-time leak detection in water distribution systems. Sensors **21**(1), 293 (2021). https://doi.org/10.3390/s21010293
17. Pope, G., Mishra, V., Lewia, S., Lowens, B., Kotz, D., Lord, S., Halter, R.: An ultra-low resource wearable EDA sensor using wavelet compression. In: 2018 IEEE 15th International Conference on Wearable and Implantable Body Sensor Networks (BSN), pp. 193–196 (2018). https://doi.org/10.1109/BSN.2018.8329691
18. Saha, R., Bera, J.N., Sarkar, G.: An alternate approach for power quality computation using sample shifting technique towards load characterization. Measurement **129**, 642–652 (2018). https://doi.org/10.1016/j.measurement.2018.07.037
19. Silva, L.R.M., Kapisch, E.B., Martins, C.H.N., Filho, L.M.A., Cerqueira, A.S., Duque, C.A., Ribeiro, P.F.: Gapless power-quality disturbance recorder. IEEE Trans. Power Deliv. **32**(2), 862–871 (2017). https://doi.org/10.1109/TPWRD.2016.2557280
20. Singh, S.P., Urooj, S.: Wavelets: biomedical applications. Int. J. Biomed. Eng. Technol. **19**(1), 1–25 (2015). https://doi.org/10.1504/IJBET.2015.071405
21. Srirangarajan, S., Allen, M., Preis, A., Iqbal, M., Lim, H.B., Whittle, A.J.: Wavelet-based burst event detection and localization in water distribution systems. J. Signal Proc. Syst. **72**(1), 1–16 (2013). https://doi.org/10.1007/s11265-012-0690-6

22. Srividhya, J.P., Sivakumar, D.: Monitoring and analysis of PQ parameters after first order derivative function to suppress the background instability in harmonic signal. J. Comput. Theoret. Nanosci. **14**(7), 3220–3226 (2017). https://doi.org/10.1166/jctn.2017.6619
23. Thirumala, K., Umarikar, A.C., Jain, T.: DSP implementation of the empirical wavelet transform for measurement of power quality indices. In: 2017 7th International Conference on Power Systems (ICPS), pp. 665–670 (2017). https://doi.org/10.1109/ICPES.2017.8387375
24. Tripathi, S., Mishra, B.: Three stage 2-D discrete wavelet transform using modified vedic multiplier. In: 2017 7th International Conference on Communication Systems and Network Technologies (CSNT), pp. 188–192 (2017). https://doi.org/10.1109/CSNT.2017.8418535
25. Trutié-Carrero, E., Cabrera-Hernández, Y., Hernández-González, A., Ramírez-Beltrán, J.: Automatic detection of burst in water distribution systems by Lipschitz exponent and Wavelet correlation criterion. Measurement **151**(107195) (2020). https://doi.org/10.1016/j.measurement.2019.107195
26. Trutié-Carrero, E., Valdés-Santiago, D., León-Mecías, Á., Ramírez-Beltrán, J.: Detección y Localización de Ruptura Súbita mediante Transformada Wavelet Discreta y Correlación Cruzada. Revista Iberoamericana de Automática e Informática industrial **15**(2), 211–216 (2018). https://doi.org/10.4995/riai.2017.8738
27. Vega-Gonzalez, I., Ramírez-Beltrán, J.: Function-oriented energy model for low power sensor nodes analysis. J. Appl. Res. Technol. **20**(2), 142–159 (2022). https://doi.org/10.22201/icat.24486736e.2022.20.2.1124
28. Zan, T.T.T., Lim, H.B., Wong, K.J., Whittle, A.J., Lee, B.S.: Event detection and localization in urban water distribution network. IEEE Sens. J. **14**(12), 4134–4142 (2014). https://doi.org/10.1109/JSEN.2014.2358842

# Poincaré Images Extracted from Vibration Signals are Useful Features for Fault Classification in a Reciprocating Compressor

**Ruben Medina, Mariela Cerrada, Diego Cabrera, and René Vinicio Sánchez**

**Abstract** Fault classification is an essential tool for Prognostics and Health Management (PHM), whose goal is the assessment of the health condition of a machine and its components. In this research, we propose using Poincaré Images (PI) as features for fault classification using Convolutional Neural Networks (CNN). The approach is validated for classifying 17 valve faults in a reciprocating compressor (RC). The proposed approach attained an averaged classification accuracy of 94.97% which represents an improvement of about 15% concerning using discrete features extracted from the Poincaré plot for classifying faults with Random Forest models. Additionally, results show that the classification performance with PI is comparable to using spectrograms as alternative input images to CNN-based fault classifiers.

## 1 Introduction

The safety and reliability of rotating machinery are critical issues of modern industrial systems, where fault detection plays a fundamental role [14]. Within this context, PHM is applied for detecting any anomaly in industrial components and predicting when a fault will occur. This prediction enables efficient maintenance schedules of maintenance [2]. In addition to anomaly detection, fault classification is important

R. Medina
Facultad de Ingeniería, CIByTEL, Universidad de Los Andes, Mérida 5101, Venezuela
e-mail: ruben.djmedina@ieee.org

M. Cerrada (✉) · D. Cabrera · R. V. Sánchez
GIDTEC, Universidad Politécnica Salesiana, Cuenca 010105, Azuay, Ecuador
e-mail: mcerrada@ups.edu.ec

D. Cabrera
e-mail: dcabrera@ups.edu.ec

R. V. Sánchez
e-mail: rsanchezl@ups.edu.ec



O. Llanes-Santiago (ed.), *Proceedings of 19th Latin American Control Congress (LACC 2022)*, Studies in Systems, Decision and Control 464,
https://doi.org/10.1007/978-3-031-26361-3_11

because it identifies the components related to the type of fault and the possible actions required for maintenance.

A positive displacement compressor, also known as an RC, is a rotating machine where pistons drive a crankshaft to transport gases at high-pressure [17]. RPs are used in the chemical industry, in refineries, and for gas transportation and storage, where reliable operation is essential. The unscheduled shutdowns must be reduced or even eliminated [20]. An important tool for fault detection and classification is monitoring different information about the machine, such as vibration, acoustic emission, or sound, using the appropriate sensors. The information collected is represented by time series that are analyzed for fault detection and classification [1, 4, 5]. Fault classification based on vibration time series has several stages. The time series are pre-processed, and useful features are extracted to feed machine learning models for fault classification. Several features have been used, including time-domain, frequency-domain, and time-frequency representation features [9]. Several classical machine learning models, such as Random Forests (RF) and Support Vector Machines (SVM), have been used for fault classification in rotating machinery. However, improvements in accuracy classification have been obtained by using deep learning models such as Long Short Term Memory (LSTM) neural networks [4] or convolutional neural networks (CNN) [18].

In previous investigations, discrete Poincaré features have been used for fault classification in mechanical rotating machines [16, 19]. Some features were proposed by the work of Brennan et al. [3] and Piskorski and Guzik [21]. However, in this research [16], a generalization of the Poincaré plot was considered where the lag between samples took an arbitrary value; this had the advantage of considering a time delay embedding system useful for modeling chaotic time series [23]. Additionally, a new feature corresponding to the area of the convex-hull of the bi-dimensional Poincaré plot was proposed.

In this research, we propose converting the Poincaré Plot of vibration signals into 2D images that are used as features for classifying faults in an RC using convolutional neural networks. Although the application of PI has been previously reported for the detection of atrial fibrillation and atrial and ventricular beats by using electrocardiogram signals [7, 8], to our knowledge, this is the first time that this approach is applied in the context of fault classification in rotating machines.

## 2 The Poincaré Plot

Non-linear dynamics of systems exhibiting chaotic behavior can be assessed using Poincaré maps [3]. Such maps can be implemented based on the orbits or trajectories in the phase space, which are the solution of a set of differential equations. Unfortunately, the set of differential equations is unknown in practical applications, and the Poincaré plot is used instead of the Poincaré map. The typical situation is a complex system where some measurements are performed using appropriate sensors. Specifically, if the system is mechanical, then accelerometers or acoustic emission

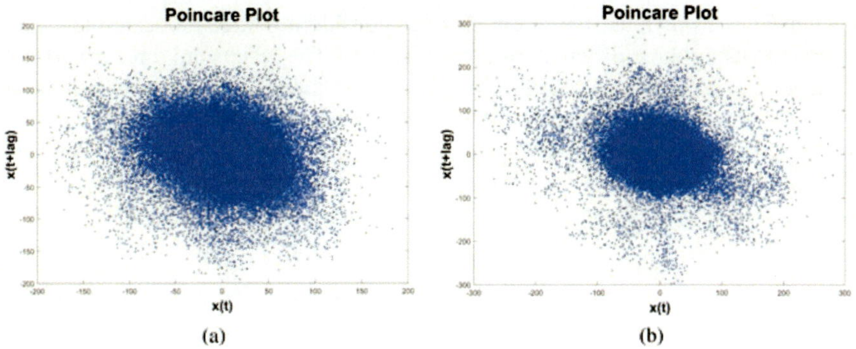


**Fig. 1** Poincaré plot for vibration signals from the RP dataset. **a** Poincaré plot for a vibration signal from the healthy class **b** Poincaré plot for a vibration signal from a faulty class

sensors would be appropriate for fault detection. In electrical motors, the appropriate sensors could be current or voltage sensors. An example of such measurements is the vibrations in mechanical systems using accelerometers. If we denote as $x(t)$ the time-series representing the vibration signal, then the Poincaré plot can simply be obtained by plotting the lagged version of the time series denoted $x(t+\tau)$ as a function of $x(t)$ [3]. Several features useful for fault classification in rotating machinery can be extracted from the Poincaré plot as was proposed in [16, 19]. An example of Poincaré plots is shown in Fig. 1 where the plots represent a vibration signal from the healthy and faulty classes, respectively, for the RC dataset.

## 3 Poincaré Images for Fault Classification Using CNN

### *3.1 Poincaré Images*

A rationale for the proposed approach is that the Poincaré plot for a uni-dimensional time-series is a 2D representation where the plot is located in a square with a side length equal to the amplitude of the signal. This fact suggests converting this representation into images that can be used as features for fault classification using convolutional neural networks. The proposed approach can be implemented using the following steps:

- (Step 1) A bi-dimensional pixel grid with the image resolution is overlaid on the square containing the Poincaré plot.
- (Step 2) In each pixel cell, the pixel value is calculated as the number of points of the Poincaré plot located within the cell space.

- (Step 3) The previous steps are performed considering three Poincaré plots constructed using three different lags: $\tau_1$, $\tau_2$ and $\tau_3$. The results are three gray-scale images.
- (Step 4) A re-scaling to enhance the images' lower pixel values is performed.
- (Step 5) The three gray-scale images are assigned to the $R$, $G$ and $B$ channels of an RGB color image.
- (Step 6) The color images can be used as features to perform the classification using a CNN.

Selection of the lag parameters ($\tau_1$, $\tau_2$ and $\tau_3$) was performed using the guidelines reported in [16].

### *3.2 Fault Classification Using Convolutional Neural Networks*

The application of CNN for the classification of faults in RCs has been reported in [24], where four-valve conditions were classified using one and multiple vibration signals. The accuracy attained with high SNR vibration signals was 100%. Research reported in [25] fed several vibration signals channels to a 2D CNN for classifying four conditions in an RC, attaining accuracies of 100%. Research reported in [10] considers the differential pressure and temperature between the compressor stages as input to a CNN for classifying seven conditions attaining accuracies of 100%. Although the accuracy attained by these investigations is 100%, only a small set of conditions is considered for classification. Additionally, some of these approaches require a complex architecture for the CNN. In contrast, we propose an approach that should be able to classify a large number of different valve fault conditions with high accuracy and using a simple architecture. We propose the CNN architecture presented in Fig. 2. The such architecture comprises three stages: a 2D convolution layer, a batch normalization layer, a rectifying unit layer, and a max-pooling layer. The last stage comprises a 2D convolution layer, a batch normalization layer, a rectifying layer unit, a dropout layer with probability 0.4, a full-connected layer for 17 classes, a softmax layer, and the final classification layer. The filter size, number of filters, and

**Fig. 2** Architecture of the convolutional neural network for fault classification using as features the PI

dropout probability were set experimentally to maximize the classification accuracy. Similarly, the rest of the hyper-parameters were also selected experimentally by varying their values around the default values set in Matlab.

## 4 Reciprocating Compressor Test Bed

The experimental test bed is composed of the RC model EGB-250. A 5.5 Hp induction motor drives this compressor to transmit mechanical rotating motion at 3470 rpm to the chamber. The RC is a two stages compressor with one inlet and one discharge valve in each stage. The gas is transferred to the tank in the second stage. The compressor is shown in Fig. 3.

The valves in the RC can be configured with four types of faults corresponding to (1) valve seat wear, (2) corrosion of the valve plate, (3) fracture of the valve

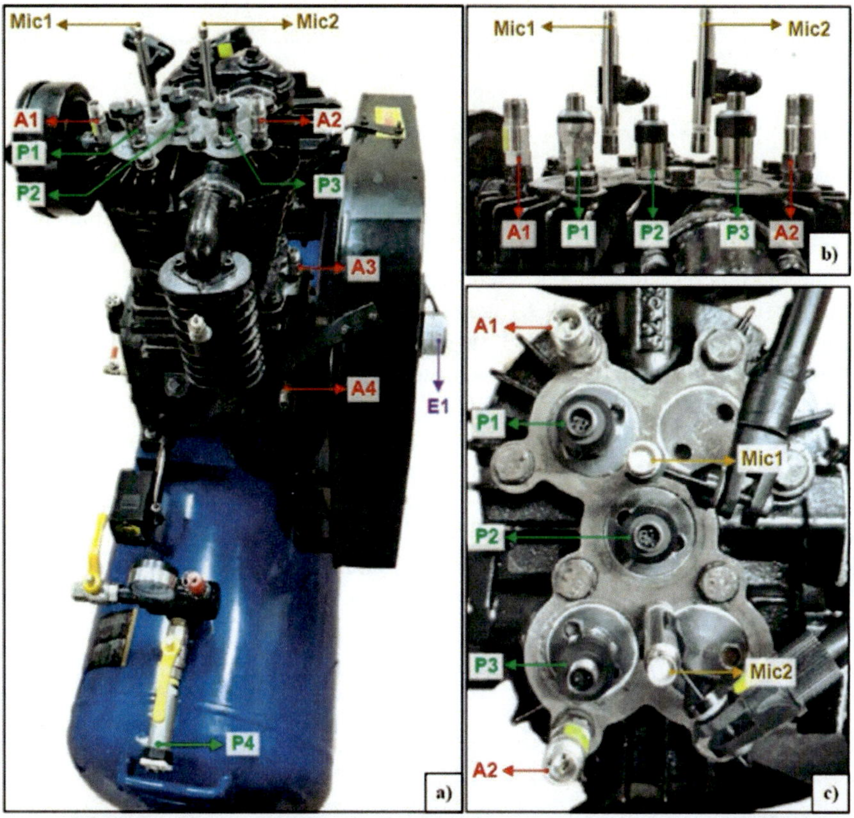


**Fig. 3** Compressor EGB-250. **a** View of the compressor with sensors location. **b** Lateral view of sensors' location. **c** Superior view of sensors location

**Table 1** Valve faults in the reciprocating compressor

| Stage and valve type | Fault type | Fault Label |
|---|---|---|
| All stages | Healthy | P1 |
| Second stage, discharge | Valve seat wear | P2 |
| Second stage, discharge | Corrosion of valve plate | P3 |
| Second stage, discharge | Fracture of valve plate | P4 |
| Second stage, discharge | Broken Spring | P5 |
| Second stage, inlet valve | Valve seat wear | P6 |
| Second stage, inlet valve | Corrosion of valve plate | P7 |
| Second stage, inlet valve | Fracture of valve plate | P8 |
| Second stage, inlet valve | Broken Spring | P9 |
| First stage, discharge | Valve seat wear | P10 |
| First stage, discharge | Corrosion of valve plate | P11 |
| First stage, discharge | Fracture of valve plate | P12 |
| First stage, discharge | Broken Spring | P13 |
| First stage, inlet valve b | Valve seat wear | P14 |
| First stage, inlet valve b | Corrosion of valve plate | P15 |
| First stage, inlet valve b | Fracture of valve plate | P16 |
| First stage, inlet valve b | Broken Spring | P17 |

plate, and (4) broken spring. Conditions in the valve fault experiment are denoted P1 representing the normal condition. P2-P17 is the set of configured valve faults studied in this research. The different fault combinations are presented in Table 1. The accelerometers PCB-603C01 denoted as $A1$, $A2$ and $A3$ were located on the compressor case as shown in Fig. 3. In general, this type of sensor should be located close to the component where we want to analyze a possible fault. Additionally, the vibration sensor should be located in a location where the vibration of the target component has a component with optimal amplitude [6, 13]. The accelerometers were located as close as possible to the valves where the different faults were implemented. Even when other sensor signals were recorded, only the vibration signals of $A1$ were considered for fault classification in this research. The vibration signals dataset is composed of 15 vibration signals with a duration of 10 seconds, sampled at a frequency of 50 kHz for each fault condition. As we have 17 fault conditions, 255 signals were recorded.

## 5 Validation of the Classification Approach

Each vibration signal has a length of 500000 time-samples, and we have considered a set of six partially overlapped signal windows of length 400000 samples with a stride of 20000 samples. The vibration signal portion in each window is used to construct the Poincaré plot and obtain a color image of size $128 \times 128$. The vibration signals dataset has 255 signals in total, which allows us to obtain 1530 images for training the CNN model. As the image dataset is relatively small, we have used a Leave-One-Out Cross-Validation (LOOCV) approach considering a partition of 85% as the training and 15% as a test set. The process is repeated ten times, and the accuracy assessment of the model corresponds to the average of the ten repetitions. The accuracy assessment was performed using several metrics reported in [12]. The metrics are calculated using the confusion matrix elements proposed in [22]. This approach is compared to a previously proposed approach where discrete features extracted from the Poincaré plot feed a classical Random Forest (RF) based fault classification model. Additionally, the results are compared using spectrograms as inputs to a CNN-based classifier.

## 6 Results and Discussion

A sensitivity analysis concerning the lag parameters ( $\tau_1, \tau_2$ and $\tau_3$) was performed by varying each parameter within the range $\tau_i - 5 \leq \tau_i \leq \tau_i + 5$, where $\tau_i$ represents the value of the parameter selected according to the guidelines reported in [16]. The average accuracy obtained during this sensitivity analysis was 94.91% with a standard deviation of 1.41%. The minimum accuracy was 93.21%, and the maximum was 97.74%. Results concerning the classification accuracy of the CNN model using PI as features are shown in Table 2. The results correspond to the cross-validation procedure of the CNN model. Results regarding using discrete Poincaré features with classical Random Forest and spectrograms inputs to a CNN are also included. The classification using discrete features was repeated ten times using different lag values for constructing the Poincaré plots. The average accuracy is 94.97% for the CNN model, while the RF model with discrete features has an average classification accuracy of 74.29%. The performance with spectrogram features is slightly better than Poincaré and reaches 96.73%. The maximum accuracy attained with discrete Poincaré Features is 80.8%, while the CNN trained with PI attained a maximum classification accuracy of 96.83%. The maximum accuracy with spectrogram inputs is 97.82%.

**Table 2** Accuracy attained during validation

| Features | ML model | Average accuracy | Std accuracy | Min accuracy | Max accuracy |
|---|---|---|---|---|---|
| Discrete features | Random forests | 74.29 | 5.33 | 67.1 | 80.8 |
| Poincaré images | CNN | 94.97 | 1.34 | 92.76 | 96.83 |
| Spectrogram | CNN | 96.73 | 0.81 | 95.12 | 97.82 |

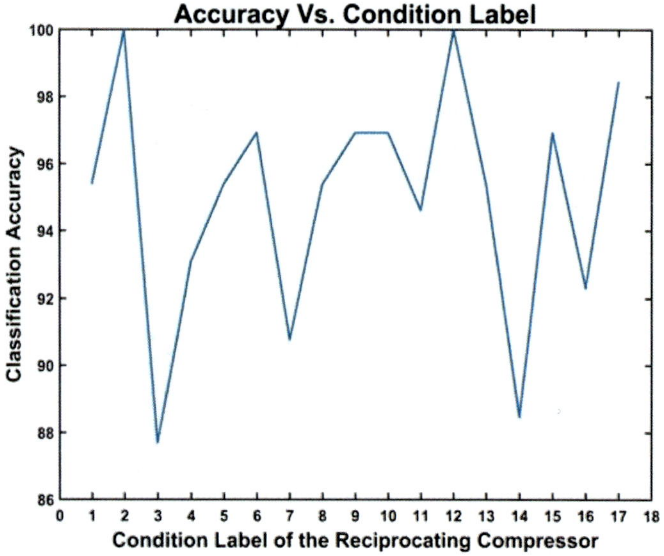


**Fig. 4** Accuracy as a function of the class label

The training was performed using Matlab 2021a with the SGDM solver. The convergence of the CNN was attained after 152 epochs, and it took 34 min and 29 s. The average accuracy as a function of the condition label in the RC is presented in Fig. 4. All classes attained an average accuracy higher than 87%. However, the lower classification accuracy is attained by classes $P3$, $P7$, $P14$, and $P16$. In contrast, $P2$ and $P12$ have the highest classification accuracy of 100%. The CNN is highly efficient for classification; however, it has a black-box nature and lacks the explanation ability [11]. In this case, it is not clear the reasons for the classes $P2$ and $P12$ for attaining the highest classification accuracy. Similarly, we cannot clearly explain the classification accuracy for the rest of the classes. The visualization of features in the softmax layer of the CNN using the t-distribution stochastic neighbor embedding (t-SNE) [15] shows that the cluster structure attained by classes $P2$ and $P12$ is easily separable from the rest of clusters. In contrast, the cluster structure for the rest of the classes presents inter-cluster intersections. This fact could explain the slightly lower classification accuracy the rest of the classes attained. In this application, we have 15 vibration signals for each condition. Although this is a small dataset, we are obtaining classification accuracies of 94.97% which is relatively high. The accuracy could be improved by increasing the dataset's size or using time-series augmentation techniques. Compared with published literature reporting deep-learning-based fault classification approaches for RCs that attain classification accuracies close to 100%. These approaches consider only a small set of faults, and the CNN architecture is generally very complex. In contrast, our approach performs the classification of 17 fault conditions with high accuracy and using a simple CNN architecture. Concerning the possible application of our approach within the context of PHM, a limitation is a fact that PI can only classify faults within a predefined set. The approach cannot be used for predicting future faults.

## 7 Conclusions

Poincaré images are valuable features that provide a 2D representation of 1D vibration time-series. These features can be fed to a CNN for classifying many valve faults in an RC. The proposed approach improves the classification accuracy by 14% for discrete features extracted from the Poincaré plot in combination with Random Forests classification models. Additionally, results show that the performance using PI is comparable to the results with classical spectrogram images, only overcome by less than 2%. Thus, PI is presented as an excellent 2D feature for fault diagnosis purposes. Future research is oriented to improve the classification accuracy and apply this approach to the problem of severity classification in a helical gearbox.

## References


1. Ali, S.M., Hui, K., Hee, L., Leong, M.S.: Automated valve fault detection based on acoustic emission parameters and support vector machine. Alex. Eng. J. **57**(1), 491–498 (2018)
2. Biggio, L., Kastanis, I.: Prognostics and health management of industrial assets: current progress and road ahead. Front. Artif. Intell. **3**, 578613 (2020)
3. Brennan, M., Palaniswami, M., Kamen, P.: Do existing measures of poincare plot geometry reflect nonlinear features of heart rate variability? IEEE Trans. Biomed. Eng. **48**(11), 1342–1347 (2001)
4. Cabrera, D., Guamán, A., Zhang, S., Cerrada, M., Sánchez, R.V., Cevallos, J., Long, J., Li, C.: Bayesian approach and time series dimensionality reduction to lstm-based model-building for fault diagnosis of a reciprocating compressor. Neurocomputing **380**, 51–66 (2020)
5. Falzone, S., Kolodziej, J.R.: Condition monitoring of a reciprocating compressor using wavelet transformation and support vector machines. In: Annual Conference of the PHM Society, vol. 9 (2017)
6. Fatima, S., Dastidar, S.G., Mohanty, A.R., Naikan, V.N.A.: Technique for optimal placement of transducers for fault detection in rotating machines. Proc. Inst. Mech. Eng. Part O: J. Risk Reliabil. **227**(2), 119–131 (2013)
7. García-Isla, G., Corino, V., Mainardi, L.: Poincaré plot image and rhythm-specific atlas for atrial bigeminy and atrial fibrillation detection. IEEE J. Biomed. Health Inform. **25**(4), 1093–1100 (2020)
8. García-Isla, G., Mainardi, L., Corino, V.D.: A poincaré image-based detector of ecg segments containing atrial and ventricular beats. In: 2021 Computing in Cardiology (CinC), vol. 48, pp. 1–4. IEEE (2021)
9. Gougam, F., Rahmoune, C., Benazzouz, D., Afia, A., Zair, M.: Bearing faults classification under various operation modes using time domain features, singular value decomposition, and fuzzy logic system. Adv. Mech. Eng. **12**(10), 1687814020967874 (2020)
10. Guo, F.y., Zhang, Y.c., Wang, Y., Wang, P., Ren, P.j., Guo, R., Wang, X.Y.: Fault detection of reciprocating compressor valve based on one-dimensional convolutional neural network. Math. Probl. Eng. **2020** (2020)

11. Hailesilassie, T.: Rule extraction algorithm for deep neural networks: a review (2016). arXiv:1610.05267
12. Hossin, M., Sulaiman, M.: A review on evaluation metrics for data classification evaluations. Int. J. Data Min. & Knowl. Manag. Proc. **5**(2), 1–11 (2015). https://doi.org/10.5121/ijdkp.2015.5201
13. ISO10816: Mechanical vibration - measurement and evaluation of machine vibration (2016)
14. Liu, R., Yang, B., Zio, E., Chen, X.: Artificial intelligence for fault diagnosis of rotating machinery: a review. Mech. Syst. Signal Process. **108**, 33–47 (2018)
15. Maaten, L.v.d., Hinton, G.: Visualizing data using t-sne. J. Mach. Learn. Res. **9**(Nov), 2579–2605 (2008)
16. Medina, R., Macancela, J.C., Lucero, P., Cabrera, D., Sánchez, R.V., Cerrada, M.: Gear and bearing fault classification under different load and speed by using poincaré plot features and svm. J. Intell. Manufact. 1–25 (2020)
17. Menon, E.S.: Transmission Pipeline Calculations and Simulations Manual. Gulf Professional Publishing (2014)
18. Monteiro, R., Bastos-Filho, C., Cerrada, M., Cabrera, D., Sánchez, R.V.: Convolutional neural networks using fourier transform spectrogram to classify the severity of gear tooth breakage. In: 2018 International Conference on Sensing, Diagnostics, Prognostics, and Control (SDPC), pp. 490–496. IEEE (2018)
19. Peña, M., Cerrada, M., Medina, R., Cabrera, D., Sánchez, R.V.: Poincaré plot features and statistical features from current and vibration signals for fault severity classification of helical gear tooth breaks. J. Comput. Inf. Sci. Eng. **23**(2), 021009 (2022)
20. Pichler, K., Lughofer, E., Pichler, M., Buchegger, T., Klement, E.P., Huschenbett, M.: Fault detection in reciprocating compressor valves under varying load conditions. Mech. Syst. Signal Process. **70**, 104–119 (2016)
21. Piskorski, J., Guzik, P.: Geometry of the poincaré plot of rr intervals and its asymmetry in healthy adults. Physiol. Meas. **28**(3), 287 (2007)
22. Sokolova, M., Lapalme, G.: A systematic analysis of performance measures for classification tasks. Inf. Proc. & Manag. **45**(4), 427–437 (2009)
23. Von Oertzen, T., Boker, S.M.: Time delay embedding increases estimation precision of models of intraindividual variability. Psychometrika **75**(1), 158–175 (2010)
24. Xiao, S., Nie, A., Zhang, Z., Liu, S., Song, M., Zhang, H.: Fault diagnosis of a reciprocating compressor air valve based on deep learning. Appl. Sci. **10**(18), 6596 (2020)
25. Yang, H.b., Zhang, J.a., Chen, L.l., Zhang, H.l., Liu, S.l.: Fault diagnosis of reciprocating compressor based on convolutional neural networks with multisource raw vibration signals. Math. Probl. Eng. **2019** (2019)

# A Novel Approach for Detection and Location of Cyber-Attacks in Water Distribution Networks

**Claudia Rodríguez-Martínez, Marcos Quiñones-Grueiro, Antônio J. Silva Neto, and Orestes Llanes-Santiago**

**Abstract** The technological development and the globalization of information allow to obtain more productivity and efficiency in industrial systems, but at the same time, these systems are more susceptible to cyber-attacks. The digital transformation process has achieved to transform the water distribution networks (WDNs) from physical systems to cyber-physical systems, making them prime targets for cyber-attacks due to their condition as critical systems for life. Several methods for detecting cyber-attacks in WDNs have been presented in the scientific literature. However, very few papers address the problem of locating them. In this paper, the Dynamic Principal Component Analysis, Adaptive Exponential Weighted Moving Average Chart, and Structural Analysis computational tools are combined to obtain an effective system of detection and location of cyber-attacks. The proposed methodology was applied to the known benchmark BATADAL. The obtained results confirm the validity of the proposal.

## 1 Introduction

New technological paradigms have changed drastically the way we live and understand technology. Cloud Computing, Ubiquitous Computing or the Internet of Things are clear examples of this new reality, increasingly latent in our daily lives.

C. Rodríguez-Martínez (✉) · O. Llanes-Santiago
Universidad Tecnológica de La Habana José Antonio Echeverría, CUJAE, Marianao, La Habana, Cuba
e-mail: claudia97rm@gmail.com

O. Llanes-Santiago
e-mail: orestes@tesla.cujae.edu.cu

M. Quiñones-Grueiro
Institute of Software Integrated Systems, Vanderbilt University, Nashville, TN, USA
e-mail: marcosqg88@gmail.com

A. J. Silva Neto
IPRJ-UERJ, Rua Bom Fin, Vila Amelia, Nova Friburgo, RJ, Brazil
e-mail: ajs_neto@iprj.uerj.br



O. Llanes-Santiago (ed.), *Proceedings of 19th Latin American Control Congress (LACC 2022)*, Studies in Systems, Decision and Control 464,
https://doi.org/10.1007/978-3-031-26361-3_12

There are already many companies that have realized the multiple benefits of implementing the paradigm Industry 4.0 in the industries. Hence, the same technologies mentioned above have been baptized again specifically for the industrial sector: they are the increasingly well-known Wireless Sensor Networks (WSN), Industrial Internet of Things (IIoT), or M2M (Machine to Machine). It is at the time of its implementation, that the concept of the 4th Industrial Revolution appears (the process by which the physical and digital worlds come together) [5].

This fact makes it necessary to rethink the security parameters in industries through the so-called Cybersecurity [9]. In the case of water, the implementation of the Industry 4.0 paradigm includes the development of new processes, new communications, and new connections as an opportunity to reduce energy consumption, mitigate unnecessary water losses and optimize advanced asset management. Also, it is essential to attend to the imminent challenge of prevention, detection, and location of cyber-attacks. However, most of the algorithms developed have focused on the detection of cyberattacks and little has been done in relation to their location [11].

In this paper, a methodology for the detection and localization of cyber-attacks in WDNs is presented which constitutes the main contribution of the paper. In the proposal, the Dynamic Principal Component Analysis, Adaptive Exponential Weighted Moving Average Chart, and Structural Analysis computational tools are combined to obtain an effective protection system.

## 2 Material and Methods

### 2.1 Dynamic Principal Component Analysis

An adequate projection of the data to low dimension space can significantly reduce the computational cost and improve the condition monitoring process in industrial systems. Principal Component Analysis (PCA) is a known computational tool very used for dimensionality reduction. The main objective of the PCA method is to obtain a new base of the data space to transform it such as the variables in the new space are uncorrelated [3].

For determining the set of transformation vectors $\mathbf{V} \in \Re^m$ from a training matrix of observations $\mathbf{X} \in \Re^{n \times m}$ where $n$ is the number of observations and $m$ the number of variables, the optimization problem $\max_{\mathbf{V} \neq 0} \frac{\mathbf{V}^T \mathbf{X}^T \mathbf{X}_\mathbf{V}}{\mathbf{V}^T \mathbf{V}}$ should be solved. To project the data in a reduce space $T$ of $m - r$ variables an operator $P$ with $m - r$ vectors is formed such that $T = XP$. The projection of T into the observation space of $m$ variables is obtained by $\check{X} = TP^T$. The difference between $X$ and $\check{X}$ form the residual matrix $E$.

PCA method assumes that the data is not time dependent. However, in industrial systems the most of variables can be classified as time series which are serially correlated. For extending PCA to the analysis of time dependent data, Dynamical PCA (DPCA) was developed [7].

DPCA implicitly assumes that the data follows an autoregressive structure. By adding lagged values of the original variables, the idea is to convert the autocorrelation in the original variables into cross-correlation, which can be modeled by PCA. To implement it, a "time shift" method is used to transform the original $X$ matrix into a new expanded $\hat{X}$ matrix, shifting the vectors in time.

$$\hat{X} = [X_t | X_{t-1} | \ldots | X_{t-h}]$$

Once the $\hat{X}$ matrix is obtained the data is scaled and the principal component analysis is performed.

## 2.2 *Adaptive Exponential Weighted Moving Average (AEWMA)*

The exponential weighted moving average (EWMA) control chart is an univariate control chart for detecting deviations in the mean $\mu$ of a signal based on current and previous observations [1]. For a signal $x(t)$, the EWMA control chart is evaluated as $z_i = \lambda \bar{x}_i + (1 - \lambda) z_{i-1}$ where $\bar{x} = \mu$, and $0 \leq \lambda \leq 1$ is a weighting factor for giving more or less importance to the recent observation. It is known that EWMA is able to detect small and large changes effectively, but not both simultaneously. To solve this drawback, Adaptive EWMA was introduced which is based on adapting the weight of previous observations based on the error, $e_i = \bar{x}_i - z_{i-1}$ such that $\lambda(e(t)) = \frac{\vartheta(e(t))}{e(t)}$ [4]. Thus, changes of various sizes could be detected in a more balanced way, while reducing the problem of inertia. In this paper $\vartheta(e(t))$ is defined as [4]:

$$\vartheta(e) = \begin{cases} e + (1-\lambda)k & if\ e < -k \\ \lambda k & if\ |e| \leq k \\ e - (1-\lambda)k & if\ e > k \end{cases}$$

For the calculation, the value of the parameters $\lambda$ and $k \geq 0$ are obtained by using the method of the least squares [8].

$$\lambda = ln(1.2219 - 0.04697 * ln(ARL) + 0.45985 * \sqrt{\delta_{\min}} - 0.02701 * \sqrt{\delta_{\max}})$$

$$k = \sqrt{4.846 + 1.5852 * ln(ARL) - 2.8697 * \sqrt{\delta_{\min}} - 1.7198 * \sqrt{\delta_{\max}}}$$

where $ARL$ is the average path length and $\delta_{\min}$ and $\delta_{\max}$ form the range of deviation to be detected.

## 2.3 Structural Analysis

Structural analysis refers to the study of the structural and functional model properties of a system [2], which provides a design approach for residual-based analytical redundancies to detect and locate cyberattacks. The number of analytic redundancy relationships (ARR) depend both on the known variables and on the structure of the system, and represent a set of operational constraints of the described model. The cyberattack is detected when at least one ARR is no longer consistent. Not needing an analytical model to analyze the detectability and isolability of different cyberattacks is the main advantage of this methodology.

A behavioral model of the system is defined by a pair $(C, V)$ where $V = v_1, v_2, \ldots, v_N$ is a set of variables and parameters, while $C = c_1, c_2, \ldots, c_M$ is a set of constraints. The system variables that make up the set V can be divided into two types: the known variables K and the unknown variables $X$, where $V = K \cup X$.

**Definition 1** (*Structural model* [2]) The structural model of a system $(C, V)$ is a bi-partite graph $(C, V, \varepsilon)$ where $\varepsilon \subset C \times V$ is a set of edges defined by: $(c_i, z_j) \in \epsilon$ if the variable $v_j$ appears in the constraint $c_i$.

A bi-partite graph can be associated with an incidence matrix. In order to set ARRs in Structural Analysis, the graph must have more constraints than unknown variables. A key tool in this case is the Dulmage–Mendelsohn decomposition [6], with which it is possible to separate the model into three parts: the over-constrained part $G^+$ with more restrictions than variables, the just-determined part $G$ with the same number of restrictions and variables, and the under-constrained part $G^-$ with fewer restrictions than variables. Regarding the diagnosis, the part of interest is the over-constrained part, since it includes redundancy. Cyberattacks belonging to $G$ and $G^-$ are not detectable.

## 2.4 Methodology for Detection and Location of Cyber-Attacks

The proposed methodology has a two-level architecture. In the first level, the detection process is developed and in the second level, the location process is executed. For its implementation, the methodology has two stages: offline and online.

In the offline stage, a dataset of the normal operation of the WDN is used. The first step is to divide the WDN into Attacked Areas (AAs). In the second step, a DPCA algorithm is used with all variables of the WDN to obtain the error matrix $E$, and an AEWMA control chart is calibrated with the mean value ($\mu_E$) and the standard deviation ($\sigma_E$) of $E$ both for the detection process in the first level of the methodology. In the third step, it is used a DPCA algorithm for each AA with the used variables in each one for obtaining the associated residual matrix and calibrating

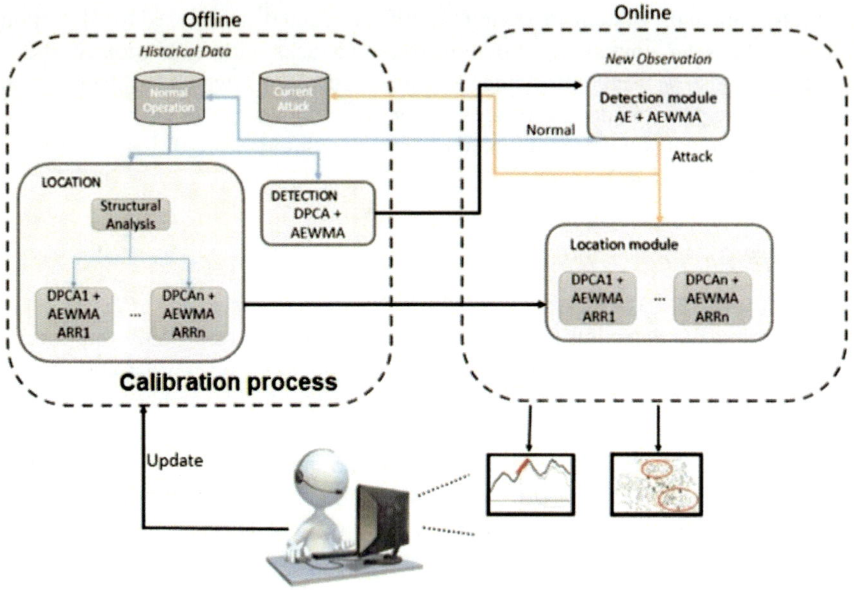


**Fig. 1** Attack detection and location system

the corresponding AEWMA control chart, all of them for the location process in the second level of the methodology.

In the online stage, each observation $\mathbf{x}$ obtained from the WDN is classified in the detection level as normal or attack by using the error $\mathbf{e} = \mathbf{x} - \hat{\mathbf{x}}$. In the first option, the observation is stored for a future recalibrate process, and in the second option, the observation is analyzed by the location module of each $AA$ to try to predict the area under attack.

A summary of the methodology is presented in the Fig. 1.

# 3 Application of the Proposed Methodology to the C-Town Case Study

## *3.1 Case Study: C-Town WDN and BATADAL Dataset*

C-Town is a mid-sized real-world water distribution network. It contains a single reservoir, 7 storage tanks, 429 pipes, 11 pumps distributed in the 5 pumping stations and 5 valves. There are 9 programmable logic controllers (PLCs) receiving information from the valves, pumps and level sensors, networked with a SCADA. The three BATADAL datasets presented in [10] contain 43 variables (32 are continuous and 12 are binary). The sampling interval is one hour. Data set 1 only contains normal operating data of the system, operating for 365 days, and it can be used for algorithm training. Dataset 2 contains information from 7 attacks with a duration of 492 h, which could also be used for training. And finally, data set 3, which has the function for testing the performance of the system for detecting and locating the DMA of a cyberattack, contains the data of 7 attacks and has a duration of 407 h.

## 3.2 Offline Stage: Detection

For preparing the detection level, a DPCA model is built using only the normal operating data of the system from Dataset 1. This data set has 43 variables, from which the discrete variables and those that are constant (do not present variability) are eliminated, so a total of 26 variables are used. In addition, the measured data is processed, normalizing their values with mean 0 and standard deviation 1. The outlier values are excluded. The 99% of the total information is chosen in terms of variance in order to eliminate only the redundant components. After some experiments, it was determined to use 3 lagged values for each used variable, so that the matrix $X$ had a dimension of $8750 \times 104$ and as a result of DPCA only 36 components are preserved. In the case of the AEWMA graph, the values of the parameters were chosen as $ARL = 900$, $\delta_{min} = 0.5$ and $\delta_{max} = 6$ according to the results present in [1], generating as a result that $\lambda = 0.1044$ and $k = 3.0107$.

## 3.3 Offline Stage: Location

For location, possible attack areas ($AAs$) must first be established. C-Town water distribution network has 5 $DMAs$ established by the physical distribution of the network, so it was initially proposed an $AA$ for each $DMA$. Since $DMA1$ is greater in terms of size and components, it was decided to have two $AAs$ in this $DMA$.

The next step is to apply the Structural Analysis to determine the $ARRs$ that characterize each $AA$. These $ARRs$ allow to locate the cyber-attack.

The variables of the model are: the level ($h\#$) of each tank ($\# : 1, 2, 3, 4, 5, 6, 7$), the flow rate ($q\#$) that passes through the pumps, the inlet and outlet pressure before($p\#a$) and after the ($p\#d$) the pumps, and the status of each pump, as well as the flow rate and valve status ($v2$). Three subsets of variables are defined: Attacks (adma&, &:$1, 2, \ldots, 6$), measured variables (identified with the letter "$y$" on the front) and system variables. Figure 2 shows the equations obtained from the structural analysis that represent the functional behavior of the system.

The results of the structural isolability analysis show the possibility to isolate all attacks. The Fig. 3a shows the Dulmage–Mendelsohn decomposition, where the isolability properties are observed. Determining the structurally over-restricted set of minimum equations (SOM) makes it possible to design the residual generators. With

| | | | |
|---|---|---|---|
| $e_1 : \{qp1, qp2, q1, qt1, h1, fdma1a\}$ | $e_11 : \{q6, p4a, p4d, qt5, h5, fdma4\}$ | $e_21 : \{yq7, q7\}$ | $e_31 : \{ypva, pva\}$ |
| $e_2 : \{q1, p1a, p1d\}$ | $e_12 : \{q6, qt5\}$ | $e_22 : \{yh1, h1\}$ | $e_32 : \{ypvd, pvd\}$ |
| $e_3 : \{q1, qt1, q2, q5\}$ | $e_13 : \{q7, p5a, p5d, h6, h7, fdma5\}$ | $e_23 : \{yh2, h2\}$ | $e_33 : \{yp2a, p2a\}$ |
| $e_4 : \{q5, q6, q7\}$ | $e_14 : \{q7, qt6, qt7\}$ | $e_24 : \{yh3, h3\}$ | $e_34 : \{yp2d, p2d\}$ |
| $e_5 : \{q2, pva, pvd, qt2, h2, fdma1b\}$ | $e_15 : \{yqp1, qp1\}$ | $e_25 : \{yh4, h4\}$ | $e_35 : \{yp3a, p3a\}$ |
| $e_6 : \{q2, qt2, q3, q4\}$ | $e_16 : \{yqp2, qp2\}$ | $e_26 : \{yh5, h5\}$ | $e_36 : \{yp3d, p3d\}$ |
| $e_7 : \{q3, p2a, p2d, qt4, h4, fdma2\}$ | $e_17 : \{yq2, q2\}$ | $e_27 : \{yh6, h6\}$ | $e_37 : \{yp4a, p4a\}$ |
| $e_8 : \{q3, qt4\}$ | $e_18 : \{yq3, q3\}$ | $e_28 : \{yh7, h7\}$ | $e_38 : \{yp4d, p4d\}$ |
| $e_9 : \{q4, p3a, p3d, qt3, h3, fdma3\}$ | $e_19 : \{yq4, q4\}$ | $e_29 : \{yp1a, p1a\}$ | $e_39 : \{yp5a, p5a\}$ |
| $e_10 : \{q4, qt3\}$ | $e_20 : \{yq6, q6\}$ | $e_30 : \{yp1d, p1d\}$ | $e_40 : \{yp5d, p5d\}$ |

**Fig. 2** Structural analysis equations

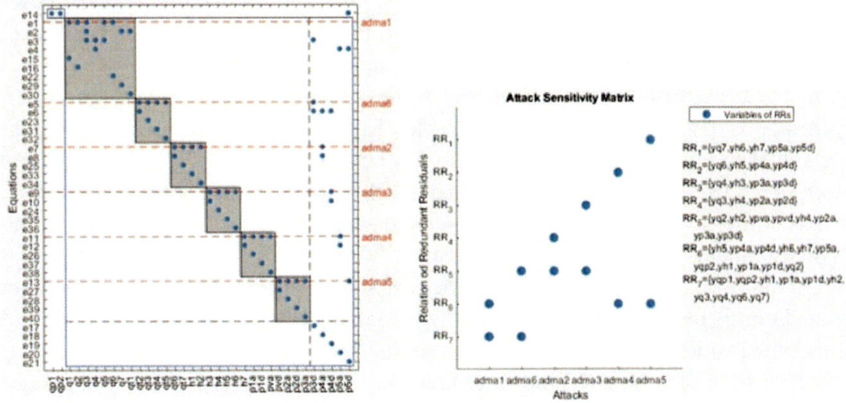


(a) Dulmage-Mendelsohn Decomposition (b) Attack Sensitivity Matrix

**Fig. 3** **a** Dulmage–Mendelson decomposition, **b** Attack sensitivity matrix

them, the set of restrictions that satisfy the isolability specifications are determined. Seven (7) $ARRs$ were obtained as shown the Attack Sensitivity Matrix in the Fig. 3b.

Similar to the detection process, a DPCA is prepared for each ARR obtained, with the samples from dataset 1 but only with the variables present in each ARR. The same number of lagged values of the variables are used, but the parameters of the AEWMA control chart associated to each DPCA algorithm are recalculated.

## *3.4 Performance Assessment*

The detection process is considered successful when it occurs early and in the presence of a true attack. The $E_{TTD}$ and $E_{CC}$ indices are used to evaluate the early detection and the correct classification respectively, in the detection process. They are defined as: $E_{TTD} = 1 - \frac{1}{n_a}\sum_{i=1}^{n_a}\frac{TTD_i}{\Delta t_i}$ and $E_{CC} = \frac{TPR+TNR}{2}$, where $n_a$ is the number of attacks, $TTD_j$ is the time to detection of the $j-th$ attack and its corresponding duration time $\Delta t_j$. True Positive Rate ($TPR$) and True Negative Rate ($TNR$) are calculated as: $TPR = \frac{TP}{TP+FN}$ and $TNR = \frac{TN}{FP+TN}$, respectively, where $TP$ is the true positive alarms, $FN$, is false negative alarms, $TN$ is true negative alarms and $FP$ is false positive alarms.

The index $E$ integrates both indices in the following form: $E = 0.5 * E_{TTD} + 0.5 * E_{CC}$.

## 4 Results and Discussion

To verify the correct functioning of the algorithm, 2089 observations were chosen from data set 3, in which 7 attacks are found. Figure 4a shows the $AA$ where each attack occurs, the duration time (DT) both in hours, and the detection time (TTD) of each attack. Figure 4b shows the results of the detection of cyberattacks where $z(t)$ corresponds to the behavior of the variable of the AEWMA control chart, and the horizontal red line represents its threshold. The true-time interval of each attack is represented by vertical lines, so that each attack is listed in order of appearance.

According to the results all attacks were detected. The Fig. 4a shows the results of the calculated indices presented above.

On the other hand, when an $ARR$ becomes inconsistent indicates that the area related to it is in presence of a cyber-attack, making possible to locate it. The Fig. 5 shows the results obtained in this regard. In the first column (I) appears the number of the attack. The second (II) indicates how long each attack takes. The third column (III) indicates the number of observations in which the attack was located in at least one area. The following four columns represent four possible alternatives that can occur when the attack is located, respectively:

IV It is only detected in the correct $AA$.
V It is detected in various $AAs$, including the actual $AA$.
VI Detected in wrong $AAs$ only.
VII Not detected in any $AA$.

To evaluate the performance of the proposed methodology, four indicators are used. The percentage location index ($LI$) is calculated as $\frac{III*100}{II}$ and represents how long it was possible to locate the attack. The percent of exact location (EXL), $\frac{IV*100}{III}$, refers to how long you were located only in the correct area of the total locate time. Finally, the percentage of effective location index (EFL) represents how long it was located in the correct area without taking into account that at the same time it could be located in another area, $\frac{(IV+V)*100}{III}$.

The most relevant results obtained from the analysis of Table 5 are:

The total average $LI$ was greater than 50%, indicating that more than half of the time, it was possible to locate the cyber-attacks. Attacks 2 and 5 have the lowest LI. The total $EXL$ was 69.3%, which is an acceptable result, since about 70 of every

| Attack | Affected Elements | $AA$ | $DT(h)$ | $TTDi(h)$ |
|---|---|---|---|---|
| 1 | Level of tank 3, flow of pump 4 and 5 | 3 | 70 | 2 |
| 2 | Level of tank 2, flow of valve 2 | 6 | 65 | 1 |
| 3 | Flow of pump 3 | 1 | 31 | 0 |
| 4 | Flow of pump 3 | 1 | 31 | 0 |
| 5 | Level of tank 2, flow of valve 2 | 6 | 100 | 0 |
| 6 | Level of tank7, flow of pumps 10 and 11 | 5 | 80 | 0 |
| 7 | Level of tank 6 | 2 | 30 | 8 |

(a) Information of Attacks

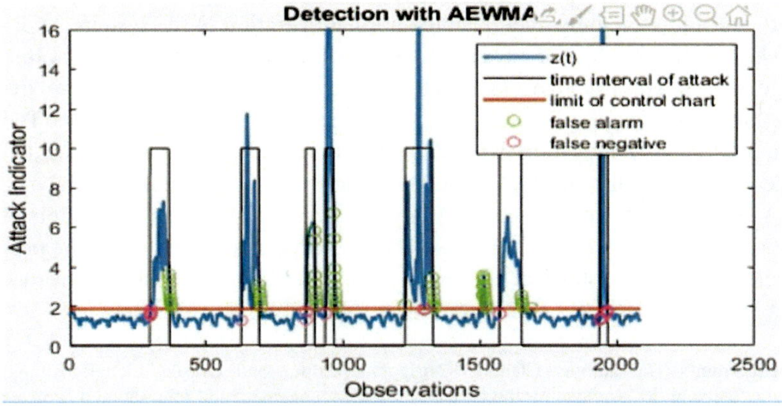


(b) Results of detection

**Fig. 4** **a** Information of attacks **b** Results of detection

| I | II | III | IV | V | VI | VII | LR(%) | ExLR (%) | EfLR(%) |
|---|---|---|---|---|---|---|---|---|---|
| 1 | 70 | 49 | 49 | 0 | 0 | 21 | 70 | 100 | 100 |
| 2 | 65 | 25 | 5 | 14 | 6 | 40 | 38.46 | 20 | 76 |
| 3 | 31 | 24 | 24 | 0 | 0 | 7 | 77.42 | 100 | 100 |
| 4 | 31 | 27 | 27 | 0 | 0 | 4 | 87.10 | 100 | 100 |
| 5 | 100 | 25 | 7 | 13 | 13 | 75 | 25 | 28 | 48 |
| 6 | 80 | 38 | 14 | 13 | 13 | 42 | 47.50 | 36.84 | 65.78 |
| 7 | 30 | 16 | 16 | 0 | 0 | 14 | 53.30 | 100 | 100 |
| Total | 407 | 204 | 142 | 30 | 32 | 203 | 57 | 69 | 84 |

**Fig. 5** Results of the cyber-attack location process

100 are localized observations which occurred only in the correct area. The most critical cases in the location process were attack 2 and 5 because of the localization was more confusing. The total $EFL$ was a satisfactory 84.25% which indicate that the correct $AA$ was identified in the most of the localization time.

## 5 Conclusions

In this article has been presented a methodology for the detection and localization of cyber-attacks in water distribution networks. For calibrating the computational tools in the offline stage only data from the normal operation of the system is necessary, which introduces a clear advantage for its practical implementation. Another advantage is the possibility to locate the attack which is a field little addressed in the scientific literature. The proposed methodology present a two level architecture. The detection process is developed in the first level and the location process is executed in the second level. In the two levels, the Dynamic Principal Component Analysis, AEWMA control chart, and Structural Analysis computational tools are combined for obtaining satisfactory results. The results of the experiments show that the proposed scheme was able to detect successfully all attacks, and exhibit satisfactory results in the location process which confirm the validity of the proposal.

**Acknowledgements** The authors Claudia Rodríguez-Martínez and Orestes Llanes-Santiago acknowledge the support provided by the Office of Fund Management and International Projects of the Ministry of Science, Technology and Environment of Cuba (CITMA) under the Project PN223LH004-23 and the authors Antônio J. Silva Neto and Orestes Llanes-Santiago acknowledge the support provided by the Coordenação de Aperfeiçoamento de Pessoal de Nivel Superior (Finance Code 001 and CAPES-PRINT Process No. 88881.311758/2018-01) from Brazil.

## References

1. Aly Aya, A., Ramadan Hamed, M., Mahmou, A.: Optimal design of the adaptive exponentially weighted moving average control chart over a range of mean shifts. Commun. Stat. - Simul. Comput. **46**(2), 37–41 (2015)
2. Blanke, M., Kinnaert, J.L., Staroswiecki, M.: Diagnosis and Fault-Tolerant Control, 3rd edn. Germany (2016). https://doi.org/10.1007/978-3-662-47943-8
3. Bro, R., Smilde, A.: Principal component analysis. Anal. Methods **6**, 2812–2831 (2014). https://doi.org/10.1039/c3ay41907
4. Capizzi, G., Masarotto, G.: An adaptive exponentially weighted moving average control chart. Technometrics **45**(3), 199–207 (2003)
5. Kamble, S.S., Gunasekaran, A., Sharma, R.: Analysis of the driving and dependence power of barriers to adopt industry 4.0 in Indian manufacturing industry. Comput. Ind. **101**, 107–119 (2018)
6. Krysander, M., Nyberg, M.: Structural analysis for fault diagnosis of DAE systems utilizing MSS sets. IFAC Proc. **35**(1), 143–148 (2002)
7. Ku, W., Storer, R.H., Georgakis, C.: Disturbance detection and isolation by dynamic principal component analysis. Chemometr. Intell. Lab. Syst. **30**(1), 179–196 (1995)

8. Mahmoud, M.A., Parker, P.A., Woodall, W.H.: A change point method for linear profile data. Qual. Reliabil. Eng. Int. **23**(247–268) (2007)
9. Moeuf, A., Lamouri, S., Pellerin, R., Tamayo-Giraldo, S., Tobon-Valencia, E., Eburdy, R.: Identification of critical success factors, risks and opportunities of Industry 4.0 in SMEs. Int. J. Prod. Res. **58**(5), 1384–1400 (2020)
10. ...Taormina, R., Galelli, S., Tippenhauer, N., Salomons, E., Ostfeld, A., Eliades, D., Aghashahi, M., Sundararajan, R., Pourahmadi, M., Banks, M., Brentan, B., Herrera, M., Rasekh, A., Campbell, E., Montalvo, I., Lima, G., Izquierdo, J., Haddad, K., Gatsis, N., Taha, A., Somasundaram, S., Ayala-Cabrera, D., Chandy, S., Campbell, B., Biswas, P., Lo, C., Manzi, D., Luvizotto, E., Jr., Barker, Z., Giacomoni, M., Pasha, M., Shfiee, M., Abokifa, A., Housh, M., Kc, B., Ohar, Z.: The battle of the attack detection algorithms Disclosing cyber attacks on water distribution networks. J. Water Res. Plan. Manag. **144**(8), 04018048 (2018). https://doi.org/10.1061/(ASCE)WR.1943-5452.0000749
11. Tuptuk, N., Hazell, P., Watson, J., Hailes, S.: A systematic review of the state of cyber-security in water systems. Water **13**(1), 81 (2021). https://doi.org/10.3390/w13010081

# Predictive Diagnosis of Isolated Photovoltaic Systems Based on PWM & MPPT Charge Regulators

**Emilio García-Moreno, Eduardo Quiles-Cucarella, Antonio Correcher-Salvador, and Francisco Morant-Anglada**

**Abstract** In this work, smart sensor networks are used to carry out predictive diagnostic techniques considering predictive symptoms obtained from the comparison of the characteristic electrical parameters of the solar panels that are part of the photovoltaic installation. The method used performs the comparative analysis of trends in each sample obtained online from the random time series generated by the solar panels, in order to detect power reduction deviations that are due to the event of partial shading on the photovoltaic installation. In this work, tests were used alternately, first with PWM-type solar charge regulators and then with MPPT-type regulators, highlighting the comparative advantages that each type of regulator provides in predictive diagnostic tasks.

## 1 Introduction

The predictive fault diagnosis (PFD) method proposed is especially suitable to be applied on isolated photovoltaic installations (IPVIs) that, due to their characteristics, allow the parallel connection of solar panels (SPs) or photovoltaic modules (PVMs).

E. García-Moreno (✉) · E. Quiles-Cucarella
Instituto de Automática e Informática Industrial, Universitat Politècnica de València, Camino de Vera, s/n, 46022 Valencia, Spain
e-mail: egarciam@isa.upv.es

E. Quiles-Cucarella
e-mail: equiles@isa.upv.es

A. Correcher-Salvador
Antonio Correcher-Salvador Instituto de Automática e Informática Industrial, Universitat Politècnica de València, Camino de Vera, s/n, 46022 Valencia, Spain
e-mail: ancorsal@ai2.upv.es

F. Morant-Anglada
Francisco Morant-Anglada Instituto de Automática e Informática Industrial, Universitat Politècnica de València, Camino de Vera, s/n, 46022 Valencia, Spain
e-mail: fmorant@isa.upv.es

 
O. Llanes-Santiago (ed.), *Proceedings of 19th Latin American Control Congress (LACC 2022)*, Studies in Systems, Decision and Control 464,
https://doi.org/10.1007/978-3-031-26361-3_13

These types of installations offer the opportunity to have clean energy, autonomously in any location and for this reason, its applications have become widespread. This type of installation, like any installation carried out outdoors, is exposed to aggressive multifactorial degradation conditions affecting the photovoltaic cells, metals, polymers and crystals that make up the solar panel. One of the most important modes of degradation is the so-called Hot-Spot, whose most common cause is due to experiencing areas of partial shading on the surface of the solar panel. This phenomenon causes the cells affected by shading to operate with reverse polarization, dissipating energy, behaving as a load that dissipates heat, instead of generating energy. This dissipation causes a progressive degradation that goes from punctual discoloration to the formation of small craters, which in a short time develop and multiply as if it were a "metastasis".

The predictive diagnosis technique used is centered on the SP, where the characteristic reduction of its electrical parameters I, V, P is undoubtedly experienced in case it is affected by partial shading. It consists of the application of a comparative analysis carried out on-line between the values of the electrical variables obtained from the time series of data corresponding to each SP. The technique makes it possible to detect and isolate predictive symptoms that precede the final failure event sufficiently in advance, thus providing a wide margin of maneuver to prevent the progression of the degrading phenomenon, applying predictive maintenance techniques for the precise analysis of the problem. In order to characterize the nature of the predictive symptoms, inductive methods of partial shading have been applied, which have made it possible to establish the threshold values of the deviations of the electrical parameters I, V, P, from small coverage shadings induced in one of the SPs of the solar installation. The importance of applying a diagnostic method of a predictive nature is due to the fact that the development of the phenomenon of Hot-Spots can occur very quickly, reaching states of visible degradation of SP irreversibly in a few hours, precursors of fires. that can be moved to adjacent panels [1–3].

This paper is organised as follows. Section 2 presents the type of smart sensor networks dedicated to the supervision of the photovoltaic installation, including also the general description of the isolated photovoltaic installation (IPVI), the symptom-based predictive diagnosis method applied and the tests carried out. Section 3 develops the tests and displays the results. In Sect. 4, the conclusions are provided, highlighting the suitability of the used PFD.

## 2 Materials and Methods

Victron GX devices can supply the photovoltaic parameters to the NMEA2000 network that is used make the predictive fault diagnosis. NMEA 2000 Network and Victron GX devices have compatible connectivity, so all sensors, battery monitors, inverters and solar chargers if they are connected to a GX controller device can send their data since GX has full connection compatibility with NMEA 2000 network bus and consequently, all the data present in the network can be jointly recorded by the Vessel Data Recorder VDR100 device and processed by N2KExtractor software installed on a PC.

In some references such as [1–3], reader can find extensive information on the used supervision and monitoring NMEA 2000 network and in Tables 1 and 2 are shown the set of specific devices, sensors and data measurement parameters as well as accuracy and range.

The photovoltaic system installed on a recreational boat. The solar plant is used as a supplemental power source for a home battery bank. Solar panels installed in the photovoltaic system of the facility are from Ecosolar company, and their characteristics can be reviewed in Table 3.

Two successive experimental studies have been done in the described PV panels. The four panels were first connected in parallel, each one of them through a Pulse Width Modulation (PWM) charge regulator (CR) and after in a second phase of experimental test using CRs of Maximum Power Point Tracking (MPPT), to feed a system of 12 V batteries bank of a total charge capacity of 750Ah (Fig. 1).

Figure 2 shows the partial shadows projected at each moment of observation on the arrangement of four parallel panels throughout the day depending on the position of the Sun. The thin dashed lines represent the projection of the shadow of the backstay in the SPs. The thick dashed line represents the projection of the mast where the weather station and other sensors are located.

To carry out the online comparative analysis of the solar panels electrical variables, the access capacity in real-world installations must be considered, as well as the symptomatic response that it can generate, correlated in general with the degradation of the PVMs, and especially with the effects of partial shading that are evidenced through the decrease in the output power that is generated in each PVM. The developed algorithm presented in detail in references [2, 3] is shown in Fig. 3.

## 3 Results

### *3.1 Results Using PWM Charge Regulators*

Figure 4 shows a comparison detail between two solar panels SP1 and SP2 installed on the port side. SP1 in its initial evolution presents a comparative deviation with respect to SP2, this is due to the greater degree of inclination that it has towards the west during sunrise and from the beginning of the day SP1, like SP2, presents a shadow effect due to the backstays and mast weather station antenna and solar radiation sensor. From 2:00 p.m., the deviation gradually decreases due to the shading of the backstay and the antenna moving towards the starboard side of the vessel where the SP3 and SP4 panels are installed (Fig. 5, Table 4).

**Table 1** PVI NMEA2000 based devices and sensors networking

| Device | Monitored variables | Parameters | Range |
|---|---|---|---|
| DCM100 | V, I | Battery sense voltage range<br>Battery sense voltage accuracy<br>Battery current range<br>Battery current accuracy | 0 to 50 VDC<br>±100 mV<br>0 to 200 A<br>±1% |
| TMP100<br>TRK3 | Solar panel temperature | Measurement range–thermistor | (−20 °C to 80 °C) |
| DCR100 | | Maximum DC switching current DC switching voltage<br>Contact resistance<br>Current sense accuracy | 10A<br><32 VDC<br>< 0 m Ω<br>±100 mA |
| USB100 | Network<br>PC connection | NMEA 2000® Connector<br>NMEA 2000®/USB Isolation<br>USB supported signals<br>USB auxiliary power<br>USB baud rate | DeviceNet Micro-C<br>Opto-Isolated<br>D+ , D−, + 5V, GND<br>+ 5 V <50 mA<br>Up to 12 Mb/s |
| VDR100 | Network data registration | USB supported signals<br>USB auxiliary power<br>USB baud rate<br>Ethernet interface | D+ , D−, + 5V, GND<br>+5 V <200 mA<br>Up to 12 Mb/s<br>100 Mb/s |
| PB200 | Outside temperatura<br>Wind speed & direction<br>GPS | NMEA input A/+<br>NMEA output A/+<br>NMEA input B/–<br>NMEA output B/–<br>12 VDC +<br>12 VDC – /ground<br>Weather station instrument must be supplied with 12 VDC (±3 VDC) at 0.5 amp | |
| IMT DiRS485TC | Solar irradiance | Irradiance measurement cell temperature measurement<br>Working temperature weight<br>Voltage supply measuring range irradiance protocol | up to 1500 W/m$^2$<br>−40 to +90 °C<br>−35 to 80 °C<br>12 to 28 VDC<br>0 to 1500 W/m2<br>MB: Modbus (RTU)<br>MT: M&T protocol |
| PWM STECA 10A | V, I | Service voltage<br>Permissible ambient temperature range<br>mA power consumption<br>Pulse width modulation (PWM) frequency | 12 V o 24 V;<br>– 10 °C y + 50 °C<br>12.5 mA<br>30 Hz |

**Table 2** PVI Victron based sensors and devices

| Device | Monitored variables | Parameters | Range |
|---|---|---|---|
| Cerbo GX | V, I | Electrical parameters V range<br>Electrical parameters I range | 0 to 70 VDC<br>1 A CDC |
| StarTech.com ports powered hub | Network MTTP chargers connection | USB ports<br>Operating temp<br>USB power<br>USB baud rate | USB 3.0<br>0 °C to 55 °C<br>CA<br>USB 3.0:5 Gbps |
| Blue Solar MPPT 75 I 15 A | V, I, P | PV electrical parameters V range<br>PV electrical parameters I range<br>Maximum PV power 12 V<br>Permissible ambient temperature range<br>mA power consuption at 12 V | 12/2 4 V<br>15 A<br>200 W<br>−30 °C to + 55 °C<br>20 mA |

**Table 3** PV Modules technical characteristics

| | | |
|---|---|---|
| Peak power($P_{max}$) | (W) | 160 |
| Production tolerance | (%) | ±3 |
| Maximum power current($I_{mp}$) | (A) | 8.88 |
| Maximum power voltage($V_{mp}$) | (V) | 18.0 |
| Short circuit current($I_{sc}$) | (A) | 9.59 |
| Open circuit voltage($V_{oc}$) | (V) | 21.6 |
| Weight | (Kg) | 2.7 |
| Dimensions | (mm) | 670 × 1510 × 3 |
| Maximum system voltage | (VDC) | 500 |
| Cell technology | Type | Monocrys |
| Cel brand | Name | Ecosolar |

All technical data at standard test conditions. AM = 1.5 E = 1000 W/m$^2$ Tc= 2 5 °C

## 3.2 *Results with MPPT Charge Regulators*

In the second experimental phase a new version of the solar installation was implemented, where not only were changes with the type of regulators. In this new phase were included the Victron devices described in Table 2 [4–8]. Also improvement operations were carried out on the solar panel installation itself to achieve, as far as possible, the orientation characteristics of the solar panels, resulting in greater convergence of the associated curves to the power series as can be seen in (Fig. 6).

**Table 4** Quantitative values obtained from PWM current outputs

| SP | Measured current (A) | | | | | | | | | | | | | | |
|---|---|---|---|---|---|---|---|---|---|---|---|---|---|---|---|
| SP1 | 5.6 | 5.9 | 6.4 | 6.8 | 7.3 | 7.4 | 7.8 | 8.1 | 7.5 | 7.3 | 6.9 | 6 | 5.8 | 5.1 | 4.2 |
| SP2 | 6.7 | 7.1 | 7.6 | 8.1 | 8.4 | 8.4 | 8.7 | 8.8 | 8.2 | 7.8 | 7.4 | 6.3 | 6 | 5.35 | 4.4 |
| SP3 | 6.4 | 6.7 | 6.9 | 7.2 | 7.4 | 7.2 | 7.2 | 7.2 | 6.4 | 5.7 | 5 | 4.1 | 3.5 | 2.4 | 0.8 |
| SP4 | 7.25 | 7.5 | 7.6 | 7.7 | 7.3 | 7.5 | 7.4 | 7.2 | 6.2 | 5.5 | 4.70 | 3.6 | 2.9 | 2.2 | 1.5 |
| DTV | SP1 | SP1 | SP1 | SP1 | SP1<br>SP3<br>SP4 | SP1<br>SP3 | SP3<br>SP4 | SP3<br>SP4 | SP3<br>SP4 | SP3<br>SP4 | SP3<br>SP4 | SP3<br>SP4 | SP3<br>SP4 | SP3<br>SP4 | SP3<br>SP4 |
| Time | 11:36 | 12:06 | 12:36 | 13:06 | 13:36 | 14:06 | 14:36 | 15:06 | 15:36 | 16:06 | 16:36 | 17:06 | 17:36 | 18:06 | 18:36 |

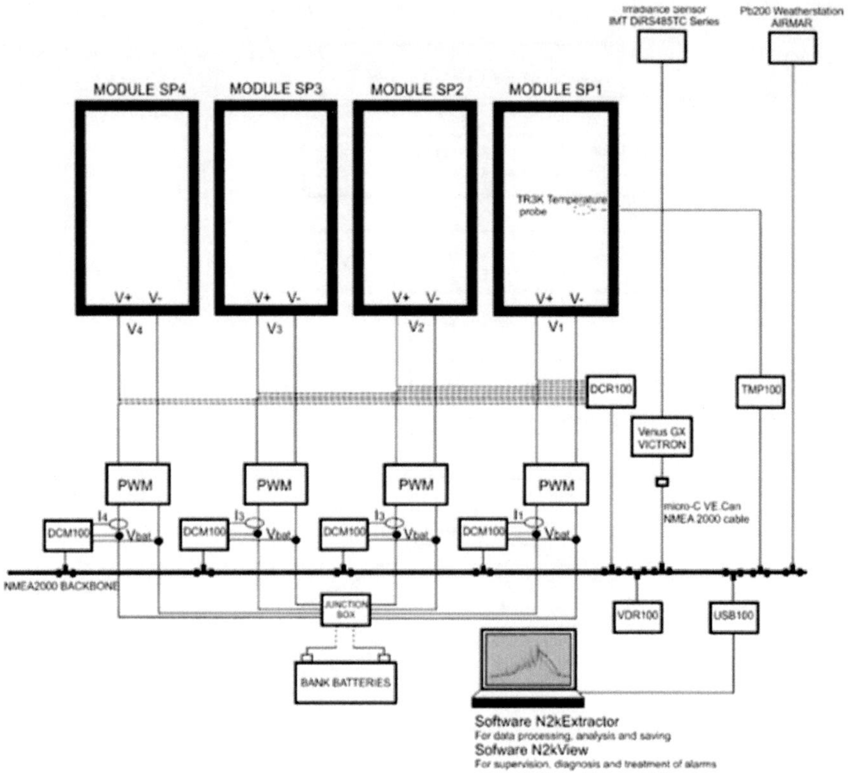


**Fig. 1** PVMs system installation

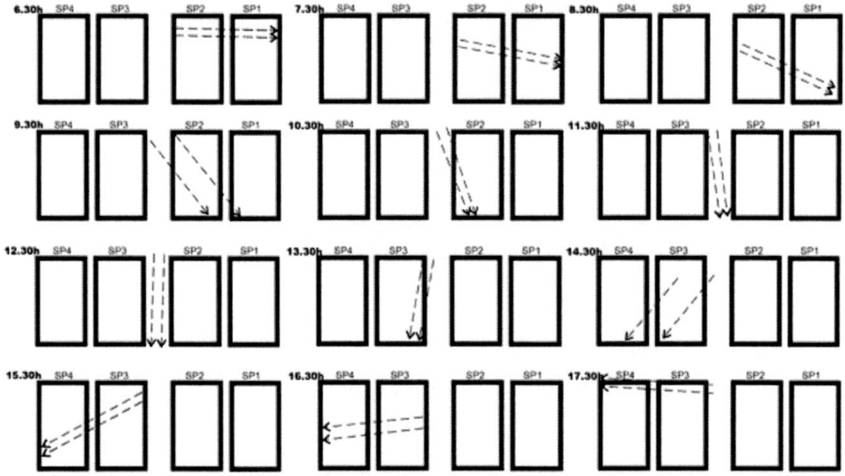


**Fig. 2** Projections of shadow lines on the PV panels

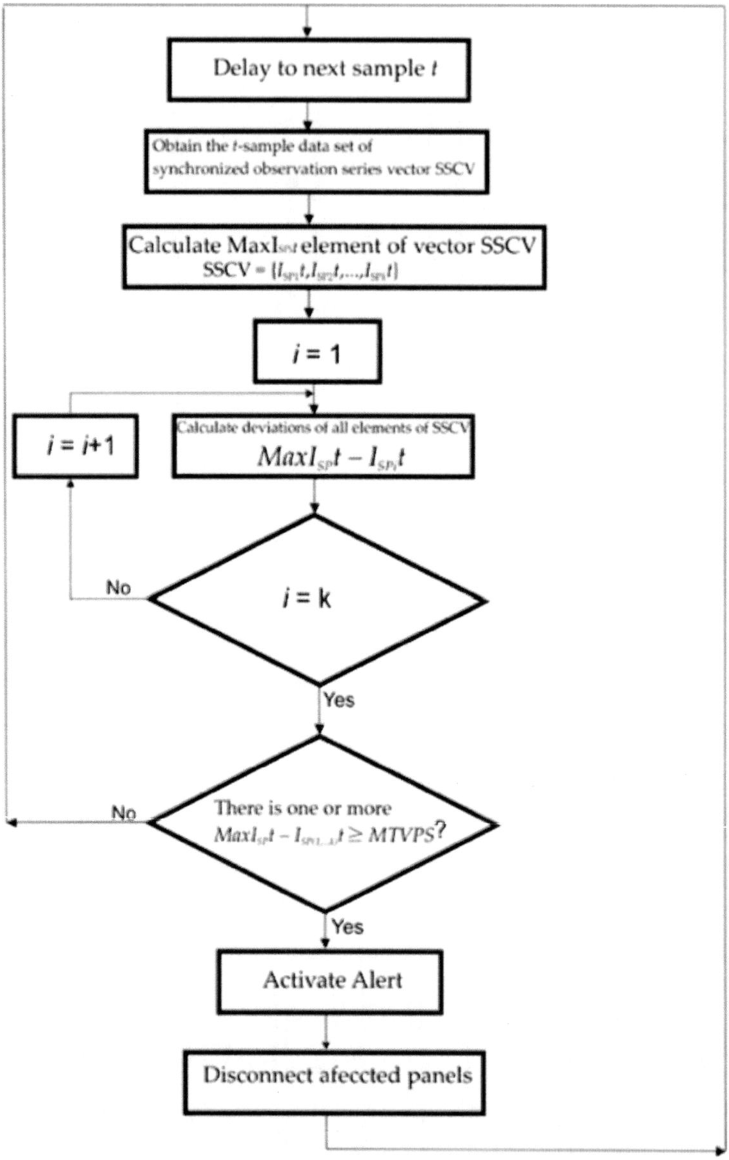


**Fig. 3** Fault diagnosis method flow chart

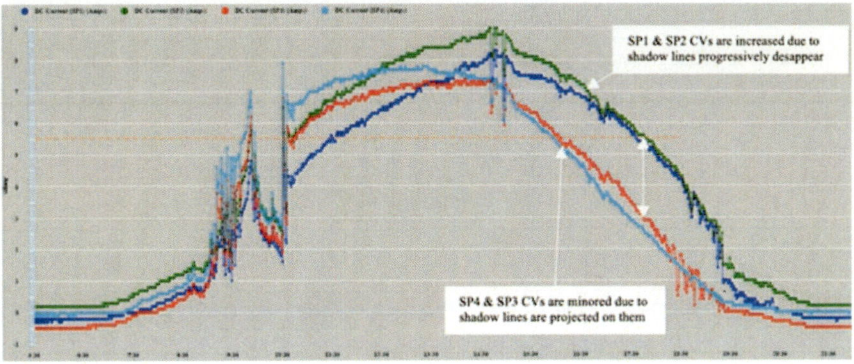


**Fig. 4** May 16th, 2021 data series power SP1, SP2, SP3 and SP4 registration

**Fig. 5** Shadow lines projections of backstay and antenna on SP1, SP2 and SP3, SP4 during the afternoon

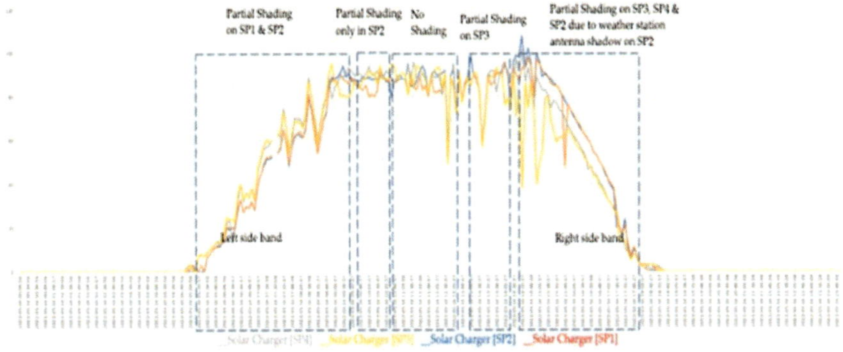


**Fig. 6** Analysis of the PVM data series

Table 5 partially shows, the values of the 4 series of generated power, for a short time in the early morning obtained from the left lateral band of Fig. 6 where the reduction of the generated power due to the partial shading that is taking place at that moment on the SP2(blue) and SP1(red) solar panels.

**Table 5** PVM data series for a sampling period of 5 min

| Time stamp UTC(+00:00) | Charge state | SP4 power | SP3 power | SP2 power | SP1 power | $P_{SPimax}t$ deviation |
|---|---|---|---|---|---|---|
| 2022–05-14 06:34:57 | Bulk | 40 | 40 | 33 | 33 | 7 |
| 2022–05-14 06:39:57 | Bulk | 32 | 33 | 27 | 27 | 6 |
| 2022–05-14 06:44:57 | Bulk | 36 | 36 | 30 | 30 | 6 |
| 2022–05-14 06:49:57 | Bulk | 34 | 34 | 29 | 29 | 5 |
| 2022–05-14 06:54:57 | Bulk | 38 | 38 | 32 | 32 | 6 |
| 2022–05-14 06:59:57 | Bulk | 33 | 33 | 28 | 28 | 5 |
| 2022–05-14 07:04:57 | Bulk | 46 | 47 | 40 | 40 | 7 |
| 2022–05-14 07:09:57 | Bulk | 51 | 52 | 44 | 44 | 8 |
| 2022–05-14 07:14:57 | Bulk | 56 | 57 | 48 | 49 | 9 |
| 2022–05-14 07:18:38 | Bulk | 57 | 60 | 51 | 52 | 9 |

## 4 Conclusions

The proposed methodology for predictive fault diagnosis in isolated photovoltaic installations allows to detect any shading process in any panel making a comparison of the deviation among the set of registration graphs of the data series corresponding to the power generated by each solar panels of the system.

For the deviation to be effective, it will be considered from a statistical point of view of the process, that is, if the deviation is punctual and random, it is not considered, while if the deviation is persistent and maintained over time, it is considered associated with an assignable no random cause as it happens in case of a partial shading occurrence on the panel. The deviation maintained over time will be considered as a SP predictor symptom that can be used for early detection and isolation of the problem that includes the immediate stoppage of the degrading process on the affected panel.

The use of CR MPPT permits to measure the predictive electrical parameters I-V-P of the solar panel part to be carried out in a specific way, unlike CR PWM where the observation of these parameters can only be carried out in the battery part, limiting the observation only to the parameter I of the output current of the PWM CR.

## References

1. García, E., Ponluisa, N., Quiles, E., Zotovic-Stanisic, R., Gutiérrez, S.C.: Solar panels string predictive and parametric fault diagnosis using low-cost sensors. Sensors **22**(1), 332 (2022). https://doi.org/10.3390/s22010332
2. García, E., Quiles, E., Zotovic-Stanisic, R., Gutiérrez, S.C.: Predictive fault diagnosis for ship photovoltaic modules systems applications. Sensors **22**(6), 2175 (2022). https://doi.org/10.3390/s22062175
3. García, E., Quiles, E., Correcher, A., Morant, F.: Marine NMEA 2000 smart sensors for ship batteries supervision and predictive fault diagnosis. Sensors **19**(20), 4480 (2019). https://doi.org/10.3390/s19204480
4. https://www.victronenergy.com/upload/documents/Cerbo_GX/GX_Device_Manual-en.pdf
5. https://www.victronenergy.com/upload/documents/Manual_SmartSolar_MPPT_75-10_up_to_100-20/MPPT_solar_charger_manual-en.pdf
6. https://vrm.victronenergy.com/installation-overview?demo-login
7. https://www.victronenergy.com/upload/documents/VictronConnect_Manual/VictronConnect_manual-en.pdf
8. https://www.victronenergy.com/upload/documents/VE.Smart_Networking/VE_Smart_Networking-en.pdf
9. https://www.startech.com/es-es/tarjetas-perifericos/hb31c4as

# Robotic and Autonomous Systems

# Path Following Control of Autonomous Tractor Using Non-linear Guidance Law

**Alexander Rodríguez Conte, Luis E. Hernández Morales, and Luis Hernández Santana**

**Abstract** The agricultural tasks of land preparation, sowing, cultivation and harvesting require a certain level of precision to achieve greater efficiency, productivity and profitability. This precision can be increased if autopilots are used in agricultural machinery such as tractors, sprayers and harvesters. This work is intended to evaluate the use of the Non-Linear Guidance Law (*NLGL*) guidance algorithm, better known as $L_1$, for the steering control of the Maxxum 110 tractor. Initially is dedicated to the dynamic modeling of the vehicle. Secondly its describes the guidance algorithm and its applicability to follow straight paths with land vehicles. Finally, a space is dedicated to present the simulation results in the tracking of a typical path in land preparation works.

**Keywords** Autonomous tractor · Dynamics model · Non-Linear Guidance Law

## 1 Introduction

Autonomous driving of heavy agricultural vehicles such as tractors, harvests and sprayers has become one of the most representative elements of precision agriculture [1]. The autopilots with which the machinery are equipped permits to perform tillage tasks with the precision required to increase productivity and input savings [2]. In this sense, the control of the lateral position of the vehicle is one of the most studied topics [3, 4]. It is essential to design guidance and control systems that allow to maintain the desired course for the fulfillment of a mission. Among the most widely used unmanned vehicle guidance algorithms are the Pure-Pursuit [5], Line of Sight (LOS)

A. R. Conte (✉) · L. E. H. Morales · L. H. Santana
Central University "Marta Abreu" of Las Villas, Santa Clara, Cuba
e-mail: arconte@uclv.cu

L. E. H. Morales
e-mail: luhmorales@uclv.cu

L. H. Santana
e-mail: luishs@uclv.edu.cu

 
O. Llanes-Santiago (ed.), *Proceedings of 19th Latin American Control Congress (LACC 2022)*, Studies in Systems, Decision and Control 464,
https://doi.org/10.1007/978-3-031-26361-3_14

[6] and Stanley Method [7]. The Non-Linear Guidance Law (*NLGL*) algorithm is an algorithm that has been mainly used in aerial [8] and marine [9] vehicles. However the mathematical expression of *NLGL* algorithm it can be used in the path-following control of ground vehicles with focus on the reduction of the Cross-track error to a straight path. For this work, the vehicle proposed to control its course is the Case IH Maxxum 110 tractor. This tractor is used in Cuba for agricultural work of different characteristics. The objective is to present the effectiveness of the *NLGL* algorithm for direction and heading control in the tracking of this tractor in the fulfillment of a certain tillage task defined by a desired path. For this purpose, the lateral dynamic model of the Maxxum 110 tractor is presented, as well as the model of the steering actuator proposed to be used. The characteristics that allow the *NLGL* algorithm to be used for the tracking of straight paths by large land vehicles are enunciated. Finally, the effectiveness of the algorithm is analyzed by simulating the tracking of two trajectories that could be used in land preparation or harvesting tasks.

## 2 Vehicle Description

The Case IH Maxxum 110 tractor is one of the vehicles used in Cuba for agricultural work. The Fig. 1 shows the reference system *OXYZ* (blue) is fixed to the vehicle, while the reference system *EXYZ* (black) is the Earth or non-mobile system. Also is assumed that the movement is made in the plane XY, therefore, $z_0 = 0m$ and the amplitude of the angles pitch and roll and it variations are zero $\theta = \phi = \dot{\theta} = \dot{\phi} \approx 0^{\underline{o}}$. In this conditions the states of the tractor are:

$$\eta = [x, y, \psi]^T; v = [u, v, r]^T \tag{1}$$

**Fig. 1** Relations of tractor

**Table 1** Characteristics of tractor Case IH Maxxum 110

| Variable | Value | Description |
|---|---|---|
| $L$ | 4.6750 m | Wheelbase distance |
| $l_f$ | 2.7470 m | Distance form COG to front axle |
| $l_r$ | 1.9280 m | Distance form COG to rear axle |
| $R$ | 5.7071 m | Turn radius of the vehicle |
| $v_{max}$ | 5 m/s | Maximum work velocity of vehicle |
| $v$ | 2 m/s | Work velocity |
| $m$ | 5820 kg | Mass of vehicle |
| $I_z$ | 30,824 kgm$^2$ | Momentum of inertia |
| $C_r$ | 182.61 kN/rad | Rear tires cornering stiffness |
| $C_f$ | 108.64 kN/rad | Front tires cornering stiffness |
| $\delta$ | 55° = 0.9599 rad | Maximum steering angle |
| $\mu$ | 1 | Friction coefficient (dry road) |

where $\eta$ denotes the position vector expressed in the ground reference frame, $v$ represents the linear and angular velocity vector with coordinates in the tractor's system. The kinematic equations can be expressed in vector form using Euler angle transformations like [10]:

$$\dot{\eta} = R(\psi)v = \begin{bmatrix} C\psi & -S\psi & 0 \\ S\psi & C\psi & 0 \\ 0 & 0 & 1 \end{bmatrix} v \tag{2}$$

The Table 1 shows the main kinematics and dynamic characteristics of the tractor [11, 12].

## 2.1 Lateral Dynamics Model

The model implemented in this paper apply the convention of the bicycle model (Fig. 2) used for modeling lateral dynamics of ground vehicles.

According to [13] the Transfer Function that relates the yaw rate ($r = \dot{\psi}$) with the steer angle ($\delta$) and has a structure with one zero and two poles:

$$G_v = \frac{r(s)}{\delta(s)} = \frac{b_1 s + b_0}{a_2 s^2 + a_1 s + a_0} \tag{3}$$

where:

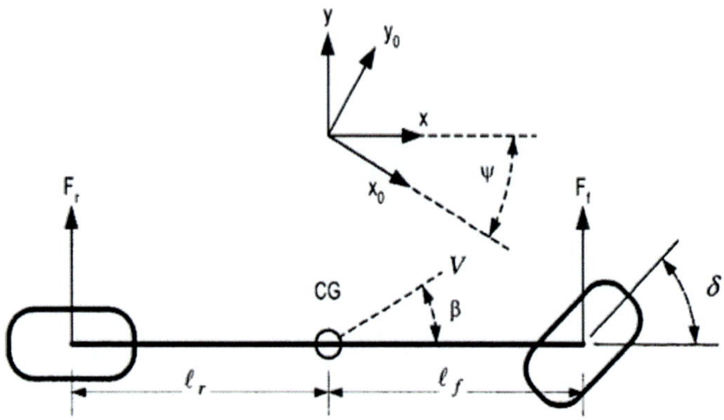


**Fig. 2** Vehicle lateral dynamic

- $b_0 = \mu^2 c_f c_r (l_f + l_r) v$
- $b_1 = \mu c_f l_f m v^2$
- $a_0 = \mu^2 c_f c_r (l_f + l_r)^2 + \mu^2 (c_r l_r - c_f l_f) m v^2$
- $a_1 = \mu \left( c_f \left( J + l_f^2 m \right) + c_r (J + l_r^2 m) \right) v$
- $a_2 = J m v^2$

$$G_v(s) = \frac{9.682s + 258.5}{s^2 + 49.33s + 3.005} \tag{4}$$

## 2.2 Steering Actuator

The steering actuator proposed to be used is similar to the DC motor used in Trimble's EZ Pilot system. The DCM50202 motor works with 50 W and 24 V power supply. Table 2 shows the motor coefficients [14].

According [15] the Transfer Function for a motor relating angular position and feeding voltage is:

**Table 2** Motor coefficients

| Variable | Value | Description |
|---|---|---|
| $K_a$ | $55*10^{-3}$ N·m/A | Torque constant |
| $K_b$ | $55*10^{-3}$ V/rad/s | Velocity constant |
| $R$ | 1,73 Ω | Resistance |
| $J_m$ | $1.62*10^{-5}$ kg·m$^2$ | Rotor inertia |

$$G_m(s) = \frac{\delta(s)}{V(s)} = \frac{Km}{s(Tms + 1)} = \frac{20.83}{s(0.01772s + 1)} \tag{5}$$

## 3 Algorithm *NLGL*

The guide algorithm *NLGL* it's based in the selection of a reference point on the desired trajectory and generate a lateral acceleration command [9, 16]. The reference point its placed to a distance ahead of the vehicle (Fig. 3). By other hand the Fig. 4 shows the general control scheme for *NLGL* algorithm, where the slip angle ($\beta$) introduced in Fig. 2 it is neglected.

| Notation | Description |
|---|---|
| $U$ | Vehicle velocity module |
| $L_1$ | A line defined from vehicle position to a VTP on a desired trajectory |

(continued)

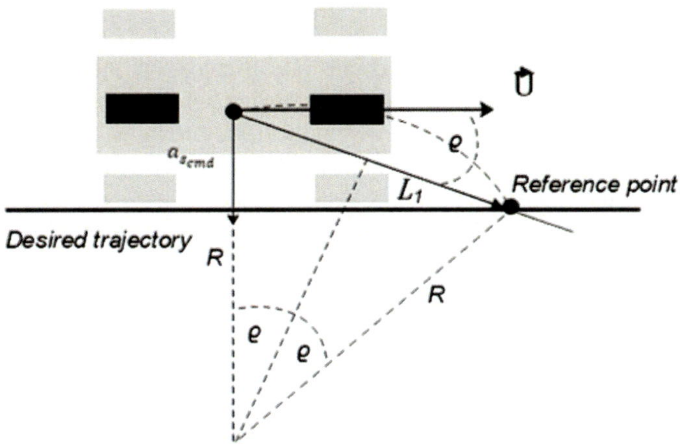


**Fig. 3** Geometric description of *NLGL* algorithm

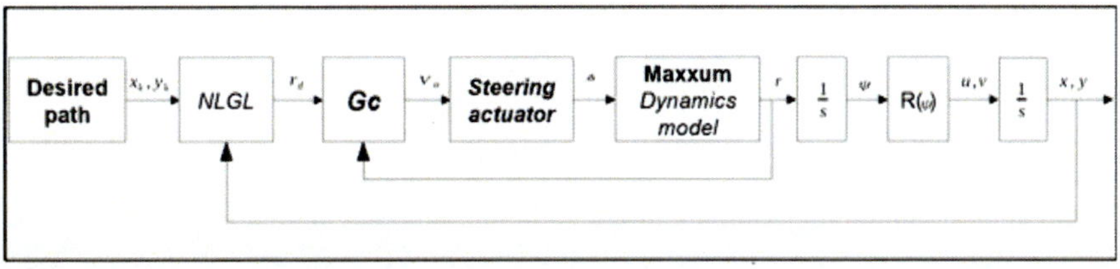


**Fig. 4** Control Scheme

(continued)

| Notation | Description |
|---|---|
| $\varrho$ | Angle created from $U$ to the line L1 |
| $a_{s_{cmd}}$ | Acceleration command sideways i.e. perpendicular to vehicle velocity direction |
| $d$ | Cross-track error |
| $R$ | Virtual circle radio |

The lateral acceleration command is the centripetal acceleration needed for follow the circular path to permits to the vehicle converge to the desired path and decreases the lateral error. It expression is an inverse relation between the square of velocity module ($U$) and the virtual circle radio ($R$):

$$a_{cen} = \frac{U^2}{R} \tag{6}$$

The relationship between the distance $L_1$ and $R$ is defined geometrically as: $L_1 = 2R sin\varrho$ (7). Therefore, the expression of the lateral command acceleration is:

$$a_{s_{cmd}} = 2\frac{U^2}{L_1} sin\varrho \tag{7}$$

Some considerations are taken account. For instance; (1) the direction of the lateral acceleration depends of the sign of the angle ($\varrho$) formed for the distance $L_1$ and $U$. (2) The vehicle is moving away from the path, the angle $\varrho$ increases and the acceleration will have a major value, and when the vehicle be on the path the value of the acceleration would be zero.

The lateral acceleration command is associated with the angular velocity. Then a desired yaw angular velocity is

$$r_d = \frac{a_{s_{cmd}}}{U} \tag{8}$$

In [9] it is used the Second *Lyapunov's* Method to analyze the stability of the nonlinear expression (8). The authors concluded that the *NLGL* algorithm is asymptotically stable when the vehicle follows a straight path. According both [9, 16], the expression (8) can be expressed linearly assuming that the vehicle is near to the trajectory. Then the value of angle $\varrho$ has small variations.

$$sin\varrho \approx \varrho_1 + \varrho_2 \tag{9}$$

The Fig. 5 shows the trigonometrical relations

$$sin\varrho \approx \varrho_1 + \varrho_2 \approx \frac{d}{L_1} + \frac{\dot{d}}{U} \tag{10}$$

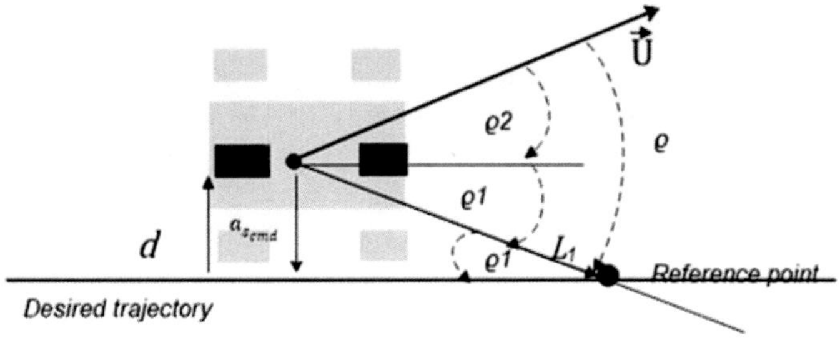


**Fig. 5** Geometric description for linearization of *NLGL* algorithm

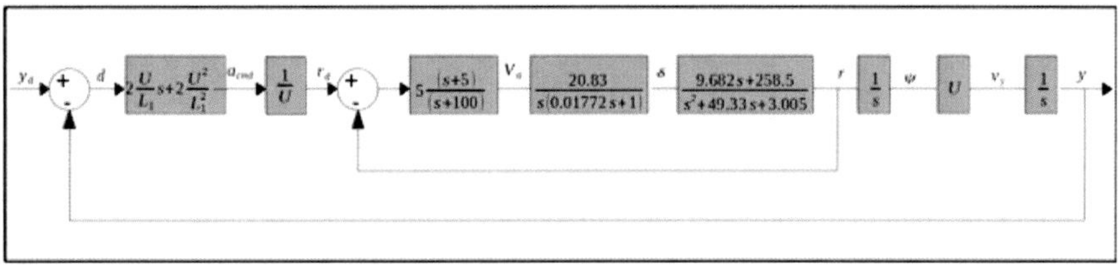


**Fig. 6** Complete linear scheme of guide and lateral control

$$a_{s_{cmd}} = 2\frac{U^2}{L_1} sin\varrho \approx \frac{2}{L_1}\left(\dot{d} + \frac{U}{L_1}\right) \tag{11}$$

$$r_d = \frac{a_{s_{cmd}}}{U} = \frac{2}{L_1}\left(\dot{d} + \frac{U}{L_1}d\right) \tag{12}$$

The above expression represents a structure of Proportional Derivative (PD) controller which acts over the lateral error to the path. The Fig. 6 presents the complete linear scheme of guide and lateral control of the vehicle in the axis $x$ with small variations in the direction $y$.

# 4 Results

## 4.1 Yaw Rate Controller

The design of the Yaw rate controller was made considering a setting time less than 2 s and rising time less than 0.5 s.

The show the response of the system formed by actuator and dynamic model. The controller designed is a Proportional-Derivative controller with $K_p = 25, K_D = 5$ and a real pole in $s = -100$. The numerical expression for the controller is:

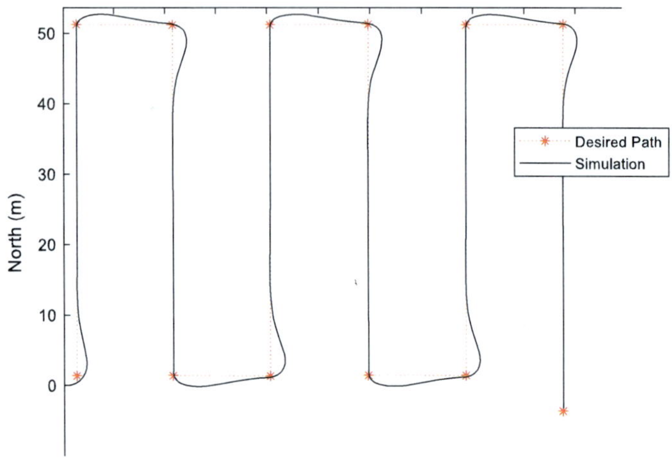


**Fig. 7** Path follow simulation results

$$G_c = 5\frac{(s+5)}{(s+100)} \tag{13}$$

## 4.2 *Simulation of Path Following*

For the simulation is used Matlab/Simulink and is performed with a constant speed of 2 m/s. The desired path represents a typical agricultural task. Also is implemented the model of the steering actuator, the tractor's lateral dynamics and the *NLGL* algorithm (Fig. 7).

The above image shows the effectiveness of the selected guidance algorithm and the stability of the implemented control systems. Also, its appreciated that the mathematical model obtained, adequately describes the dynamics of the vehicle. Since the wheelbase of the tractor is large, the above image shows the drift that occurs when the vehicle changes the direction of motion. When the path is straight and have sufficient large the trajectory converges to the road and the Cross-track error is reduced to zero. The next figure shows the behavior of the lateral error (Fig. 8).

# 5 Conclusions

The present research contributes to the development of a national autopilot to be used in sugarcane agriculture in Cuba. A mathematical model describing the lateral

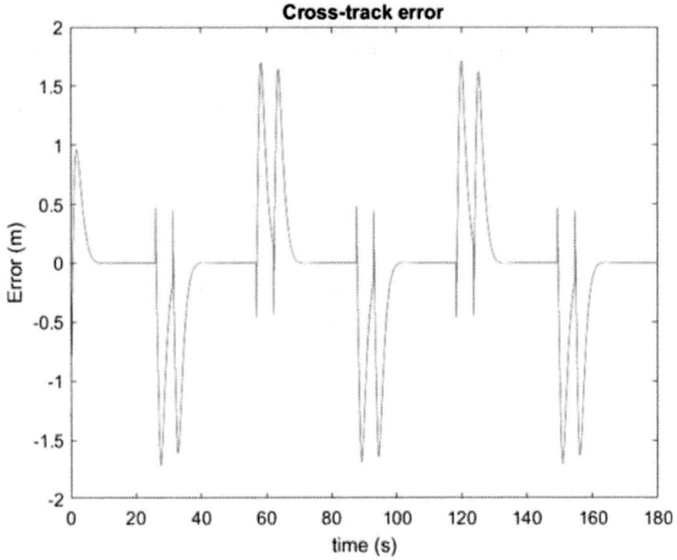


**Fig. 8** Cross-track error

dynamics of the Maxxum 110 tractor was obtained using the bicycle-type configuration. The results of the simulations carried out demonstrate the validity of the model found, as well as the usefulness of the *NLGL* algorithm for its use in the autonomous driving applied to this vehicle in straight paths follow.

## References


1. Mahmud, M.S.A., Abidin, M.S.Z., Emmanuel, A.A., Hasan, H.S.: Robotics and automation in agriculture: present and future applications. Appl. Model. Simul. **4**, 130–140 (2020)
2. Rehman, A., Jingdong, L., Khatoon, R., Hussain, I., Iqbal, M.S.: Modern agricultural technology adoption its importance, role and usage for the improvement of agriculture. Life Sci. J. **14**(2), 70–74 (2016)
3. Alberto-Rodriguez, A., et al.: Review of control on agricultural robot tractors. Int. J. Comb. Optim. Probl. Inform. **11**(3), 9–20 (2020)
4. Matholiya, C.S., Gupta, P., Agravat, V.V., Patel, U.V., Balas, P.R.: Automatic guidance systems in agricultural autonomous robotic machine: a review (2022)
5. Li, H., Luo, J., Yan, S., Zhu, M., Hu, Q., Liu, Z.: Research on parking control of bus based on improved pure pursuit algorithms. In: 2019 18th International Symposium on Distributed Computing and Applications for Business Engineering and Science (DCABES), pp. 21–26 (2019)
6. Jin, X., Dai, S.-L., Liang, J., Guo, D., Tan, H.: Constrained line-of-sight tracking control of a tractor-trailer mobile robot system with multiple constraints. Am. Control Conf. (ACC) **2021**, 1046–1051 (2021)

7. Wang, L., Zhai, Z., Zhu, Z., Mao, E.: Path Tracking Control of an Autonomous Tractor Using Improved Stanley Controller Optimized with Multiple-Population Genetic Algorithm, vol. 11, no. 1, p. 22 (2022)
8. He, M., Wang, X., Cui, N.: Modified vector field and nonlinear guidance law for low-cost UAV path following. Aircr. Eng. Aerosp. Technol. no. ahead-of-print (2022)
9. Hernández-Morales, L., Valeriano-Medina, Y., Hernández-Santana, L., Mesa-Suarez, E.: Nonlinear guidance law algorithm applied to a small unmanned surface vehicle. Proc. Inst. Mech. Eng. Part M J. Eng. Marit. Environ. **234**(3), 623–633 (2020)
10. Jazar, R.N.: Vehicle planar dynamics. Advanced Vehicle Dynamics. Springer International Publishing, Cham, pp. 115–213 (2019).https://doi.org/10.1007/978-3-030-13062-6_2
11. Report on test in accordance with OECD STANDARD CODE II for the Official Testing of Agricultural Tractor Performance, England (1997). https://pruefberichte.dlg.org/filestorage/CASE-IH_MA_110_MAXXUM-Nr1680-1997-englisch.pdf
12. Marugán, J.: Los nuevos CASE. Serie MX Maxxum, Agrotécnica. Cuadernos de Agronomía y Tecnología, no. 8, Ocutbre 1998. https://www.mapa.gob.es/ministerio/pags/Biblioteca/Revistas/pdf_Agrotec%2FAgrotec_1998_8_94_100.pdf
13. Ackermann, J. et al.: Robust Control: the Parameter Space Approach, vol. 2. Springer (2002)
14. GOWE DC Servo Motor dcm50202 - 02d-1000 24 VDC 50W 3500rpm 1000-line. https://www.amazon.com/Servo-DCM50202-02D-1000-3500rpm-1000-Line-Encoder/dp/B078CYDYXH
15. Ogata, K.: Modern control Engineering, vol. 5. Prentice Hall, Upper Saddle River, NJ (2010)
16. Park, S., Deyst, J., How, J.: A New Nonlinear Guidance Logic for Trajectory Tracking, p. 4900 (2004)

# Integral Line-of-Sight Controller for 3-D Path Following in Presence of Marine Currents

**Yunier Valeriano Medina, Omar Milián Morón, Leidys Miranda Jiménez, and Anailys Hernández Julián**

**Abstract** The development of autonomous underwater vehicles is a topic of great interest. The guidance system plays a vital role in fulfilling missions without human intervention. This paper presents the design of a guidance scheme formed by two uncoupled I-LOS controllers, one for the horizontal plane and one for the vertical plane. The scheme achieves following straight paths in three dimensions and reduces tracking errors caused by marine currents. Adjusting the controllers' gains is done according to the vehicle's characteristics and the path's geometry. The proposed guidance scheme is validated through simulations, demonstrating the feasibility of its use in the underwater vehicle HRC-AUV.

## 1 Introduction

The marine environment represents a challenging environment for exploiting techniques and technologies related to automatic control. Fortunately, the technological development achieved in recent years has facilitated the implementation of automatic control systems in marine vehicles. An example is the Autonomous Underwater Vehicles (AUV), which carry their power source and computing units, where software and control solutions are executed, allowing them to fulfill predefined tasks without human intervention [1]. Marine vehicles' motion control system comprises three fundamental modules: guidance, control, and navigation [8]. The guidance and control blocks have a high impact on the autonomous behavior of the vehicle. Guidance

Y. Valeriano Medina · O. Milián Morón (✉) · L. Miranda Jiménez · A. Hernández Julián
Central University "Marta Abreu" of las Villas, Carretera a Camajuaní Km. 5 y 1/2, Santa Clara, Villa Clara, Cuba
e-mail: omilian@uclv.cu

Y. Valeriano Medina
e-mail: yunierv@uclv.edu.cu

L. Miranda Jiménez
e-mail: lemiranda@uclv.cu



O. Llanes-Santiago (ed.), *Proceedings of 19th Latin American Control Congress (LACC 2022)*, Studies in Systems, Decision and Control 464,
https://doi.org/10.1007/978-3-031-26361-3_15

determines the control module's desired position, velocity, and acceleration values. Control deals with determining the necessary forces and moments to be applied to the vehicle to satisfy a given control objective [10]. Path following without time constraints is a motion control scenario, where the control objective is to converge and follow a previously defined geometric path without considering any time constraints. When the vehicle is under-actuated [9], numerous guidance and control solutions are proposed for this type of scenario, ranging from the PID controller [14] to others that include complex algorithms [2, 15]. However, in many works, the guidance strategy known as the Line Of Sight (LOS) is used [4]. The LOS strategy cannot ensure accuracy during path following in the presence of ocean currents. A reasonable way to solve the following error is to incorporate an integral action in the guidance law. This variant, known as I-LOS [7], has been mainly used in surface vehicles [5, 11, 13]. In those operating underwater, the guidance scheme must be able to lead the vehicle to a path at a specific depth . In this case, sea currents cause tracking errors in horizontal and vertical planes, which must be counteracted with control actions.

This research is based on the autonomous underwater vehicle HRC-AUV [12, 16], for which a decoupled guidance scheme is proposed, composed of two I-LOS controllers, one to correct the error in the horizontal plane and the another to correct it in the vertical plane. The contribution of this work lies in proposing a guidance scheme formed by a horizontal I-LOS controller and another I-LOS controller for the vertical plane, capable of ensuring the convergence of a 3D path and reducing tracking errors; despite the effects caused by sea currents, the non-linearities that characterize the dynamics of the vehicle and the fact that the vehicle has fewer actuators than degrees of freedom to control.

## 2 Dynamic Models of HRC-AUV

The HRC-AUV is a vehicle equipped with three actuators located aft: a thruster, a horizontal steering rudder, and a vertical stabilizer rudder. HRC-AUV can safely navigate depths of 10 m for up to 6 h. The geometric and inertial data of the vehicle are shown in Table 1 [16, 17]. In order to analyze the movement of marine vehicles in the six degrees of freedom (6 DOF), it is convenient to define two coordinate systems, as shown in Fig. 1. The mobile coordinate system is fixed on the vehicle, and its origin is made to coincide with the center of buoyancy. The ground-fixed reference system is considered inertial and with origin at OE. The following vectors describe the movement of a marine vehicle in the 6 DOFs [17]: $\boldsymbol{\upsilon} = [u, v, w, p, q, r]^T$, $\boldsymbol{\eta} = [x, y, y, \phi, \theta, \psi]^T$ and $\boldsymbol{\tau} = [X, Y, Z, K, M, NN]^T$. The position and orientation vector is denoted as $\boldsymbol{\eta}$, $\boldsymbol{\upsilon}$ represents the linear and angular velocity vector, and $\boldsymbol{\tau}$ is used to represent the forces and moments acting on the vehicle. The kinematic equations can be expressed in vector form as shown in Eq. 1, where $\mathbf{J}(\boldsymbol{\eta})$ represents the transformation matrix as a function of the Euler angles [9].

$$\dot{\boldsymbol{\eta}} = \boldsymbol{J}(\boldsymbol{\eta})\boldsymbol{\upsilon} \tag{1}$$

**Table 1** Geometric and inertial values of the HRC-AUV

| Parameters | Description | Value | Parameters | Description | Value |
|---|---|---|---|---|---|
| $m$ | Mass | 4094.56 kg | $n$ | Propeller revolutions | 52.36 rad/s |
| $u_0$ | Cruise speed | 1.9 m/s | **BG** | Distance between CG and CB | $[0, 0, 22\text{mm}]^T$ |
| $L$ | Length | 9.46 m | $\delta_T$ | Tail rudder deflection angle | $\pm 30\,^\circ$S |
| $R$ | Radius | 0.4 m | $\delta_E$ | Tail stabilizer deflection angle | $\pm 30\,^\circ$S |
| $I_{xx}$ | Moment of inertia | 450.1 kgm$^2$ | $I_{zz}$ | Moment of inertia | 20816 kgm$^2$ |
| $I_{yy}$ | Moment of inertia | 21010.4 kgm$^2$ | $I_{xz}$ | Moment of inertia | 275.44 kgm$^2$ |

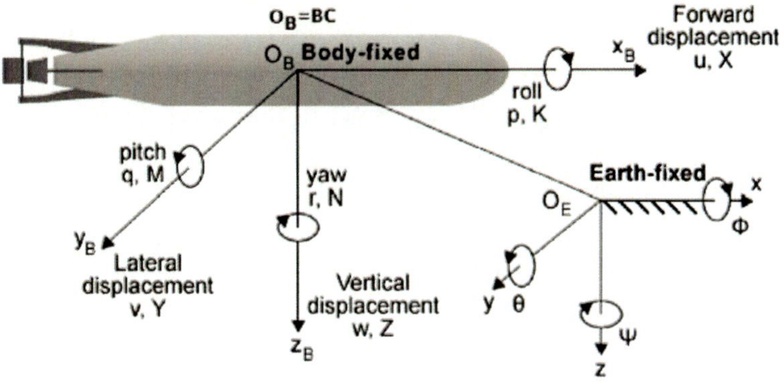


**Fig. 1** Coordinate systems used in the modeling of marine vehicles

The equations of motion for a marine vehicle can be expressed as Eq. 2.

$$\boldsymbol{M_{RB}\dot{\upsilon}_r} + \boldsymbol{C_{RB}(\upsilon_r)\upsilon_r} + \boldsymbol{M_A\dot{\upsilon}_r} + \boldsymbol{C_A(\upsilon_r)\upsilon_r} + \boldsymbol{D(\upsilon_r)\upsilon_r} + \boldsymbol{g(\eta)} = \boldsymbol{\tau} + \boldsymbol{\tau_{waves}} \quad (2)$$

where $\boldsymbol{M} = \boldsymbol{M_R B} + \boldsymbol{M_A}$ is the inertia matrix, $\boldsymbol{C(\upsilon_r)} = \boldsymbol{C_{RB}(\upsilon_r)} + \boldsymbol{C_A(\upsilon_r)}$ is the Coriolis matrix, $\boldsymbol{D(\upsilon_r)}$ groups the damping terms, $\boldsymbol{g(\eta)}$ is the force vector gravitational and buoyancy forces, finally, $\boldsymbol{\tau}$ is the vector of controlling forces and moments. Equation 2 incorporates the terms related to waves ($\boldsymbol{\tau_{waves}}$) and marine currents ($\boldsymbol{\upsilon_r}$).

The JONSWAP spectrum is used to represent the waves generated by the wind in areas where the sea region is limited, and the depth is considered finite [9]. The second-order transfer function in Eq. 3 represents the effects of waves, which can be included in the control loops.

$$y(s) = \frac{K_w s}{s^2 + 2\zeta\omega_0 s + \omega_0^2}. \quad (3)$$

The gain is defined as $K_w = 2\zeta\omega_0\sigma$, where $\zeta_\omega$ is a constant that describes the intensity of the waves, $\zeta$ is the damping coefficient, whose value is 0.1 for the JONSWAP spectrum, and $\omega_0$ is the fundamental frequency of the waves.

Sea currents are defined in the OE inertial reference system. They are considered constant, irrotational, and bounded [6]. For underwater vehicles, they are defined by the vector $\boldsymbol{V_c} = [V_x,\ V_y,\ V_z)]^T$, whose magnitude value is determined as $V_{ca} = \sqrt{V_x^2 + V_y^2 + V_z^2}$. In addition to the magnitude, the components of the vector $\boldsymbol{V_c}$ depend on two angles: the lateral direction angle ($\beta_c$) and the vertical direction angle ($\alpha_c$).

$$\begin{aligned} V_x &= V_{ca}\cos\alpha_c\cos\beta_c \\ V_y &= V_{ca}\sin\beta_c \\ V_z &= V_{ca}\sin\alpha_c\cos\beta_c \end{aligned} \tag{4}$$

The speed of sea currents of an irrotational fluid can be represented in the vehicle reference system as:

$$\boldsymbol{\upsilon}_c = \begin{bmatrix} u_c, & v_c, w_c, 0, 0, 0 \end{bmatrix}^T \tag{5}$$

where $u_c$, $v_c$ and $w_c$ represent the linear components of the speed of currents. The effect of sea currents is added to the non-linear 6 DOF model of the vehicle by the term $\boldsymbol{\upsilon_r}$ (relative velocity). Relative velocity is defined in the OB reference frame as:

$$\boldsymbol{\upsilon_r} = \boldsymbol{\upsilon} + \boldsymbol{\upsilon_c} \tag{6}$$

The vector $\boldsymbol{\upsilon_c}$ is related to $\boldsymbol{V_c}$ by the following equation:

$$\begin{bmatrix} u_c \\ v_c \\ w_c \end{bmatrix} = \boldsymbol{J}_1(\boldsymbol{\eta}_2)^T \begin{bmatrix} V_x \\ V_y \\ V_z \end{bmatrix} \tag{7}$$

where $\boldsymbol{J}_1(\boldsymbol{\eta}_2)$ represents the translation matrix from OB to OE using the Euler angles [9]. The model expressed by Eq. 2 is used for the simulation. In [17] the values of the parameters that are part of it and that have been determined from the geometric and inertial data of the vehicle, as well as the result of experimental tests, are presented.

## 3 Decoupled Steering and Pitch Controllers

The vehicle control system is responsible for supplying the forces and moments necessary to satisfy a specific control objective. The control objective is often defined in conjunction with the guidance system. When the desired values come from a guidance algorithm, the controllers' functions would be to steer the vehicle to a certain point or follow a path.

Equation 2 is not useful for decoupled controller design. One solution is to divide the system into several subsystems with little interaction. The three subsystems are velocity, lateral, and longitudinal. An underwater vehicle that navigates at a constant speed requires a heading and pitch controller to follow a path at a certain depth. The heading controller must be able to follow the desired values corresponding to the yaw angle, while the pitch controller is in charge of following the references of the angle $\theta$.

## 3.1 Heading Controller

The vehicle dynamics for the lateral subsystem can be described as:

$$\begin{bmatrix} \dot{v} \\ \dot{r} \\ \dot{\psi} \end{bmatrix} = \begin{bmatrix} \frac{Y_v}{m-Y_{\dot{v}}} & \frac{-mu_0}{m-Y_{\dot{v}}} & 0 \\ 0 & \frac{N_r}{I_{zz}-N_{\dot{r}}} & 0 \\ 0 & 1 & 0 \end{bmatrix} \begin{bmatrix} v \\ r \\ \psi \end{bmatrix} + \begin{bmatrix} \frac{b_2}{m-Y_{\dot{v}}} \\ \frac{b_5}{I_{zz}-N_{\dot{r}}} \\ 0 \end{bmatrix} \delta_T \tag{8}$$

where $Y_{\dot{v}}$ and $N_{\dot{r}}$ are added mass coefficients, $Y_v$ and $N_r$ represent the damping terms, while $b_2$ and $b_5$ are the gains related to the control input. From this model it is possible to obtain the transfer function between the yaw angle and the deflection angle of the horizontal rudder:

$$\frac{\psi(s)}{\delta_T(s)} = \frac{b_5}{(I_{zz} - N_{\dot{r}})\, s^2 - N_r s} \tag{9}$$

The proposed steering control solution consists of a P-D controller, whose output equation is given by:

$$\delta_T = K_p(\psi_d - \psi) - K_d r \tag{10}$$

The gain values are adjusted from Eq. 9, using the pole placement technique.

## 3.2 Pitch Controller

The vehicle dynamics for the longitudinal subsystem is defined by Eq. 11.

$$\begin{bmatrix} \dot{z} \\ \dot{\theta} \\ \dot{q} \end{bmatrix} = \begin{bmatrix} 0 & -u_0 & 0 \\ 0 & 0 & 1 \\ 0 & -\frac{WBG_z}{I_{yy}-M_{\dot{q}}} & \frac{M_q}{I_{yy}-M_{\dot{q}}} \end{bmatrix} \begin{bmatrix} z \\ \theta \\ q \end{bmatrix} + \begin{bmatrix} 0 \\ 0 \\ \frac{b_4}{I_{yy}-M_{\dot{q}}} \end{bmatrix} \delta_E \tag{11}$$

where $W$ represents the weight of the vehicle $M_{\dot{q}}$ and $M_q$ constitute the moments related to the added masses and damping, respectively. For its part, $b_4$ represents the gain relative to the control input. From this model we arrive at the transfer function that relates the pitch angle to the tail stabilizer deflection, which is used to adjust the controller.

$$\frac{\theta(s)}{\delta_E(s)} = \frac{b_4}{\left(I_{yy} - M_{\dot{q}}\right)s^2 - M_q s + WBG_z} \quad (12)$$

To control the vertical position of the vehicle a pitch controller is needed. In this work it is proposed to use a P-D structure, whose output equation is the following:

$$\delta_E = G_1(\theta_d - \theta) - G_2 q \quad (13)$$

The gains $G_1$ and $G_2$ are determined using the pole placement technique.

## 4 I-LOS Controllers

Figure 2 shows the decoupled guidance scheme proposed in this work for 3-D path tracking. The scheme is made up of an I-LOS controller that generates the appropriate heading references for the vehicle to converge to a straight path in the x-y plane. Similarly, an I-LOS controller generates the pitch references for convergence to the path in the x-z plane. The modification of the LOS algorithm, introducing an integral action, responds to the need to effectively counteract the deviations caused by sea currents in the horizontal and vertical planes while tracking a 3D path. The scheme's structure for the horizontal plane is similar to that used for road following in surface vehicles [3]. Dividing the schematic into two parts makes the guidance solution easier to understand and easier to use for 3-D roads.

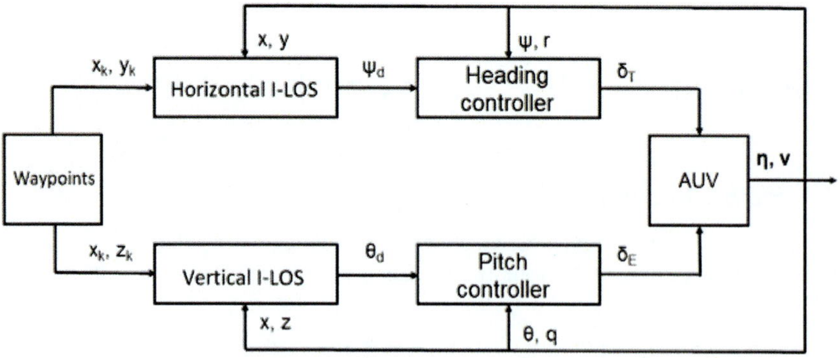


**Fig. 2** Decoupled guidance scheme

### 4.1 I-LOS Controller for the Horizontal Plane

The lookahead distance-based guidance law is used to steer the vehicle onto the path in the direction of the LOS vector [7]. The LOS vector is oriented from the vessel to a point located on a line tangential to the desired path at a lookahead distance ($\Delta_h > 0$) from the projection of the vehicle's position on the path [10].

The x-axis of the path has been rotated by a positive and constant angle $\alpha_K = atan2(y_{k+1} - y_k,\ x_{k+1} - x_k)$ relative to the $x$-axis of OE.

The expressions for the tracking error along the path $x_e(t)$ and for the tracking error perpendicular to the path $y_e(t)$ are stated as:

$$x_e(t) = (x_t - x_k)\cos\alpha_k + (y_t - y_k)\sin\alpha_k \tag{14}$$

$$y_e(t) = -(x_t - x_k)\sin\alpha_k + (y_t - y_k)\cos\alpha_k \tag{15}$$

This law determines the course angle of the vehicle as the sum of two angles:

$$\chi = \chi_r(y_e) + \chi_p \tag{16}$$

where $\chi_r(y_e)$ is a correction angle, which ensures that the speed of the vehicle is in the direction of the point on the path towards which the vessel is heading, and $\chi_p = \alpha_K$. The control objectives in the horizontal plane for the path following scenario are:

$$\lim_{t\to\infty} y_e(t) = 0, \qquad \lim_{t\to\infty} \psi(t) = \psi_d \tag{17}$$

In the presence of sea currents, a difference between $\psi$ and $\chi$ is produced. The desired heading angle must then be calculated as:

$$\psi_d = \chi + \beta_r \tag{18}$$

where $\beta_r$ is known as the sideslip angle. When an I-LOS controller is used, $\beta_r \approx 0$ is considered and the guidance algorithm based on the lookahead distance is modified, being expressed as follows [13]:

$$\psi_d = \alpha_k - \arctan\left(\frac{y_e + \sigma_h y_{int}}{\Delta_h}\right) \tag{19}$$

$$\dot{y}_{int} = \left(\frac{\Delta_h y_e}{(y_e + \sigma_h y_{int})^2 + {\Delta_h}^2}\right) \tag{20}$$

where $\sigma_h > 0$ and $\Delta_h$ are the fit parameters. In the case of $\Delta_h$, its value depends on the length of the vehicle.

## 4.2 I-LOS Controller for the Vertical Plane

In the vertical plane, the $(x, z)$ coordinates of the path points are used. For this case, the tangential angle to the path is determined as $\gamma_K = atan2(-(z_{k+1} - z_k), (x_{k+1} - x_k))$. The lookahead distance-based guidance algorithm is adapted to achieve convergence and path following in the vertical plane.

In this plane, both the tracking error along the path and the vertical tracking error to the path are produced, which are determined by Eqs. 21 and 22. The sum of angles to ensure that the vehicle navigates towards the next point on the path is stated as shown in Eq. 23, where $\chi_r(z_e)$ is the correction angle for the vertical plane and in this case $\chi_p = \gamma_K$. The control objectives for this plane are represented in Eq. 24.

$$x_e(t) = (x_t - x_k)\cos\gamma_k - (z_t - z_k)\sin\gamma_k \tag{21}$$

$$z_e(t) = (x_t - x_k)\sin\gamma_k + (z_t - z_k)\cos\gamma_k \tag{22}$$

$$\chi = \chi_r(z_e) + \chi_p \tag{23}$$

$$\lim_{t\to\infty} z_e(t) = 0, \qquad \lim_{t\to\infty} \theta(t) = \theta_d \tag{24}$$

The sea currents cause a deviation in the pitch angle during the vertical movement of the vehicle. According to equation (25), this effect is mathematically represented by the angle of attack $\alpha_r$. To counteract the effect caused by marine currents, a vertical I-LOS driver is used. Then, it is considered $\alpha_r \approx 0$ and an integral action is included in Eq. 25, as shown Eqs. 26 and 27, where $\sigma_v > 0$ and $\Delta_v$ are the fit parameters. Therefore, the control objectives expressed in Eq. 24 are met.

$$\theta_d = \chi + \alpha_r \tag{25}$$

$$\theta_d = \gamma_k + \arctan\left(\frac{z_e + \sigma_v z_{int}}{\Delta_v}\right) \tag{26}$$

$$\dot{z}_{int} = \left(\frac{\Delta_v z_e}{(z_e + \sigma_v z_{int})^2 + {\Delta_v}^2}\right) \tag{27}$$

If the path to follow is made up of n straight line segments connected by $n + 1$ points, it is necessary to use a criterion to know when the vehicle has overcome a point on the path. This makes it possible to change direction to the next point. If for each point a sphere is established, whose radius will be of value $R_{k+1} > 0$ for the point of the path $k + 1$, then it could be defined that the vehicle goes to the next point when:

$$(x_{k+1} - x(t))^2 + (y_{k+1} - y(t))^2 + (z_{k+1} - z(t))^2 \le {R^2}_{k+1} \tag{28}$$

## 5 Results of the Simulations

The scheme shown in Fig. 2 is used to achieve accuracy during 3-D path following. The I-LOS controllers are responsible for ensuring convergence to the path that is generated from a set of points. The nonlinear 6-DOF model, which includes the effect of waves and ocean currents, is used in numerical simulation tests with MATLAB software to evaluate the scheme's performance. The numerical values of the model matrices and vectors are presented in [17]. The model outputs affected by marine disturbances are not processed by any filtering algorithm, which constitutes an additional challenge for the proposed guidance solution.

The first test aims to identify the most appropriate values for the lookahead distances. The parameters $\Delta_h$ and $\Delta_v$ are the most influential when it comes to ensuring the convergence of the vehicle to the road. For this, a straight path is defined formed by the points: $p1 = (300, 400, 3)$ and $p2 = (1000, 0.10)$. The vehicle is initially located at the following coordinates $(x0, y0, z0) = (0, 400, 0)$. The speed at which the motor rotates is 500 rpm, and the vehicle is affected by a constant sea current: $V_{ca} = 0.1$ m/s, $\beta_c = 45\,^{\circ}$S and $\alpha_c = 20\,^{\circ}$S. Wave effects are represented with values of $w_0 = 3, 0$ rad/s and $\varsigma_w = 0.5$. The gains for the direction and pitch controllers are: $K_p = 1.54$, $K_d = 1$ and $G_1 = -1.8$, $G_2 = 1$, respectively. On the other hand, the values selected for the integral action of the I-LOS controllers are: $\sigma_y = \sigma_z = 0.5$

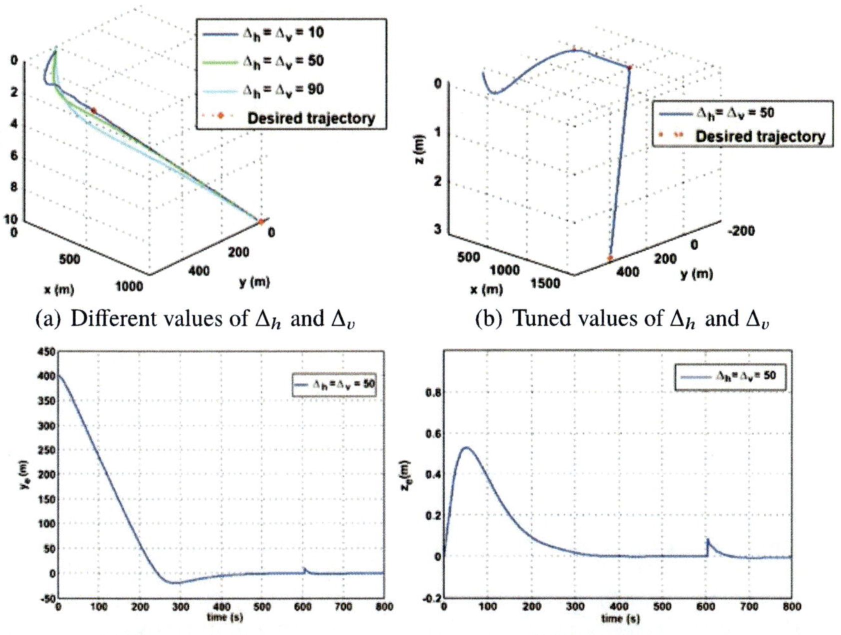


(a) Different values of $\Delta_h$ and $\Delta_v$ (b) Tuned values of $\Delta_h$ and $\Delta_v$

(c) Perpendicular error (d) Vertical error

**Fig. 3** Performance obtained while following a 3-D path

m/s. The radius of the sphere that is used to consider a point on the path as expired is 20 m. Under these conditions, three simulations are performed, each with different values of $\Delta_h$ and $\Delta_v$ and proportional to the length of the vehicle. Figure 3a shows the results achieved during the path following $\Delta_h = \Delta_v = 10$ m, $\Delta_h = \Delta_v = 50$ m, and $\Delta_h = \Delta_v = 90$ m. If the values of the lookahead distances are selected small, oscillations are produced that can affect the convergence. On the other hand, high values of $\Delta_h$ and $\Delta_v$ cause the convergence to be slow. That is why it is suggested to use $\Delta_h = \Delta_v = 50$ m as lookahead distance values for the case of the HRC-AUV vehicle. Another numerical simulation is analyzed using a more complex 3-D straight path, which is defined with the points: $p1 = (200, 0, 0)$, $p2 = (900, 0, 0)$ and $p3 = (1600, 400, 3)$ . The conditions of the experiment are the same as for the previous case. According to Fig. 3b, the HRC-AUV successfully follows the road segments. Figure 3c and d show how the I-LOS controllers manage to reduce the tracking errors that arise in each plane due to the presence of sea currents. The spikes that occur in the error signals are caused by the I-LOS controllers, which immediately try to steer the vehicle onto the next leg of the path.

## 6 Conclusion

The decoupled guidance scheme ensures the convergence and following of a 3-D path by an underwater and sub-actuated vehicle, such as the HRC-AUV vehicle. The horizontal I-LOS controller is responsible for reducing the tracking error perpendicular to the path. In contrast, the vertical I-LOS controller acts the same way with the vertical error to the path, despite the presence of marine currents, dynamic complexities, and model uncertainties. The results obtained in the numerical simulations show that the decoupled guidance scheme based on I-LOS controllers constitutes a feasible solution for following paths in 3-D without considering time restrictions.

## References


1. Antonelli, G., Fossen, T.I., Yoerger, D.R.: Modeling and Control of Underwater Robots, pp. 1285–1306. Springer International Publishing, Cham (2016)
2. Assunção, M., Teodoro, P., Marat-Mendes, R., Franco, V.: Design of an underactuated usv catamaran. In: Brito Palma, L., Neves-Silva, R., Gomes, L. (eds.) CONTROLO 2022, pp. 656–666. Springer International Publishing, Cham (2022)
3. Bibuli, M., Caharija, W., Pettersen, K.Y., Bruzzone, G., Caccia, M., Zereik, E.: Ilos guidance - experiments and tuning. IFAC Proc. Vol. **47**(3), 4209–4214 (2014). 19th IFAC World Congress
4. Breivik, M., Fossen, T.I.: Guidance laws for autonomous underwater vehicles (Chap. 4). In: Inzartsev, A.V. (ed.) Underwater Vehicles. IntechOpen, Rijeka (2009)
5. Caharija, W., Pettersen, K.Y., Bibuli, M., Calado, P., Zereik, E., Braga, J., Gravdahl, J.T., Sreønsen, A.J., Milovanović, M., Bruzzone, G.: Integral line-of-sight guidance and control of underactuated marine vehicles: theory, simulations, and experiments. IEEE Trans. Control Syst. Technol. **24**(5), 1623–1642 (2016)

6. Caharija, W., Pettersen, K.Y., Gravdahl, J.T.: Path following of underactuated surface vessels in presence of unknown constant environmental forces: preliminary results. IFAC Proc. Vol. **46**(33), 85–90 (2013). 9th IFAC Conference on Control Applications in Marine Systems
7. Caharija, W., Sørensen, A.J., Pettersen, K.Y., Greco, M., Gravdahl, J.T.: Path following control of underactuated surface vessels in the presence of multiple disturbances. In: 2019 18th European Control Conference (ECC), pp. 2551–2558 (2019)
8. Cruz, J.M., Aranda, J.G.J.: Tutorial automática marina: una revisión desde el punto de vista del control. Revista Iberoamericana de Automática e Informática Industrial **9**(3), 205–218 (2012)
9. Fossen, T.I.: Handbook of Marine Craft Hydrodynamics and Motion Control, 2nd edn. Wiley, Hoboken (2021)
10. Lekkas, A.M.: Guidance and path-planning systems for autonomous. Norwegian University of Science and Technology, Noruega, Tesis de doctorado (2014)
11. Lekkas, A.M., Fossen, T.I.: Integral los path following for curved paths based on a monotone cubic hermite spline parametrization. IEEE Trans. Control Syst. Technol. **22**(6), 2287–2301 (2014)
12. Martníez, A., Rodriguez, Y., Hernández, L., Guerra, C., Lemus, J., Sahli, H.: Diseño de auv.arquitectura de hardware y software. Revista Iberoamericana de Automática e Informática industrial **10**(3), 333–343 (2013)
13. Milián, O., Garcia, D., Valeriano, Y.: Autopilot for a robotic boat based on an open hardware configuration. Int. J. Embed. Syst. **14**(6), 553–562 (2021)
14. Mišković, N., Triska, Z., Na đ, Đ., Vukić, Z.: Guidance of a small-scale overactuated marine platform - experimental results. In: 2011 Proceedings of the 34th International Convention MIPRO, pp. 684–689 (2011)
15. Rego, R.C., de Araújo, F.M.U.: Lyapunov-based continuous-time nonlinear control using deep neural network applied to underactuated systems. Eng. Appl. Artif. Intell. **107**, 104519 (2022)
16. Valeriano, Y., Fernandez, A., Hernandez, L., Prieto, P.: Yaw controller in sliding mode for underwater autonomous vehicle. IEEE Latin Amer. Trans. **14**(3), 1213–1220 (2016)
17. Valeriano-Medina, Y., Martínez, A., Hernández, L., Sahli, H., Rodríguez, Y., Cañizares, J.: Dynamic model for an autonomous underwater vehicle based on experimental data. Math. Comput. Modell. Dyn. Syst. **19**(2), 175–200 (2013)

# Battery Health-Aware MPC Planning for Autonomous Racing Vehicles

**Sergio E. Samada, Vicenç Puig, and Fatiha Nejjari**

**Abstract** A battery health-aware trajectory planner for autonomous racing vehicles is developed in this work. The proposed scheme is based on a two-layer robust control strategy which the outer layer defines a zonotopic-tube-based model predictive control (MPC), meanwhile, the inner layer combines a linear quadratic zonotopic controller (LQZ) with a zonotopic Kalman filter (ZKF). In addition, a data-driven Takagi-Sugeno (TS) model is trained using an adaptive neuro-fuzzy inference system for control and estimation purposes. For assessing the proposed control scheme a simulated 1/10 Scale RC car is employed.

## 1 Introduction

The autonomous driving issue has reached huge interest due to the potential impacts that can be generated on society. Autonomous driving is mainly addressed by solving trajectory tracking and trajectory planning problems. In fact, each problem constitutes a challenging task from the control point of view due to the uncertainty which the vehicle interacts with, that is, modelling errors or exogenous disturbances. Nevertheless, planning and trajectory tracking in racing scenarios can lead to a more challenging task since vehicles are subjected to higher levels of acceleration and velocity because of steep curves, hence it implies faster reaction times. In this sense, for achieving the goal both dynamic and kinematic models should be considered.

Particularly, motion planning is grounded on the guidance of the vehicle through unknown environments. In the literature, motion planning approaches are mainly classified as geometric-based, interpolation-based, sampling-based and optimization-

S. E. Samada · V. Puig (✉) · F. Nejjari
Polytechnic University of Catalonia (UPC), Rambla Sant Nebridi 22, 08222 Terrassa, Spain
e-mail: vicenc.puig@upc.edu

S. E. Samada
e-mail: sergio.emil.samada@upc.edu

F. Nejjari
e-mail: fatiha.nejjari@upc.edu

 
O. Llanes-Santiago (ed.), *Proceedings of 19th Latin American Control Congress (LACC 2022)*, Studies in Systems, Decision and Control 464,
https://doi.org/10.1007/978-3-031-26361-3_16

based [7]. Optimization strategies, among them, model predictive control (MPC) have gained great popularity due to their capability to handle hard constraints. So, several works can be consulted, for instance, [3] provides an MPC path planning strategy based on sequential linearization. This reduced complexity model allows implementing real applications but their quality could be insufficient in racing environments. In turn, an MPC trajectory planner based on a non-linear (NL) vehicle model and Pacejka's magic tire formula for racing is presented in [13]. Moreover, sequential convex programming (SCP) is employed for convexifying the objective function, meanwhile, both dynamic and kinematic expressions are considered. Likewise, [1] develops an MPC planning algorithm in racing scenarios where the NL dynamics of the vehicle are reformulated into a linear parameter varying (LPV) representation. Furthermore, a planning performance comparison between NL-MPC and LPV-MPC is carried out which results that the computational cost being significantly reduced when the LPV model is assumed. However, regarding the developed trajectories, both cases are quite similar.

The aforementioned works employ dependent dynamics from physical parameters, some of which are difficult to determine in practice. In addition, the uncertainty in the vehicle's model, as well as the health management capability, are not considered for planning purposes. In this sense, this work proposes a deterministic two-layer control methodology using a data-driven TS model that is trained from a neuro-fuzzy inference algorithm. Particularly, the outer layer synthesises a zonotopic-tube-based MPC with health management capability by means of the battery state of charge. Meanwhile, a zonotopic Kalman filter (ZKF) and a linear quadratic zonotopic (LQZ) feedback controller are combined in the inner layer. It is worth noting that the proposed modelling approach improves the controller performance without increasing the computational burden. Moreover, in the existing literature related to robust MPC (tube-based), this two-layer control scheme is suitably employed since without using it the controller performance could be more conservative or even could violate the control inputs constraints' satisfaction.

The remainder of this paper is structured as follows: Sect. 2 details the vehicle modelling and problem definition. Section 3 addresses the TS model learning whereas Sect. 4 presents the designs of the state-feedback controller and estimator. In the same way, the predictive controller design is provided in Sect. 5. Finally, results and conclusions are presented in Sects. 6 and 7 respectively.

## 2 Vehicle Modelling and Problem Definition

### *2.1 Non-linear Model Description*

In this work, the standard bicycle model [12] for simulating the non-linear dynamics as for the time-reference velocity rates evolution is assumed:

$$\begin{aligned}
\dot{v}_x &= a - \frac{F_{yF}\sin\delta}{m} - \mu_s g + w_z v_y \\
\dot{v}_y &= \frac{F_{yF}\cos\delta}{m} + \frac{F_{yR}}{m} - w_z v_x \\
\dot{w}_z &= \frac{F_{yF} l_f \cos\delta - F_{yR} l_r}{I} \\
F_{yF} &= C_f\left(\delta - \frac{v_y}{v_x} + \frac{l_f w_z}{v_x}\right) \\
F_{yR} &= C_r\left(-\frac{v_y}{v_x} - \frac{l_r w_z}{v_x}\right)
\end{aligned} \tag{1}$$

where $v_x$, $v_y$, and $w_z$ represent the linear velocities, and the angular velocity in the body frame, respectively. In turn, $\delta$ and $a$ are the steering angle at the front wheels, and the longitudinal acceleration on the rear wheels, respectively, which are defined as control inputs. $F_{yF}$, $F_{yR}$, $C_f$ and $C_r$ denote the lateral forces and tire stiffness coefficients produced in front and rear wheels. In addition, $m$ and $I$ are the vehicle mass and inertia whereas $l_f$ and $l_r$ represent the distances from the vehicle center of mass toward the front and rear wheel axes. Finally, $\mu_s$, and $g$ stand for the static friction coefficient and the acceleration due to gravity, respectively.

The kinematic model, on the other hand, describes the evolution of the lateral and orientation errors respect to time based on the curvilinear [15] reference frame:

$$\begin{aligned}
\dot{y}_e &= v_x \sin\theta_e + v_y \cos\theta_e \\
\dot{\theta}_e &= w_z - \frac{v_x \cos\theta_e - v_y \sin\theta_e}{1 - y_e \gamma_r}\gamma_r
\end{aligned} \tag{2}$$

where $y_e$ is the lateral distance from the centerline of the road, $\theta_e$ is the heading angle error, and $\gamma_r$ denotes the track reference curvature.

Regarding the vehicle health management capability, the state of charge (SoC) of the integrated electric battery on the vehicle is considered in the model. This fact contributes to the lengthening of its useful life and, hence, avoids additional costs. The available energy in the battery is usually quantified in percentage points, i.e. $SoC \in [0,\ 1]$, such that 0 indicates the battery is discharged (0%), and 1 indicates the battery is totally charged (100%). The discharge rate representation [9] is employed to include the $SoC$ as a state variable:

$$\begin{aligned}
S\dot{o}C &= -\frac{1}{C_\alpha} p_\alpha \\
p_\alpha &= \frac{1}{2} C_d \rho A_r v_x^2 + \mu_s m g v_x
\end{aligned} \tag{3}$$

where $p_\alpha$ and $C_\alpha$ are the power and the capacity of the battery, respectively. Moreover, $p_\alpha$ is expressed as a function of the longitudinal velocity $v_x$, such that $C_d$, $A_r$ and $\rho$ are the wheels drag coefficient, the vehicle frontal cross-sectional area and the air density at 25 °C, respectively.

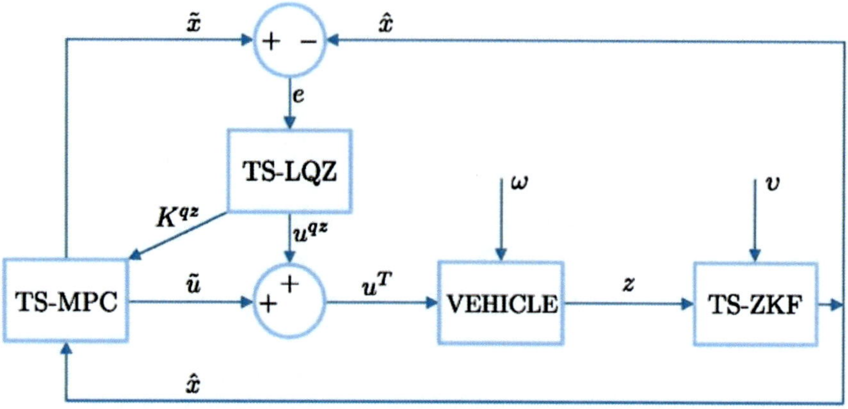


**Fig. 1** Robust control diagram. TS; Takagi-Sugeno; MPC, Model Predictive Controller; LQZ, Linear Quadratic Zonotopic Controller; ZKF, Zonotopic Kalman Filter

## 2.2 *Problem Definition*

Figure 1 depicts the proposed control strategy considering that (1)–(3) can be expressed as the uncertain and discrete-time TS representation using the sector non-linearity [16] approach:

$$\begin{aligned} x_{k+1} &= \sum_{i=1}^{n_r} \bar{\omega}_k^i(\zeta_k)\big(A_k^i x_k + B_k^i u_k\big) + E_k \omega_k \\ z_k &= C_k x_k + F_k \upsilon_k \end{aligned} \tag{4}$$

where $x_k \in \mathbb{R}^{n_x}$, $u_k \in \mathbb{R}^{n_u}$, and $z_k \in \mathbb{R}^{n_z}$ denote the state, input and output vectors, respectively. $\zeta_k \in \mathbb{R}^{n_\zeta}$ is the vector of premise variables with $\zeta = [x_k^T \;\; u_k^T]^T$, and $n_\zeta = n_x + n_u$, whereas $n_r$ is the number of IF-THEN fuzzy rules defined as $n_r = 2^{n_\zeta}$. $\bar{\omega}_k^i(\zeta_k)$ represents the membership functions set, such that $\bar{\omega}_k^i(\zeta_k) > 0$, and $\sum_{i=1}^{n_r} \bar{\omega}_k^i(\zeta_k) = 1$. $A_k^i \in \mathbb{R}^{n_x \times n_x}$, $B_k^i \in \mathbb{R}^{n_x \times n_u}$, and $C_k \in \mathbb{R}^{n_z \times n_x}$ are the system matrices. $\omega_k \in \mathbb{R}^{n_w}$ and $\upsilon_k \in \mathbb{R}^{n_\upsilon}$ denote the state and measurement uncertainties which are assumed unknown but bounded. $E_k$ and $F_k$ are diagonal matrices with fixed dimensions.

In addition, (4) is subject to the following constraints:

$$\begin{aligned} x_k \in \mathcal{X}_k \subset \mathbb{R}^{n_x}, \quad u_k \in \mathcal{U}_k \subset \mathbb{R}^{n_u} \\ \omega_k \in \mathcal{W}_k \subset \mathbb{R}^{n_\omega}, \quad \upsilon_k \in \mathcal{V}_k \subset \mathbb{R}^{n_\upsilon} \end{aligned} \tag{5}$$

with $\mathcal{X}_k$, $\mathcal{U}_k$, $\mathcal{W}_k$, and $\mathcal{V}_k$ representing convex sets. Then, recalling (1)–(3) the state and input vectors are defined as follows:

$$x = [v_x \;\; v_y \;\; w_z \;\; y_e \;\; \theta_e \;\; SoC]^T$$
$$u = [\delta \;\; a]^T \tag{6}$$

According to Fig. 4, the MPC strategy provides a feasible trajectory planning, i.e., ensures the constraint satisfactions using the nominal system,

$$\tilde{x}_{k+1} = \sum_{i=1}^{n_r} \bar{\omega}_k^i(\zeta_k)\big(A_k^i \tilde{x}_k + B_k^i \tilde{u}_k\big), \tag{7}$$

in contrast, TS-LQZ is in charge of compensating for the mismatch between the nominal and real systems. In this sense, the control action is computed as:

$$u_k^{qz} = -K_k^{qz}(\tilde{x}_k - \hat{x}_k) \tag{8}$$

where $K_k^{qz} \in \mathbb{R}^{n_u \times n_x}$ stands the state-feedback gain. Moreover, the overall control input applied to the vehicle lead to:

$$u_k^T = \tilde{u}_k + u_k^{qz} \tag{9}$$

## 3 TS Neuro-Fuzzy Model Learning

The Adaptive Neuro-Fuzzy Inference System (ANFIS) is employed in this work for learning the vehicle dynamic model (1). This is due to the fact that some parameters are not easy to obtain in practice. In contrast, the kinematic model (2) only includes geometric parameters that can be easily determined which implies that it is not incorporated into the learning approach. Similar situation applies for $SoC$ in (3).

### *3.1 Structure and Training Algorithms*

Roughly speaking, ANFIS is a five-layer feed-forward network with fixed and adaptive nodes [8]. Furthermore, ANFIS generates a weighted summation of linear models able to accurately represent non-linear systems. The methodology for adjusting the parameters of the adaptive nodes, also called training methods, has evolved into several variants. For instance, these approaches can be classified as derivative-based, heuristics or hybrid. This work employs the gradient descent and recursive least-squares algorithms for adjusting the adaptive parameters and guaranteeing a fast convergence toward their optimal values.

### 3.2 TS Space-State Representation

Obtaining the TS-ANFIS representation of (4) implies merging the dynamic model learning processes with the kinematic and battery models. Particularly, the learning phase is developed for each multiple-input single-output subsystem, that is, the states of motion $v_x$, $v_y$, and $w_z$. Thus, the premise variables vector $\zeta = [x_k^T \;\; u_k^T]^T$, according to (4), is defined as:

$$\zeta_k = [v_{x_k} \;\; v_{y_k} \;\; w_{z_k} \;\; \delta_k \;\; a_k]^T \tag{10}$$

Once that ANFIS has been properly trained, the TS space-state representation for three dynamic states $v_x$, $v_y$, and $w_z$ can be obtained using the procedure presented in [2], as follow:

$$\bar{x}_{k+1} = A_k^{fis}(\zeta_k)\bar{x}_k + B_k^{fis}(\zeta_k)u_k + h_k^{fis}(\zeta_k) \tag{11}$$

with

$$\begin{gathered} A_k^{fis}(\zeta_k) = \sum_{i=1}^{n_r} \begin{bmatrix} \bar{\omega}_k^{i,1}(\zeta_k)\bar{A}_k^{i,1} \\ \bar{\omega}_k^{i,2}(\zeta_k)\bar{A}_k^{i,2} \\ \bar{\omega}_k^{i,3}(\zeta_k)\bar{A}_k^{i,3} \end{bmatrix}, \quad B_k^{fis}(\zeta_k) = \sum_{i=1}^{n_r} \begin{bmatrix} \bar{\omega}_k^{i,1}(\zeta_k)\bar{B}_k^{i,1} \\ \bar{\omega}_k^{i,2}(\zeta_k)\bar{B}_k^{i,2} \\ \bar{\omega}_k^{i,3}(\zeta_k)\bar{B}_k^{i,3} \end{bmatrix} \\ h_k^{fis}(\zeta_k) = \sum_{i=1}^{n_r} \begin{bmatrix} \bar{\omega}_k^{i,1}(\zeta_k)\bar{h}_k^{i,1} \\ \bar{\omega}_k^{i,2}(\zeta_k)\bar{h}_k^{i,2} \\ \bar{\omega}_k^{i,3}(\zeta_k)\bar{h}_k^{i,3} \end{bmatrix} \end{gathered} \tag{12}$$

Likewise, utilizing (4), the unknown but bounded uncertainties that are related with the three learning process can be determined in (14), s.t, the $l$th uncertainty interval [5] $\bar{E}_k^l$ is computed by solving the following optimization problem:

$$\begin{aligned} \bar{E}_k^l &= \max(y_k^l - \hat{y}_k^l - \sigma, \hat{y}_k^l - y_k^l - \sigma) \\ \forall k &= 1, \ldots, N \\ \forall l &= 1, \ldots, 3 \end{aligned} \tag{13}$$

where $y_k^l$ is the available measurement, $\hat{y}_k^l$ stands the estimated output by ANFIS, $\sigma \in \mathbb{R}$ bounds the additive uncertainty, and $N$ refers to the size of the trained data set.

$$E_k^{fis} = \text{diag}(\bar{E}_k^1, \;\; \bar{E}_k^2, \;\; \bar{E}_k^3) \tag{14}$$

At this point the characterization of the non-linear dynamics (1) by means of the ANFIS learning approach, and a state-space representation has been determined (11). However, kinematic model (2) and the battery state of charge (3) should be considered for control and estimation purposes, as stated before. Then, using (11), and (14), model (4) can be reformulated as:

$$\begin{aligned} x_{k+1} &= A_k x_k + B_k u_k + h_k + E_k \omega_k \\ z_k &= C_k x_k + F_k \upsilon_k \end{aligned} \quad (15)$$

with

$$A_k = \begin{bmatrix} A_k^{fis}(\zeta_k) & 0_3 \\ A_k^{kin}(\zeta_k^e) & I_3 \end{bmatrix}, B_k = \begin{bmatrix} B_k^{fis}(\zeta_k) \\ 0_{3\times2} \end{bmatrix}, h_k = \begin{bmatrix} h_k^{fis}(\zeta_k) \\ 0_{3\times1} \end{bmatrix}, E_k = \begin{bmatrix} E_k^{fis} & 0_3 \\ 0_3 & 0_3 \end{bmatrix} \quad (16)$$

where $x_k$ and $u_k$ come from (6), and $\zeta_k$ comes from (10). In turn, matrix $A_k^{kin}(\zeta_k^e)$ (17), such that $\zeta_k^e = [y_{e_k} \ \ \theta_{e_k} \ \ \gamma_{r_k} \ \ v_{x_k} \ \ v_{y_k}]^T$ is calculated by discretizing (2), (3), and using the Euler method with sampling time $T_s$. It is worth pointing out that the time discretization is introduced into learning stages, so the TS model (11) is already in discrete-time.

$$A_k^{kin}(\zeta_k^e) = \begin{bmatrix} T_s sin\theta_{e_k} & T_s cos\theta_{e_k} & 0 \\ -\frac{T_s\gamma_{r_k}}{1-y_{e_k}\gamma_{r_k}} cos\theta_{e_k} & \frac{T_s\gamma_{r_k}}{1-y_{e_k}\gamma_{r_k}} sin\theta_{e_k} & T_s \\ -\frac{T_s}{C_\alpha}(\frac{1}{2}C_d\rho A_r v_{x_k} + \mu_s mg) & 0 & 0 \end{bmatrix} \quad (17)$$

## 4 State-Feedback Controller and Estimator Designs

The zonotopic linear quadratic controller (LQZ) and Kalman filter (ZKF) designs, according to the deterministic system (4) and Fig. 1 are addressed in this section. ZKF provides a robust convergence to the minimum zonotopic set that contains all the admissible states, as well as, the estimation of the unmeasured lateral velocity $\upsilon y$. In turn, LQZ [14] compensates for the deviation between nominal $\tilde{x}$ and estimated states $\hat{x}$, respectively.

Synthesizing the LQZ controller implies assuming the positive definite weighting matrices $Q_k^{qz} \in \mathbb{R}^{n_x \times n_x}$, $R_k^{qz} \in \mathbb{R}^{n_u \times n_u}$. Furthermore, $Q_k^{qz} = H_k H_k^T$, such that $H_k \in \mathbb{R}^{n_q \times n_x}$, $n_q = \text{rank}(Q_k^r)$, and $\gamma_k \in \mathbb{R}$. So, the LMI-based design of the LQZ controller leads to solve the following convex optimization problem [11]:

$$\min_{\gamma_k, Y_k = Y_k^T, W_k} \quad \gamma_k$$

subject to:

$$\begin{bmatrix} \gamma_k I_{n_x} & I_{n_x} \\ I_{n_x} & Y_k \end{bmatrix} \succ 0, \quad \begin{bmatrix} -Y_k & Y_k A_k^T - W_k^T B_k & Y_k H_k^T & W_k^T \\ \star & -Y_k & 0 & 0 \\ \star & \star & -I_{n_q} & 0 \\ \star & \star & \star & -R_k^{r^{-1}} \end{bmatrix} \prec 0 \quad (18)$$

such that $Y_k = P_k^{qz^{-1}} \in \mathbb{R}^{n_x \times n_x}$, $P_k^{qz} \succ 0$ denotes the common Lyapunov matrix, whereas $A_k$ and $B_k$ represent the TS fuzzy model matrices. Moreover, the LQZ control gain $K_k^{qz} \in \mathbb{R}^{n_u \times n_x}$ results in:

$$K_k^{qz} = W_k Y_k^{-1} \tag{19}$$

On the other hand, recalling (15) and following a similar methodology presented in [6], also assuming the initial states vector $x_0 = \langle c_0^z, R_0^z \rangle$, $\omega_k = \langle 0, I_{n_\omega} \rangle$ and $\upsilon_k = \langle 0, I_{n_\upsilon} \rangle$ as centered zonotopes, and $x_k = \langle c_k^z, R_k^z \rangle$ is satisfied $\forall k \geq 0$. Then, the center $c_{k+1}^z$ and generator matrix $R_{k+1}^z$ are recursively forward predicted as:

$$\begin{aligned} c_{k+1}^z &= (A_k - L_k^z C_k) c_k^z + B_k u_k + h_k + L_k^z z_k \\ R_{k+1}^z &= [(A_k - L_k^z C_k) \bar{R}_k^z,\ E_k,\ -L_k^z F_k] \\ L_k^z &= A_k \bar{P}_k^z C_k^T (C_k \bar{P}_k^z C_k^T + Q_{\upsilon_k}^z)^{-1} \\ \bar{R}_k^z &= \downarrow_q (R_k^z) \end{aligned} \tag{20}$$

such that $\bar{P}_k^z = \bar{R}_k^z \bar{R}_k^{z^T}$, $Q_{\omega_k}^z = E_k E_k^T$, and $Q_{\upsilon_k}^z = F_k F_k^T$ are covariance matrices, the reduction operator $\downarrow_q$, fulfills $\bar{R}_k = \downarrow_q (R)$, and $\hat{x}_k \in \langle c_k, R_k \rangle \subset \langle c_k, \bar{R}_k \rangle$. Finally, $L_k^z$ is the optimal observer gain matrix.

**Remark 1** The definitions related to sets such as, e.g., zonotopes, centred zonotopes, the reduction operator as well as the Minkowski sum $\oplus$ and linear image $\odot$ operations can be consulted in [6] due to the space limitation.

## 5 Planner Design Based on Model Predictive Control

The main objective of the planning algorithm is to guarantee a feasible response in terms of the trajectory and energy consumption that allows the vehicle not to leave the track despite perturbations. In particular, the MPC strategy provides a trade-off by maximizing both the longitudinal velocity and the battery state of charge while minimizes the control actions variation and the lateral velocity.

Following the above idea, it is deduced that the solution conduces to a multiple-objective optimization problem that is non-convex because of the maximization pair $v_x$ and $SoC$. So, an approximated reformulation that transforms the non-convex problem into convex finite-horizon optimization problem is proposed. The proposed function is defined as $r^T Q r + q^T r$ where $q \in \mathbb{R}^2$, and $Q \in \mathbb{R}^{2\times 2}$ denotes a symmetric weighting matrix, such that $Q \prec 0$. It is worth noting that the last constraint guarantees the concavity criterion.

On the other hand, to ensure the overall stability, both a terminal cost function and a terminal constraint set are introduced. The terminal cost function results in $V(x) = x_{H_p}^T P^c x_{H_p}$, such that $H_p$ stands the prediction horizon and $P^c = Y^{-1}$ is taken as the Lyapunov matrix in (18). In contrast, the terminal constraint set $\mathcal{X}_f$

is defined as the minimal robustly positively invariant (mRPI) set [4] where states converge, i.e., $\mathcal{X}_f \subseteq \mathcal{X}$.

Hence, taking into account the uncertain close-loop dynamic (15) and using the previous analysis, the mRPI set can be explicitly determined if and only if $\exists K \in \mathbb{R}^{n_u \times n_x}$, such that $(A - KB)$ is stable, so $\Omega_\infty = \oplus_{i=1}^{\infty}(A - KB)^i \oplus (E \odot \mathcal{W})$. However, due to the infinite iterations needed to compute $\Omega_\infty$, a finite recursive method based on an outer-approximation set with a given precision is considered [10]. In the following context, the outer-approximation set $\Omega$ is utilized instead of $\Omega_\infty$ which implies that the terminal constraint set $\mathcal{X}_f = \Omega$.

At this point, the convergence of states toward a stable region is ensured. Then, by means of the zonotopic notation, a reachable sets sequence $\Phi$ can be computed for propagating the uncertainty along the prediction horizon $H_p$, and afterwards, refining the trajectory tubes of states $\mathcal{X}$ and control inputs $\mathcal{U}$ in the TS-MPC problem (22). Therefore, taking the initial reachable set as the uncertainty provided by ZKF, i.e., $\Phi_0 = \langle 0, R_k^z \rangle$, such that $R_k^z$ comes from (20), the reachable sets sequence at each sampling period $k$ results in:

$$\begin{aligned} &\Phi_{k+i+1} = (A_{k+i} - K_{k+i}^{qz} B_{k+i}) \odot \Phi_{k+i} \oplus (E \odot \mathcal{W}) \\ &\forall i = 0, \ldots, H_p, \quad i \in \mathbb{N} \end{aligned} \tag{21}$$

such that $\mathcal{W} = \langle 0, R_{\mathcal{W}} \rangle$ and $K_{k+i}^{qz}$ is determined in (19).

Finally, using (15) and considering the states $\hat{x}$ and past predicted control inputs $\tilde{u}_{k-1}$ as known, also assuming the previous section results, the following quadratic optimization problem is defined at each time $k$:

$$\begin{aligned} \min_{\Delta\tilde{u}_k, \tilde{x}_k} \; & - \sum_{i=0}^{H_p-1} \left(\tilde{x}_{k+i}^T Q_{k+i}^c \tilde{x}_{k+i} + q_{k+i}^{c^T} \tilde{x}_{k+i}\right) + \sum_{i=0}^{H_p-1} \Delta\tilde{u}_{k+i}^T R_{k+i}^c \Delta\tilde{u}_{k+i} + \tilde{x}_{k+H_p}^T P_{k+H_p}^c \tilde{x}_{k+H_p} \\ \text{subject to:} \; & \\ & \tilde{x}_{k+i+1} = A_{k+i}\tilde{x}_{k+i} + B_{k+i}\tilde{u}_{k+i} + h_{k+i} \\ & \tilde{x}_{k+0} = \hat{x}_k \\ & \tilde{u}_{k+i} = \tilde{u}_{k+i-1} + \Delta\tilde{u}_{k+i} \\ & \tilde{x}_{k+i} \in \mathcal{X}_{k+i} \ominus \Phi_{k+i} \\ & \tilde{u}_{k+i} \in \mathcal{U}_{k+i} \ominus (K_{k+i}^{qz} \odot \Phi_{k+i}) \\ & \tilde{x}_{k+H_p} \in \mathcal{X}_f \end{aligned} \tag{22}$$

with $\tilde{x}_{k+i} \in \mathbb{R}^{n_x}$, and $\tilde{u}_{k+i} \in \mathbb{R}^{n_u}$ denoting the predictions of states and control inputs, respectively, $\Delta\tilde{u}_{k+i} \in \mathbb{R}^{n_u}$ represents the control actions variation, $Q_{k+i}^c \in \mathbb{R}^{n_x \times n_x}$, $q_{k+i}^c \in \mathbb{R}^{n_x}$, $R_{k+i}^c \in \mathbb{R}^{n_u \times n_u}$ stand weighting expressions, and $A_{k+i}$, $B_{k+i}$, and $h_{k+i}$ are the TS system (15) matrices and vector, respectively. The proposed methodology is summarized in Algorithm 1.

**Algorithm 1** Online phase of the planning strategy

1: $\Phi_0 \leftarrow \langle 0, R^z \rangle$ using (20).
2: **for** $j = 1$ **to** $H_p$ **do**
3: Instantiate $A$, $B$, and $h$ (15) with the previous premise variables $\zeta$ and $\zeta^e$.
4: Compute the state-feedback gain $K^{qz}$ using (19).
5: Compute the reachable zonotopic sets sequence $\Phi$ using (21).
6: Compute the states $\mathcal{X} \ominus \Phi$ and control inputs $\mathcal{U} \ominus (K^{qz} \odot \Phi)$ constraint sets.
7: **end for**
8: Compute the Tube-based MPC predictions $\tilde{x}$, $\tilde{u}$ solving (22).
9: Compute the LQZ control inputs $u^{qz}$ using (8).
10: Apply the overall control input $u^T$ using (9).
11: Compute the state estimations $\hat{x} = \langle c^z, R^z \rangle$ using (20).

## 6 Results

The performance of the developed planning strategy is assessed in this section by means of a racing L-shaped circuit. Particularly, Matlab, and SeDuMi through YALMIP are employed for solving the optimization problems (19) and (22). Likewise, exogenous disturbances in states and measurements are considered in order to simulate an almost real environment. Specifically, a friction force is generated due to the high levels of acceleration and velocity to which the vehicle is subjected. This effect is assumed unknown but bounded and for modelling, purposes are contained in matrix $E$ (15). In the same way, measurements are also contaminated with bounded noises which are contained in matrix $F$ (15).

On the other hand, the weighting matrices $Q^c$, $q^c$, $R^c$, $Q^{qz}$, and $R^{qz}$ are iteratively calculated until achieving the desired behaviour. Meanwhile, the initial constraint zonotopic sets for the states $\mathcal{X}$ and control inputs $\mathcal{U}$ are calculated taking into consideration the physical limits of the vehicle dynamic. Finally, the feedback controller LQZ is run at a higher frequency (50 Hz) than the predictive controller MPC (10 Hz).

The test consists in comparing the trajectory developed by the vehicle when the battery state of charge is controlled and without controlling it. Figure 2a and b depict both scenarios in the longitudinal velocity profile. As can be seen, the response of the vehicle is quite good. As expected, the longitudinal velocity decrease when the $SoC$ objective is taken into account, but instead, the saving energy is considerable. For instance, after finishing the lap, $SoC$ remains over 70% whereas without considering the $SoC$ it remains closed to 60% according to Fig. 2c. Regarding the time lap, the vehicle finishes lap 2.2 s later when the $SoC$ is prioritized, whereas, concerning its position, it tends to approach the inner border on the steep curves and the outer border on straight segments as the velocity increases.

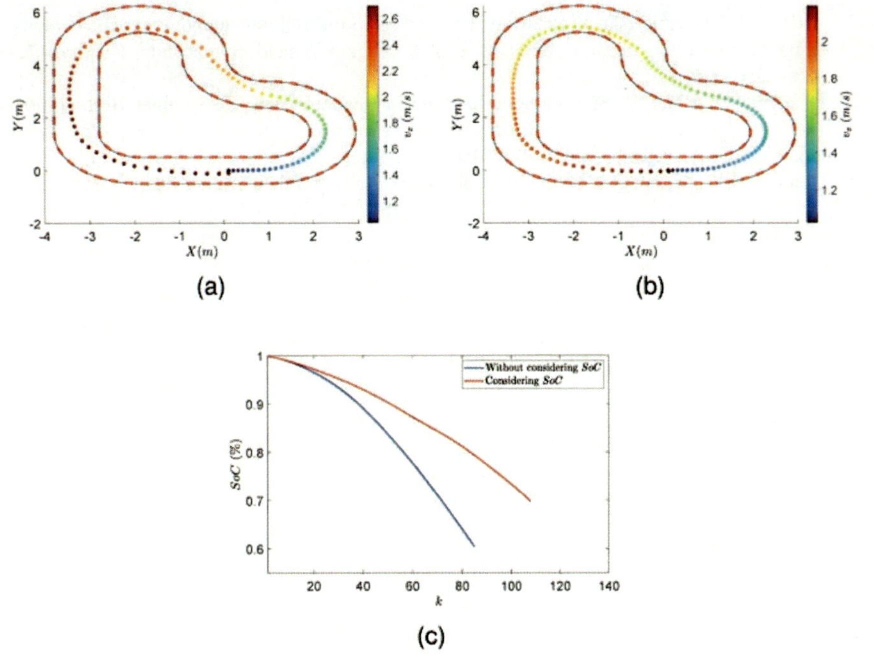


**Fig. 2** Planned trajectory in terms of the longitudinal velocity $v_x$. **a** Without considering the $SoC$ objective. **b** Considering the $SoC$ objective. **c** $SoC$ comparison

## 7 Conclusions

This work has presented a planning strategy for autonomous vehicles in racing scenarios. The proposed control scheme includes a robust MPC controller that determines the optimal trajectory by computing a trade-off between the longitudinal velocity and the battery state of charge. In turn, an LQZ controller compensates for the deviation between the real system and the estimation provided by ZKF. As can be checked, the planner guarantees the constraint satisfactions even when the vehicle dealing with uncertainties, as well as, improves the energy consumption significantly. On the other hand, the learned TS model suitably represents the non-linear vehicle dynamics.

## References


1. Alcalá, E., Puig, V., Quevedo, J.: Lpv-mp planning for autonomous racing vehicles considering obstacles. Robot. Auton. Syst. **124**, 103392 (2020)
2. Alcalá, E., Sename, O., Puig, V., Quevedo, J.: TS-MPC for autonomous vehicle using a learning approach. IFAC-PapersOnLine **53**(2), 15110–15115 (2020). 21th IFAC World Congress

3. Alrifaee, B., Maczijewski, J.: Real-time trajectory optimization for autonomous vehicle racing using sequential linearization. In: 2018 IEEE Intelligent Vehicles Symposium (IV), pp. 476–483. IEEE (2018)
4. Blanchini, F., Miani, S.: Set-Theoretic Methods in in Control, vol. 78. Springer, Berlin (2008)
5. Blesa, J., Puig, V., Bolea, Y.: Fault detection using interval lpv models in an open-flow canal. Control Eng. Pract. **18**(5), 460–470 (2010)
6. Combastel, C.: Zonotopes and kalman observers: gain optimality under distinct uncertainty paradigms and robust convergence. Automatica **55**, 265–273 (2015)
7. González, D., Pérez, J., Milanés, V., Nashashibi, F.: A review of motion planning techniques for automated vehicles. IEEE Trans. Intell. Trans. Syst. **17**(4), 1135–1145 (2015)
8. Jang, J.S.: Anfis: adaptive-network-based fuzzy inference system. IEEE Trans. Syst. Man Cybern. **23**(3), 665–685 (1993)
9. Nair, U.R., Costa-Castelló, R.: A model predictive control-based energy management scheme for hybrid storage system in islanded microgrids. IEEE Access **8**, 97809–97822 (2020)
10. Olaru, S., De Doná, J.A., Seron, M.M., Stoican, F.: Positive invariant sets for fault tolerant multisensor control schemes. Int. J. Control **83**(12), 2622–2640 (2010)
11. Ostertag, E.: Mono-and Multivariable Control and Estimation: Linear, Quadratic and LMI Methods. Springer Science & Business Media, Berlin (2011)
12. Pacejka, H.: Tire and Vehicle Dynamics. Elsevier, Amsterdam (2005)
13. Scheffe, P., Henneken, T.M., Kloock, M., Alrifaee, B.: Sequential convex programming methods for real-time optimal trajectory planning in autonomous vehicle racing. IEEE Trans. Intell. Vehicles (2022)
14. Trapiello, C., Puig, V.: Replay attack detection using a zonotopic kf and lq approach. In: 2020 IEEE International Conference on Systems, Man, and Cybernetics (SMC), pp. 3117–3122. IEEE (2020)
15. Verschueren, R., De Bruyne, S., Zanon, M., Frasch, J.V., Diehl, M.: Towards time-optimal race car driving using nonlinear mpc in real-time. In: 53rd IEEE Conference on Decision and Control, pp. 2505–2510. IEEE (2014)
16. Wang, H.O., Tanaka, K.: Fuzzy Control Systems Design and Analysis: A Linear Matrix Inequality Approach. Wiley, New York (2004)

# Leader-Follower Formation Tracking Strategies Using Heterogeneous Robots for Monitoring and Remediation Applications

**Jesus Santiaguillo-Salinas, Hiram N. Garcia-Lozano, and Rafael F. Gonzalez-Zarate**

**Abstract** This work presents two formation tracking strategies using a leader-follower scheme. The robots of the system are heterogeneous. In one of the strategies, formation tracking using a time-varying position vector is considered, while in the other one, formation tracking with time-gap separation is considered. The strategies presented in this work are intended for crop monitoring and remediation applications in precision agriculture. The theoretical results are validated by numerical simulations and real-time experiments.

## 1 Introduction

Precision agriculture (PA) consists in applying new technologies for control and management of farmland production, in order to increase productivity [1]. Unlike traditional agricultural machinery, which includes automatic irrigation and fertigation systems [2], PA is able to maximize productivity on each stage of the crop cycle, optimizing use efficiency of water and natural resources [3].

PA aims for decision making based on farmland real data, rather than on intuition. Sensors, data acquisition systems and Big Data are some techniques used to gather crop information like soil parameters, crop data or field humidity [4]. From such information, decisions can focus on the most suitable action to solve a given issue. For instance, unmanned aerial and ground robots can spray pesticides, apply fertilizers or harvest fruits [5]. Multi-agent systems (MAS) can be used to assist all agricultural work. When multi-agents are used it is possible to split a given agricultural task into

J. Santiaguillo-Salinas (✉) · H. N. Garcia-Lozano · R. F. Gonzalez-Zarate
Universidad del Papaloapan, Av. Ferrocarril S/N, Col. Ciudad Universitaria, 68400, Loma Bonita, Oaxaca, Mexico
e-mail: jsantiaguillo@outlook.com

H. N. Garcia-Lozano
e-mail: garcia.hiram@gmail.com

R. F. Gonzalez-Zarate
e-mail: rfgz_unpa@gmail.com

 
O. Llanes-Santiago (ed.), *Proceedings of 19th Latin American Control Congress (LACC 2022)*, Studies in Systems, Decision and Control 464,
https://doi.org/10.1007/978-3-031-26361-3_17

smaller labours for faster completion. An agent is an entity that has the capability to perceive the environment and interact with it by means of sensors and actuators. In this way, the PA uses MAS to improve farmland profitability and production.

In the present work, two formation tracking following strategies using a leader-follower scheme are presented. The proposed MAS is a heterogeneous system based on two robots with different capabilities (different mathematical models). Whereas the leader robot follows a path, the follower robot adjusts to a desired formation. Two strategies are implemented for formation tracking, the first uses a time-varying position vector [6–9] and the second is drawn by a time-gap separation [10–13]. The strategies presented in this work are intended for crop monitoring by applying special treatment applications in precision agriculture, i.e., the leader follows a path in the crop while collecting agrotechnical information of each farmland section; this information is processed and sent to the follower, which follows the leader's trajectory while performing a variety of operations including weed control, watering, fertilization, and sprinkling pesticides. This is why the precision on the formation tracking is a key point in these tasks.

## 2 Preliminaries

This section presents the mathematical models of the two mobile robots that will be used, the car-like robot and the omnidirectional robot, and the problem to be solved.

### 2.1 *Kinematics Model of the Car-Like Robot (Leader Robot)*

Consider a car-like robot like the one shown in Fig. 1a. This type of mobile robot is similar to a car, where the front wheels can be steered while the rear wheels are fixed. Based on [14], the kinematics model considering rear-wheel drive and a point $P_l$ as output of the system, is given by

$$\begin{bmatrix} \dot{p}_{xl} \\ \dot{p}_{yl} \\ \dot{\theta}_l \\ \dot{\phi}_l \end{bmatrix} = \begin{bmatrix} \cos\theta_l - \tan\phi_l(\sin\theta_l + \frac{\Delta \sin\delta_l}{\ell}) & -\Delta\sin\delta_l \\ \sin\theta_l - \tan\phi_l(\cos\theta_l + \frac{\Delta\cos\delta_l}{\ell}) & \Delta\cos\delta_l \\ \frac{\tan\phi_l}{\ell} & 0 \\ 0 & 1 \end{bmatrix} \begin{bmatrix} v_l \\ w_l \end{bmatrix} = \begin{bmatrix} A_l(\theta_l, \phi_l) \\ \frac{\tan\phi_l}{\ell} \quad 0 \\ 0 \quad 1 \end{bmatrix} \begin{bmatrix} v_l \\ w_l \end{bmatrix} \tag{1}$$

where $\left[p_{xl}, p_{yl}\right]^T$ are the cartesian coordinates of the point $P_l$, $\theta_l$ is the orientation angle of the robot body with respect to the $X$ axis, $\phi_l$ is the steering angle, $\ell$ is the distance between the front and rear wheels, $v_l$ and $w_l$ are the longitudinal velocity of the robot and the steering angular velocity, respectively, with $\Delta \neq 0$ and $\delta_l = \theta_l + \phi_l$.

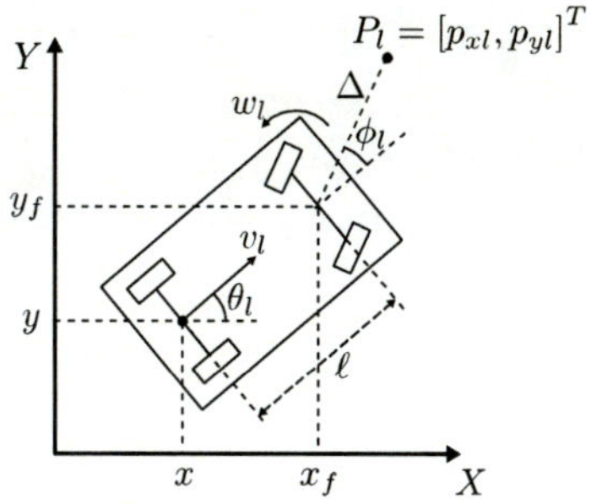


(a) Car-like Robot.

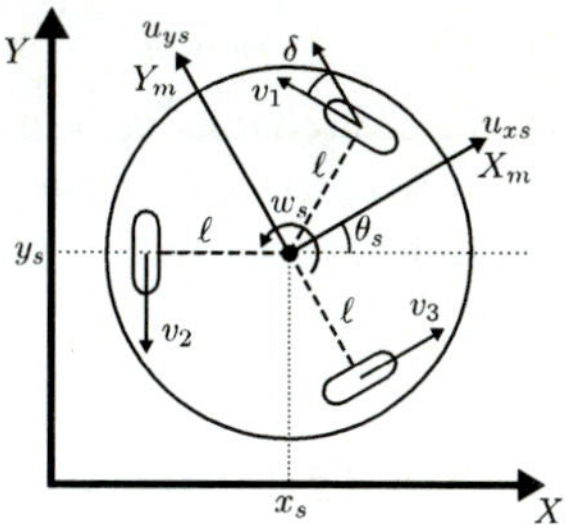


(b) Omnidirectional Robot.

**Fig. 1** Mobile robots schemes

## 2.2 *Kinematics Model of the Omnidirectional Robot (Follower Robot)*

Consider the omnidirectional mobile robot shown in Fig. 1b. This robot has three omnidirectional wheels. The wheel mechanism consists of small rollers located on the outside diameter of the wheel, giving the robot free movement. These wheels are separated by an angle of 120 degrees from each other. Based on [15] the kinematic model of this robot is given by

$$\begin{bmatrix} \dot{x}_s \\ \dot{y}_s \\ \dot{\theta}_s \end{bmatrix} = \begin{bmatrix} \cos\theta_s & -\sin\theta_s & 0 \\ \sin\theta_s & \cos\theta_s & 0 \\ 0 & 0 & 1 \end{bmatrix} \begin{bmatrix} u_{xs} \\ u_{ys} \\ w_s \end{bmatrix} \tag{2}$$

with $\xi_s = [x_s, y_s]^T$ the position of the robot in the plane, $u_{xs}$ and $u_{ys}$ the longitudinal velocities in $X$ and $Y$ axes, respectively, $w_s$ its angular velocity and $\theta_s$ the orientation angle of the robot body.

## 2.3 *Problem Statement*

Let $m(t) = \left[m_x(t), m_y(t)\right]^T$ be a continuously differentiable preestablished trajectory. The goal of this work is to design a decentralized control strategy that ensures

- Asymptotic tracking of a prescribed trajectory by the leader (tracking control)

$$\lim_{t\to\infty} (P_l(t) - m(t)) = 0.$$

- Asymptotic formation by the follower (formation control)

$$\lim_{t \to \infty} (\xi_s(t) - \xi_s^*(t)) = 0.$$

with $\xi_s^*(t)$ the desired relative position of the follower within the desired formation. In the latter, two formation strategies will be implemented, time-varying position vector and time-gap separation.

## 3 Formation Tracking Strategies

In this section we explain the formation tracking strategies for the leader and follower robots, time-varying position vector and time-gap separation.

### 3.1 Control Law for the Leader

By defining the auxiliary control variables $u_l = \left[u_{xl}, u_{yl}\right]^T$ it is possible to obtain a strategy to control the position of the point $P_l$ (however, the orientation angle is not controllable) in the car-like robot as

$$\begin{bmatrix} v_l \\ w_l \end{bmatrix} = A_l(\theta_l, \phi_l)^{-1} \begin{bmatrix} u_{xl} \\ u_{yl} \end{bmatrix} \tag{3}$$

where $A_l(\theta_l, \phi_l)^{-1}$ is the inverse of the decoupling matrix. To solve the trajectory tracking problem for the leader, we propose an auxiliary bounded control strategy

$$\begin{bmatrix} u_{xl} \\ u_{yl} \end{bmatrix} = \begin{bmatrix} -k_{1l} \tanh\left(p_{xl} - m_x\right) + \dot{m}_x \\ -k_{2l} \tanh\left(p_{yl} - m_y\right) + \dot{m}_y \end{bmatrix} \tag{4}$$

where $\left[\dot{m}_x, \dot{m}_y\right]^T$ is the velocity of the trajectory, $k_{1l}$ and $k_{2l}$ are the control gains.

### 3.2 Time-Varying Position Vector

To maintain a formation (by the follower agent) oriented to the direction of the leader agent, we use a time-varying position vector [8] given by

$$C_{ls}(t) = \begin{bmatrix} \cos\theta_l & -\sin\theta_l \\ \sin\theta_l & \cos\theta_l \end{bmatrix} c_{ls} = R(\theta_l) c_{ls} \tag{5}$$

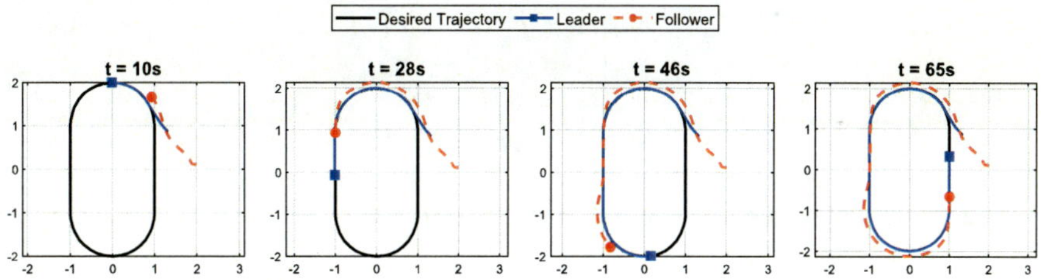


**Fig. 2** Simulation results with time-varying position vector

where $\theta_l$ is the leader angle and $c_{ls}$ is a static position vector corresponding to the desired formation and it can be arbitrarily chosen, as long as there is no collision between robots.

For formation tracking using a time-varying position vector, we propose a bounded control law for the follower agent given by

$$\begin{aligned} u_{xs} &= -k_{1s}\tanh(x_s - x_s^*) + \dot{x}_s^* \\ u_{ys} &= -k_{2s}\tanh(y_s - y_s^*) + \dot{y}_s^* \\ w_s &= -k_{3s}\tanh(\theta_s - \theta_s^*) + \dot{\theta}_s^* \end{aligned} \tag{6}$$

where $k_{1s}, k_{2s}$ and $k_{3s}$ are the control gains, $\xi_s^* = \left[x_s^*, y_s^*\right]^T = \xi_l + C_{ls}(t)$ and $\theta_s^* = \theta_l$ the orientation angle of the leader robot body. The results obtained from a numerical simulation are shown in Fig. 2. This figure shows the positions of the robots and their paths at 10, 28, 46 and 60s. It can be seen how the leader follows the trajectory while the follower remains at a fixed distance from the leader. The formation of the follower is oriented according to the steering angle of the leader, so it is possible to see that in the curves, the trajectory of the follower is displaced.

### 3.3 Time-Gap Separation

A formation in which the follower robot performs the same trajectory as the leader robot but with a time-gap of $\tau$ units of time [13] is addressed in this section.

The leader performs a pre-established trajectory with bounded longitudinal and angular velocities. It is assumed that only the pose of the leader is known and that the desired position and velocity of the delayed trajectory for the follower is obtained from a delayed state observer. In this way, the follower will performs the delayed trajectory of the leader from the delayed observer using a strategy based on the fundamental consensus algorithm. Consider the kinematics model of the leader (1) and the following delayed variables

$$
\begin{aligned}
w_{1l}(t) &= p_{xl}(t-\tau)\\
w_{2l}(t) &= p_{yl}(t-\tau)\\
w_{3l}(t) &= \theta_l(t-\tau)\\
w_{4l}(t) &= \phi_l(t-\tau)
\end{aligned} \tag{7}
$$

The virtual delayed representation of the system (1) is

$$
\begin{aligned}
\dot{w}_{1l}(t) &= \left(\cos w_{3l}(t) - \tan w_{4l}(t)(\sin w_{3l}(t) + \tfrac{\Delta \sin \delta_{wl}}{\ell})\right) v_l(t-\tau)\\
&\quad - (\Delta \sin \delta_{wl})\, w_l(t-\tau)\\
\dot{w}_{2l}(t) &= \left(\sin w_{3l}(t) - \tan w_{4l}(t)(\cos w_{3l}(t) + \tfrac{\Delta \cos \delta_{wl}}{\ell})\right) v_l(t-\tau)\\
&\quad + (\Delta \cos \delta_{wl})\, w_l(t-\tau)\\
\dot{w}_{3l}(t) &= \tfrac{\tan w_{4l}(t)}{\ell} v_l(t-\tau)\\
\dot{w}_{4l}(t) &= w_l(t-\tau)
\end{aligned} \tag{8}
$$

with $\delta_{wl} = w_{3l}(t) + w_{4l}(t)$. The previous system allows to propose a delayed observer

$$
\begin{aligned}
\dot{\hat{w}}_{1l}(t) &= \left(\cos \hat{w}_{3l}(t) - \tan \hat{w}_{4l}(t)(\sin \hat{w}_{3l}(t) + \tfrac{\Delta \sin \delta_{\hat{w}l}}{\ell})\right) v_l(t-\tau)\\
&\quad - (\Delta \sin \delta_{\hat{w}l})\, w_l(t-\tau) + \lambda_{1l} e_{w_{1l}}(t)\\
\dot{\hat{w}}_{2l}(t) &= \left(\sin \hat{w}_{3l}(t) - \tan \hat{w}_{4l}(t)(\cos \hat{w}_{3l}(t) + \tfrac{\Delta \cos \delta_{\hat{w}l}}{\ell})\right) v_l(t-\tau)\\
&\quad + (\Delta \cos \delta_{\hat{w}l})\, w_l(t-\tau) + \lambda_{2l} e_{w_{2l}}(t)\\
\dot{\hat{w}}_{3l}(t) &= \tfrac{\tan \hat{w}_{4l}(t)}{\ell} v_l(t-\tau) + \lambda_{3l} e_{w_{3l}}(t)\\
\dot{\hat{w}}_{4l}(t) &= w_l(t-\tau) + \lambda_{4l} e_{w_{4l}}(t)
\end{aligned} \tag{9}
$$

with $\delta_{\hat{w}l} = \hat{w}_{3l}(t) + \hat{w}_{4l}(t)$, $\lambda_{1l}$, $\lambda_{2l}$, $\lambda_{3l}$, $\lambda_{4l}$ are positive gains and where $e_{w_{il}}(t) = w_{il}(t) - \hat{w}_{il}(t)$, $i = 1, 2, 3, 4$ are the observation errors. The desired trajectory for the follower agent is obtained from the trajectory of the leader delayed $\tau$ unit of time, the delayed observer from system (9). Based on the latter, we can propose

$$
\begin{aligned}
u_{xs} &= -k_{1s} \tanh(e_{1s}(t)) + \dot{\hat{w}}_{1l}(t)\\
u_{ys} &= -k_{2s} \tanh(e_{2s}(t)) + \dot{\hat{w}}_{2l}(t)\\
w_s &= -k_{3s} \tanh(e_{3s}(t)) + \dot{\hat{w}}_{3l}(t)
\end{aligned} \tag{10}
$$

where $e_{is}(t) = x_s(t) - \hat{w}_{il}(t)$, $i = 1, 2, 3$ are the formation tracking errors. The results obtained from a numerical simulation are shown in Fig. 3. This figure shows the positions of the robots and their paths at 10, 28, 46 and 60s. It can be seen how the leader robot follows the trajectory while the follower robot maintains a formation with respect to the leader, following the same path as the leader, but delayed, with a constant time-gap separation. The distance of the follower with respect to the leader is not fixed. It varies according to the velocity of the trajectory, maintaining a distance during the curve and another on the straight due to the change in the velocities in said segments. However, its trajectory is not displaces, but remains the same unlike the formation control strategy based on the time-varying position vector.

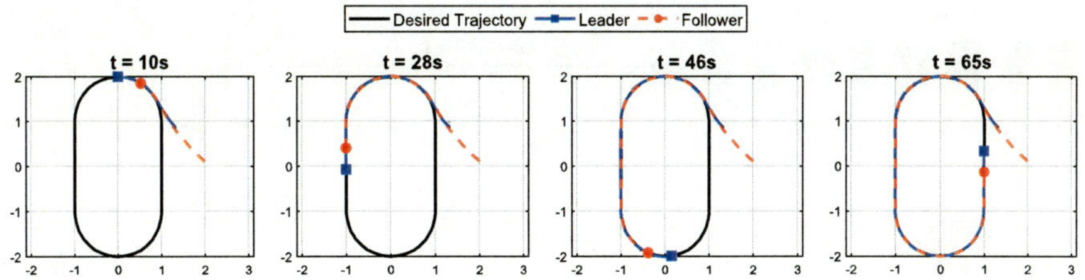


**Fig. 3** Simulation results with time-gap separation

# 4 Experimental Results

This section presents the experimental platform used to validate the theory, as well as the experimental results obtained.

## 4.1 *Experimental Platform*

The experimental platform includes an Optitrack vision system to detect the position and orientation of each robot. The vision system consists of 8 Flex 13 cameras and software Motive designed to track objects in 6 degrees of freedom, with high precision and real-time processing. The leader is an AutoMiny version 4.0, this robot is an autonomous vehicle model (1:10 scale) developed at the Freie Universität Berlin, for educational and research purposes. The follower is an omnidirectional capable of moving in any direction, without the need to perform extra maneuvers, this robot was developed at the Universidad del Papaloapan.

The experimental platform used is shown in Fig. 4. The vision system is responsible for the location of the robots (position and orientation). This information is sent to a computer with the Motive software. Once having the position and orientation in Motive, these dates are sent to the computers that control the robots, one computer per robot, through a Wi-Fi network. These computers perform the control law calculations for formation tracking and send them wirelessly to each robot.

## 4.2 *Experimental Results*

The experimental results obtained using the leader-follower control strategies for formation tracking presented in Sect. 3 are shown below. In both strategies, the leader robot (AutoMiny) follows the desired trajectory, while the follower robot (omnidirectional) maintains a formation with respect to the leader.

Figures 5a–6b present the results obtained from experiment 1 using the formation strategy with time-varying position vector. The parameters used in the experiement

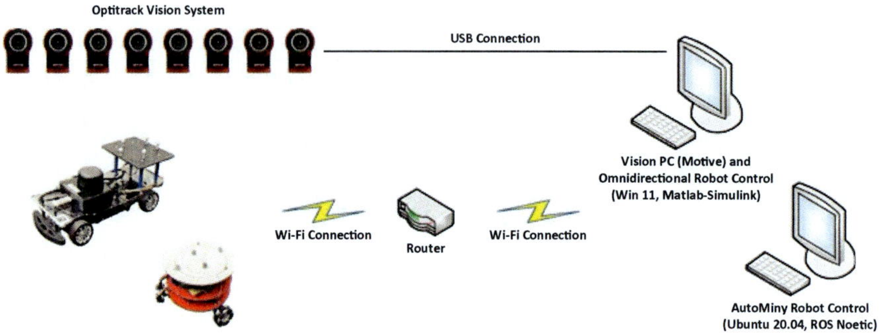


**Fig. 4** Experimental platform

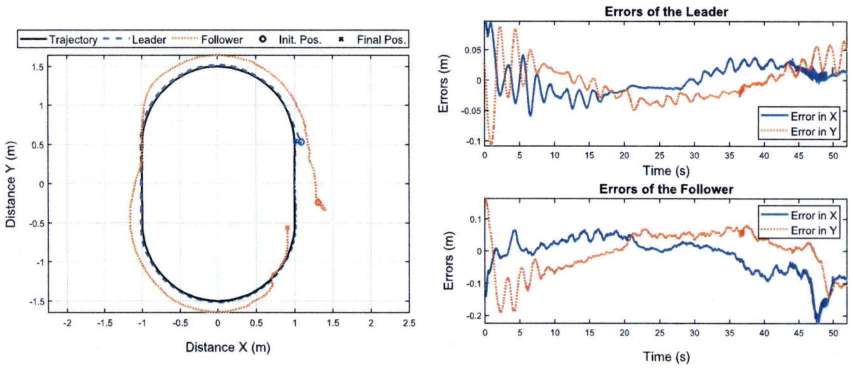


(a) Trajectories on the plane. (b) Tracking errors.

**Fig. 5** Trajectories and position tracking errors of the robots, experiment 1

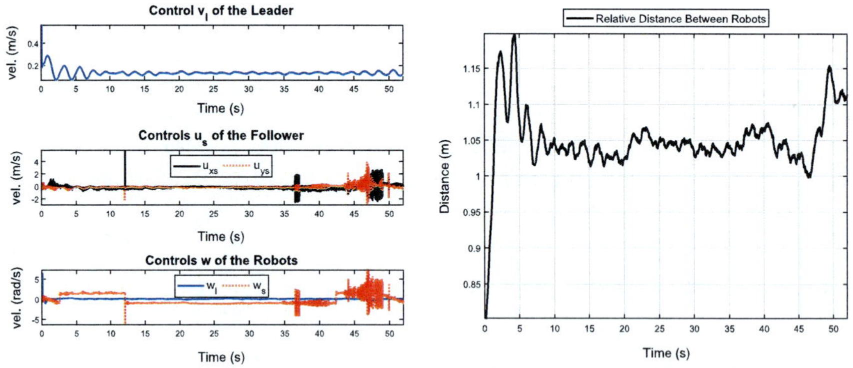


(a) Control signals. (b) Relative distance

**Fig. 6** Control signals and relative distance between robots, experiment 1

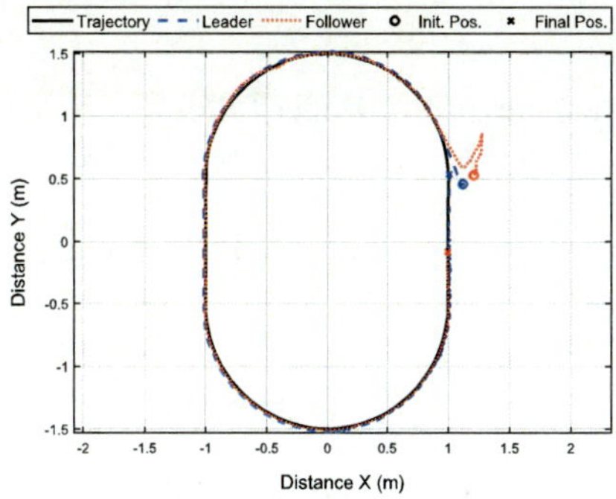


(a) Trajectories on the plane.

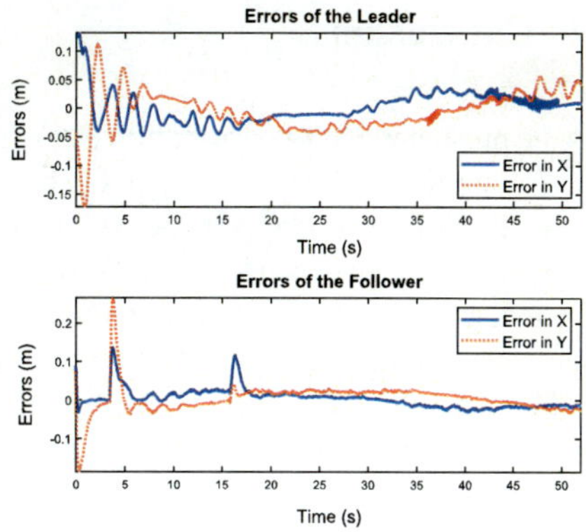


(b) Tracking errors.

**Fig. 7** Trajectories and position tracking errors of the robots, experiment 2

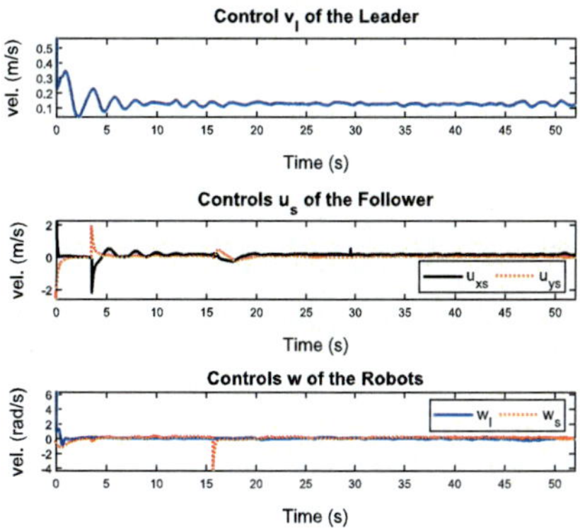


(a) Control signals.

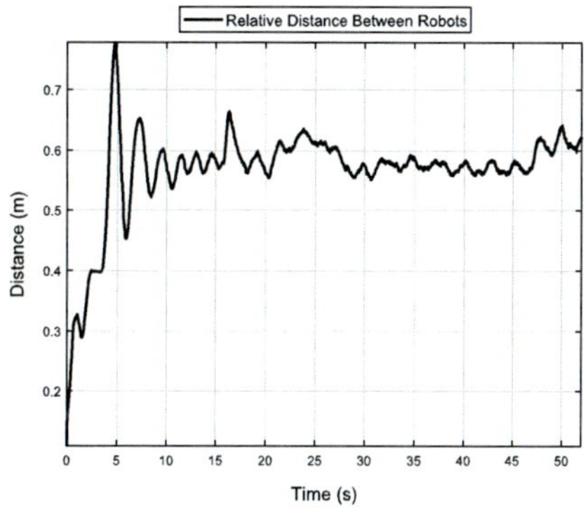


(b) Relative distance

**Fig. 8** Control signals and relative distance between robots, experiment 2

are $k_{1l} = k_{2l} = 1$, $k_{1s} = k_{2s} = k_{3s} = 1.2$ and $c_{ls} = [-1, 0]^T$. Figure 5a presents the trajectories of the robots in the plane. The position tracking errors are shown in Fig. 5b. Figure 6a presents the control signals sent to each robot. Relative distance between robots are shown in Fig. 6b.

Figures 7a–8b present the results obtained from experiment 2 using the formation strategy with time-gap separation. The parameters used in the experiemnt are $k_{1l} = k_{2l} = 1$, $k_{1s} = k_{2s} = k_{3s} = 1$, $\lambda_{1l} = \lambda_{2l} = \lambda_{3l} = \lambda_{4l} = 7$ and $\tau = 3.5s$. Figure 7a presents the trajectories of the robots in the plane. The position tracking errors are shown in Fig. 7b. Figure 8a presents the control signals sent to each robot. Relative distance between robots are shown in Fig. 8b.

## 5 Conclusions

In the present work, two formation tracking strategies using a leader-follower scheme for PA applications are presented. On both strategies the leader follows the trajectory while the follower adapts its behavior for each strategy.

On the strategy of time-varying position vector the follower robot remains at a fixed distance from the leader but must be displaced from the leader, since both robots share the orientation angle. On the strategy of time-gap separation, the follower robot hold formation with respect to the leader, then it follows the same path. The distance between them is not fixed, but varies according to the velocity.

An analysis of the experimental results allows us to conclude that both strategies are carried out with an acceptable error for the type of application, therefore the type of strategy to use depends on the available spaces at the cultivation area.

The formation tracking strategies can be applied to a real process as long as they are implemented in a MAS designed specifically for agricultural applications.

## References

1. Dwivedi, A.: Precision agriculture. In: E. Ahmad, E. (ed.) Agri-Horti Technological Innovations, pp. 83–105. Washington State University Extension (2017)
2. Krishnan, S., Lakkanige, K., Ananthakrishnan, R., Dhaneesh, V., Laungani, V.: Automated irrigation system. Int. J. Eng. Res. Tech. **9**, 915–918 (2020)
3. Weddell, B.: Precision Agriculture. In: G. Yorgey. (ed.) Advances in Dryland Farming in the Inland Pacific Northwest, pp. 319–352. Washington State University Extension (2017)
4. Jurišić, M., Plaščak, I., Barač, Ž., Radocaj, D., Domagoj., Z.: Sensors and their application in precision agriculture. Tehnički Glasnik. (2021). https://doi.org/10.31803/tg-20201015132216
5. Rajmane, R., Gitay, N., Yadav, A., Patil, Y., Sonawane, A.: Precision agriculture and robotics. Int. J. Eng. Res. Tech. **9**, 17–22 (2020)
6. Rendon-Benitez, F., Santiaguillo-Salinas, J., Gonzalez-Sierra, J., et al.: Marching control of multi-agent systems with orientation to the marching angle of the leader (in Spanish). In: Proceedings on the XV Congreso Latinoamericano de Control Automatico (2012)
7. Briñon-Arranz, L., Seuret, A., Canudas-de Wit, C.: Cooperative control design for time-varying formations of multiagent systems. IEEE Trans. Autom. Control **59**, 2283–2288 (2014)
8. Santiaguillo-Salinas, J., Aranda-Bricaire, E.: Containment problem with time-varying formation and collision avoidance for multiagent systems. Int. J. Adv. Robot. Syst. (2017). https://doi.org/10.1177/1729881417703929
9. Santiaguillo-Salinas, J., Aranda-Bricaire, E.: Time-varying formation tracking with collision avoidance for multi-agent systems. In: IFAC-PapersOnLine, vol. 50, pp. 309–314 (2017)
10. Swaroop, D., Rajagopal, K.R.: A review of constant time headway policy for automatic vehicle following. In: Proceedings on the 2001 IEEE Intelligent Transportation Systems, pp. 65–69. IEEE (2001)
11. Klancar, G., Matko, D., Blazic, S.: A control strategy for platoons of differential drive wheeled mobile robot. Robot. Auton. Syst. (2011). https://doi.org/10.1016/j.robot.2010.12.002
12. Milanes, V., Shladover, S.E., Spring, J., et al.: Cooperative adaptive cruise control in real traffic situations. IEEE Trans. Int. Trans. Syst. (2014). https://doi.org/10.1109/TITS.2013.2278494
13. Cruz-Morales, R.D., Velasco-Villa, M., Rodriguez-Angeles, A.: Chain formation control for a platoon of robots using time-gap separation. Int. J. Adv. Robot. Syst. (2018). https://doi.org/10.1177/1729881418770858

14. De Luca, A., Oriolo, G., Samson, C.: Feedback control of a nonholonomic car-like robot. In: Laumond, J.P. (ed.) Robot Motion Planning and Control. Lectures notes in control and information sciences, vol. 229, pp. 171–253. Springer, Verlag/Berlin/Heidelberg (1998)
15. Li, W., Yang, C., Jiang, Y., et al.: Motion planning for omnidirectional wheeled mobile robot by potential field method. J. Adv. Transp. (2017). https://doi.org/10.1155/2017/4961383

# Modeling, Identification and Delayed Systems

# Model-Based Efficient Selection of Centrifugal Pumps for Simple Branch Pumping Systems

**Cláudio C. Pellegrini, Jacqueline Pedrera-Yanes, Orestes Llanes-Santiago, and Guillermo Vilalta-Alonso**

**Abstract** The efficient selection of centrifugal pumps to operate in a system is a subject that needs the utmost attention due to its high energy consumption. Nowadays, a significant percentage of the pumps work outside of the preferred operating region (POR). In the present paper, a mathematical model is developed for identifying if a pump operating in a single branch pumping system, is inside of the POR, which indicates that it was selected correctly and, therefore, operates efficiently. The developed model was implemented in MatLab®, through a discrete algorithm. For the study, a pump database was created with the digitalization of the $H(Q)$ and $\eta(Q)$ curves of 43 pumps from different manufacturers and operating ranges. Seven different sizes systems were defined for model validation. Each system was tested for all pumps belonging to the database. The results show that the model correctly identifies which pumps work inside and outside the POR, using system-pump non-dimensional coefficients, without calculating the operating flow rate. In the case that the analyzed pump is outside the POR, the method also indicates how far a pump is from the POR limits. This allows to the pipeline system designers to choose a strategy for efficient pump selection.

**Keywords** Efficient selection of centrifugal pumps · Preferred operating region (POR) · Energy efficiency · Efficient operation of centrifugal pumps

C. C. Pellegrini · J. Pedrera-Yanes · G. Vilalta-Alonso (✉)
Federal University of São João del Rei, Praça Frei Orlando 170, Centro SJDR, Brazil
e-mail: gvilalta@ufsj.edu.br

C. C. Pellegrini
e-mail: pelle@ufsj.edu.br

J. Pedrera-Yanes
e-mail: jpedrera@ufsj.edu.br

O. Llanes-Santiago
Universidad Tecnológica de La Habana José Antonio Echeverría, CUJAE, Calle 114 No. 11901, Marianao, La Habana, Cuba
e-mail: orestes@tesla.cujae.edu.cu



O. Llanes-Santiago (ed.), *Proceedings of 19th Latin American Control Congress (LACC 2022)*, Studies in Systems, Decision and Control 464,
https://doi.org/10.1007/978-3-031-26361-3_18

## 1 Introduction

Currently, centrifugal pumps are the most widely used type of turbomachinery in industrial and domestic applications. Therefore, the electrical energy consumption by those equipments represents an important parcel of the overall energy budget and, consequently, accounts for higher values of $CO_2$ emitted into the atmosphere [1].

Statistics associated with energy consumption in different industries and countries, reflect the need for the proper selection, maintenance, and operation of pumps [2]. A centrifugal pump is designed to operate at its best efficiency point (BEP).

Taking into account the different applications where the pump may work, the pump manufacturers indicate a permissible deviation region around the BEP named preferred operation region (POR) where the operation of the pump is considered satisfactory. In ANSI/HI 9.6.3-2017 standard [3], the POR is defined as a range of rates of flow to either side of the predicted BEP within which the hydraulic efficiency and the operational reliability of the pump are not substantially degraded. The operation of centrifugal pumps away from the POR for long periods of time leads to adverse results, reducing the life expectancy of the pump and its components [4]. It is a well-known fact that a significant quantity of pumps operates outside of the POR.

For this reason, intensive research efforts focused on the development of more efficient centrifugal pumps and on a better understanding of the factors that affect the efficient operation of the pumping systems, has been made in the last decades employing numerical techniques, experimental studies, and optimization methods, as it is summarized. The impeller is the pump's component that has received more attention because it has the most influence on the pump performance. Flow guidance inside the vane, the number and thickness of blades [5, 6], development of the secondary flows [7], flow pattern [8, 9], and models for characterizing the losses inside the pumps [10] are the most investigated factors in order to enhance the energy conversion efficiency with respect to conventional configurations. The obtained results in all these works, support strategies for the better hydraulic performance of the new pump's models.

Regarding water pumping systems, the water distribution networks (WDNs) have been widely studied because they constitute one of the vital infrastructures of modern cities, which consumes a lot of energy [11]. Due to the variable water demand in WDNs, pumps operate with variable efficiency and thus, necessarily out of the BEP. Therefore, optimizing the operating condition of the pumps is very important for minimizing costs. According to [12], the main factors assessed are pump operation, size, and location and its results show that pump operation is the most relevant factor. Selection of the pumps operating at the BEP significantly reduces the energy expenses in a WDNs.

The above-cited research compilation has mainly focused on the efficient design of new centrifugal pump models and on some strategies for the proper selection of pumps when operating in WDNs. However, no research centered on the criteria that should be used for the efficient selection of a pump for a simple branch pumping system was found.

The main objective of the present paper is to develop a mathematical model to know if a pump is suited for a given system. In the proposed model, which was implemented in MatLab® utilizing a discrete algorithm, the centrifugal pumps are characterized by the set of coefficients of head and efficiency performance curves, and the pipeline system is defined by pipes of different lengths, diameters, materials, absolute roughness, and different accessories and fittings. The model allows determining, for a given system, which pumps from a database of pumps operate inside of the POR. This supports the decision-making process on the most appropriate selection of a pump for a specific system.

## 2 Mathematical Analysis

### 2.1 *Mathematical Model*

Consider a simple branch hydraulic system, as the one depicted in Fig. 1, comprising two reservoirs connected by a pipes, a centrifugal pump and a number of pipe fittings. Points A and B are located at the fluid's free surfaces at rest, and the line connecting those points is a streamline.

The energy conservation equation for steady-state, incompressible flow, with uniform velocities between points A and B situated on the same streamline is

$$\frac{P_B - P_A}{\gamma} + (z_B - z_A) + h - H_P = 0 \tag{1}$$

where $P$ is the pressure at the suction and discharge tanks, $\gamma$ is the fluid specific gravity, $z$ is the vertical position of points A and B, $h$ is the total loss due to friction and $H_P$ is the energy added by the pump, or the pump's head.

In Eq. 1, the three first terms at the left hand side represent the energy required by the flow (the system's head), denoted by $H_S$. Thus,

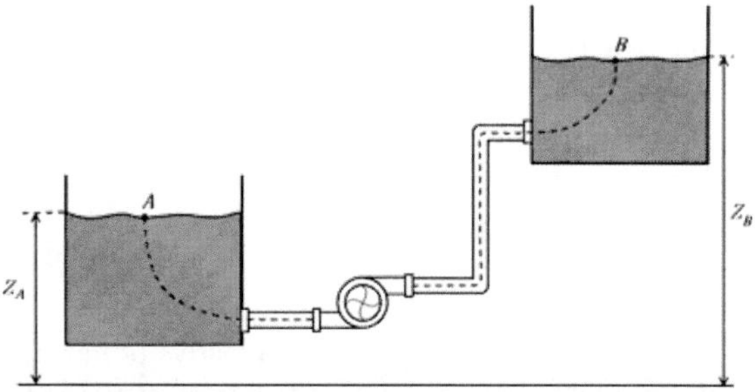


**Fig. 1** Simple branch hydraulic system: two reservoirs and one centrifugal pump

$$H_S = \frac{P_B - P_A}{\gamma} + (z_B - z_A) + h \tag{2}$$

indicating that at the operating point $H_P = H_S$.

The pressure and elevation terms in Eq. 1 are known to the system operator, but the pump and the losses terms must be modelled. The distributed losses, $h_f$, in Newtonian fluid flows through circular pipes can be modelled using the Darcy-Weisbach equation [13] as

$$h_f = f_1 \frac{L_1}{D_1}\frac{V_1^2}{2g} + f_2 \frac{L_2}{D_2}\frac{V_2^2}{2} \tag{3}$$

In this equation and hereafter, subscripts 1 and 2 represent properties associated with the piping system between reservoir 1 and the suction nozzle and between the discharge nozzle and reservoir 2 respectively. Thus, $f_1$ and $f_2$ are the friction factors of the suction and discharge tubes and, similarly, $L_1$ and $L_2$ are the total tube lengths, and $D_1$ and $D_2$ are the diameters. The velocities, $V_1$ and $V_2$ are related to the flow rate in the system through $V_1 A_1 = V_2 A_2 = Q$, where $A_1 = \pi D_1^2/4$ and $A_2 = \pi D_2^2/4$ are the cross-sectional area of the pipes

The localized losses, $h_L$, can be modelled according to K theory [13] as

$$h_L = \sum_{i=1}^{n_s} K_{1,i} \frac{V_1^2}{2g} + \sum_{i=1}^{n_d} K_{2,i} \frac{V_2^2}{2g} \tag{4}$$

In this equation, $K_{1,i}$ and $K_{2,i}$ are the localized loss coefficients of the individual pipe fittings. The summations run from 1 to $n_s$ and $n_d$, which are the numbers of pipe fittings in the suction and discharge lines respectively.

To model the pump, polynomial approximations are frequently used [14]. For centrifugal pumps, a second degree equation for the pump's head is generally appropriate:

$$H_P = aQ^2 + bQ + H_{P0} \tag{5}$$

where $H_P$ is the energy required by the pump, i.e. the pump's head, and $a$, $b$ and $H_{P0}$ are constants that must be obtained experimentally. $H_{P0}$ is the static head of the pump, which is the energy required by the pump at $Q = 0$. Hereafter, the subscript '0' is used to indicate variables at $Q = 0$.

The $H_P(Q)$ dependence is illustrated in Fig. 2 by the black curve. The blue, green and gray curves in Fig. 2 represent the behaviour of the system in three distinct cases to be explained soon.

In centrifugal pumps, due to internal losses, the concavity of the $H_P(Q)$ curve is always down, as shown in Fig. 2 and, therefore, $a$ must be negative. The constant $b$, on the other hand, may be zero, positive or negative depending on the sign of the first derivative of $H_P(Q)$ at $Q = 0$, i.e., $H'_P(0)$. As $H'_P = 2aQ + b$, then $H'_P(0) = b$. Thus, $b > 0$ implies $H'_P(0) > 0$ (Fig. 2), $b < 0$ implies $H'_P(0) < 0$ and $b = 0$ implies

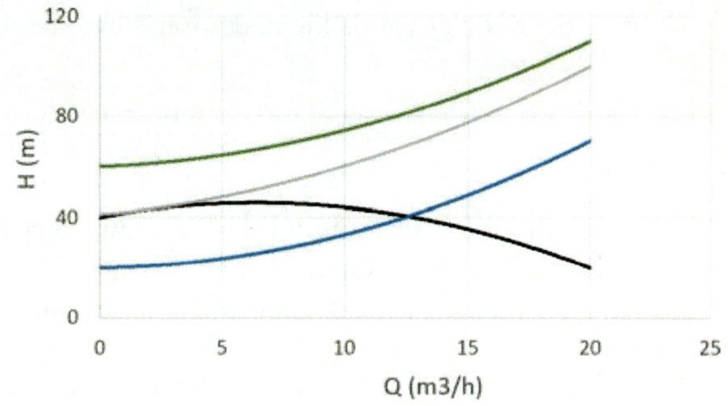


**Fig. 2** Relative positions of the pump and the system's curves in simple branch hydraulic systems. Pump curve (—), system curves (—, —, —).

$H'_P(0) = 0$. Ideally, $H_P(Q)$ curves should decrease monotonically for $Q > 0$ to avoid operation instabilities at low values of $Q$ and, therefore, should have $b > 0$.

Another important characteristic curve of centrifugal pumps is related to its efficiency, which can be described by an incomplete second degree polynomial [14]:

$$\eta = AQ^2 + BQ \tag{6}$$

where $A$ and $B$ are constants that must be obtained experimentally.

Differentiating the preceding equation in relation to $Q$ shows that the efficiency and flow rate at the BEP are

$$\eta_{max} = -B^2/4A \quad \text{and} \quad Q_{BEP} = -B/2A \tag{7}$$

Substituting Eqs. 3–5 into Eq. 1 yields

$$(\alpha - a)\,Q^2 - bQ + H_{S0} - H_{P0} = 0 \tag{8}$$

where

$$H_{S0} = \frac{P_B - P_A}{\gamma} + z_B - z_A \tag{9}$$

is the static head of the system, i.e., $H_S$ for $Q = 0$ and

$$\alpha = \frac{8}{g\pi^2}\left[\left(\frac{f_1 L_1}{D_1^5} + \frac{f_2 L_2}{D_2^5}\right) + \left(\frac{\sum K_1}{D_1^4} + \frac{\sum K_2}{D_2^4}\right)\right] \tag{10}$$

where $\alpha Q^2$ is the term that represents the total energy losses by friction.

To obtain the characteristic curve for the system, Eqs. 3, 4 and 9 may be substituted into Eq. 2 to yield

$$H_S = \alpha Q^2 + H_{S0} \tag{11}$$

Solving Eq. 8 for $Q$ yields the operation flow rate of the system:

$$Q_{\text{op}} = \frac{b \pm \sqrt{b^2 + 4\,(\alpha - a)\,(H_{P0} - H_{S0})}}{2(\alpha - a)} \tag{12}$$

All possibilities for the roots of Eq. 12 are illustrated in Fig. 2, for one hypothetical pump and three different systems.

The first possibility is the existence of two real and distinct roots, represented by the blue and black curves. Two cases are possible here: one of the roots is negative and the other is positive or both are positive. In the first case, the negative root must be discarded; in the second, the smaller root is located to the left of the point of maximum of the $H_P(Q)$ curve and thus lies in a region to be avoided, because it causes unstable operation of the pump. In any case, the result to be considered here is the larger root, and therefore, the positive sign in Eq. 12 must be chosen.

Another possibility is the existence of one repeated real root, illustrated by the black and grey curves in Fig. 2. This solution is never recommended in practice, because the system is physically unable to operate at any other point except at the exact value given by that root. For the case of two complex and conjugate roots, corresponding to the green and black curves in Fig. 2, where it is evident the no intersection between the curves and, thus, there is not a real value for $Q_{\text{op}}$ as a solution of Eq. 12. This indicates that a negative value of the discriminant in Eq. 12 may be used as a flag to exclude that particular pump of the analysis.

It is important to notice that $\alpha$ in Eq. 10 depends on $f$ that depends on $Q$, as can seen in the most recommend empirical formula available for $f$ [15] in turbulent flows

$$f = \frac{1}{1.8^2}\left\{\log_{10}\left[\frac{6.9}{\text{Re}} + \left(\frac{\epsilon}{3,7D}\right)^{1.11}\right]\right\}^{-2} \tag{13}$$

where $\epsilon$ is the absolute roughness of the pipe. The dependence with $Q$ is embedded in the Reynolds number, $\text{Re} = 4\rho Q/\pi d$. For this reason, Eq. 12 is implicit in $Q$. As Eq. 13 is algebraically a bit complicated, analytical solutions to Eq. 12 with variable $f$ have never been obtained.

In fact, the use of variable $f$ does not make sense in a proposal of method for optimizing the choice of a pump without performs the full calculations. Indeed, obtaining $f$ means calculating $Q_{\text{op}}$, which is what the method seeks to avoid. Therefore, the limiting value of $f$ for large Re number in the Moody diagram ($\text{Re} = 10^8$) is used here as an approximation, by assuming the system is operating at the fully rough region.

The most commonly used definition for the POR [3] establishes that the operation flow rate must fall between certain limits, $Q_1$ and $Q_2$, related to the flow rate at BEP:

$$Q_1 \leq Q_{\text{op}} \leq Q_2 \tag{14}$$

where $Q_1 = \kappa_1 Q_{\text{BEP}}$ and $Q_2 = \kappa_2 Q_{\text{BEP}}$, with $0 < \kappa_1 < 1$ and $\kappa_2 > 1$. The most accepted values are $\kappa_1 = 0.7$ and $\kappa_2 = 1.2$ [3]. Thus, from Eq. 12, the POR limits are given by

$$-\frac{\kappa B}{A} = \frac{b + \sqrt{b^2 + 4(\alpha - a)\Delta H}}{(\alpha - a)} \tag{15}$$

where $\kappa = \kappa_1$ or $\kappa_2$ and $\Delta H = H_{P0} - H_{S0}$

After some straightforward algebraic manipulation, Eq. 15 may be rewritten in terms of system-pump non-dimensional coefficients $C_{\text{SP0}}$ and $C_{\text{SP}}$ as

$$-\kappa = C_{\text{SP}} + \sqrt{C_{\text{SP}}^2 + 4C_{\text{SP0}}} \tag{16}$$

where $C_{\text{SP0}} = A^2\Delta H/B^2(\alpha - a)$ and $C_{\text{SP}} = Ab/B(\alpha - a)$. Note that, by definition, $C_{\text{SP0}} > 0$, then the square root is always a real number. Squaring both sides of Eq. 16 yields the POR limit relations:

$$C_{\text{SP}} = \frac{2C_{\text{SP0}}}{\kappa} - \frac{\kappa}{2} \tag{17}$$

for $\kappa = \kappa_1, \kappa_2$. Clearly, taking $\kappa = 1$ gives the relation for the BEP, $C_{\text{SP}} = 2C_{\text{SP0}} - 1/2$. This relation and Eq. 17 represent the locus of the pumps operating at flow rates $Q_{\text{BEP}}$, $Q_1$ and $Q_2$ respectively. Plotting them on a graph provides a simple way to decide if a set of pumps is well suited for a given system. Calculating the values of $C_{\text{SP}}$ and $C_{\text{SP0}}$ for a number of pumps, it is possible to decide graphically if they are inside or outside of the POR, comparing their positions to the straight lines representing Eq. 17 for $\kappa = \kappa_1$ and $\kappa_2$, without calculating $Q_{\text{op}}$. This will be illustrated and discussed in the next section.

## 2.2 Discrete Algorithm for the Model Implementation

The mathematical model obtained in this paper was implemented in MatLab® through the Algorithm 1. The basic premise was to make it as simple as possible: all needed information for identifying if a particular centrifugal pump operates inside the POR for a given pumping system must be available in an integrated environment that includes all the resources needed to obtain the results.

The model is formed by Eqs. 7, 16 and 17. They depend on the pump's and system's information, provided as input data. The pump information is defined by the coefficients $a$, $b$ $H_{P0}$, $A$ and $B$ in Eqs. 5 and 6, respectively. These coefficients were obtained from the digitalization of the head and efficiency curves of 43 centrifugal pumps from different manufacturers and operational ranges. For coefficients, $a$, $b$, and $H_{P0}$ the head curves were fitted by second degree polynomials using the MatLab built-in standard least square method. For coefficients $A$ and $B$, the efficiency curves were fitted in the same way by incomplete second degree polynomials. Each system is characterized by its layout, pipe diameters, lengths and roughnesses, fittings, and pressure data. The information is concentrated on terms $\alpha$ and $\Delta H$. To obtain the solution, the values of chosen POR limits $\kappa_1$ and $\kappa_2$ are also needed.

**Algorithm 1** Pump's selection

**Input** pumps and systems data
**Output** results showing pumps in the POR
**for** $j = 1$ to $N_S$ **do** ▷ $N_S$ is the number of systems
  **for** $i = 1$ to $N_P$ **do** ▷ $N_p$ is the number of pumps
    **if** $\Delta H > 0$ **then** ▷ Pump is a valid candidate
      Calculate $C_{\text{SP0}}$ and $C_{\text{SP}}$ for each pump (Eq. 17)
      Generate theoretical $C_{\text{SP0}} \times C_{\text{SP}}$ lines
      Calculate $H_P$, $H_S$ and $\eta$ (Eqs. 5, 6 and 11)
      Calculate $Q_{\text{OP}}$ to check the POR condition ($Q_1 \le Q_{\text{OP}} \le Q_2$) ▷ Not mandatory
    **end if**
  **end for**
**end for**

## 3 Results and Discussion

As mentioned before, plotting Eq. 17 together with the calculated values of $C_{\text{SP}}$ and $C_{\text{SP0}}$ shows which pumps operate inside and outside the POR, that is, pumps correctly selected and operating efficiently and pumps that should not be used for this application, respectively. The lines $S_1$ and $S_2$ correspond to $\kappa = \kappa_1$ and $\kappa = \kappa_2$ respectively and define the POR envelope. Line $S_{\text{BEP}}$ corresponds to pumps operating exactly on the BEP, i.e., with $\kappa = 1$. Figure 3 illustrates the process.

To validate the proposed model, seven systems were considered. Each one was paired to all the pumps on the database. The results of the application of the model is presented for one system and three pumps from the respective databases. The selected system was the number 2 whose information is shown in Table 1. In the case

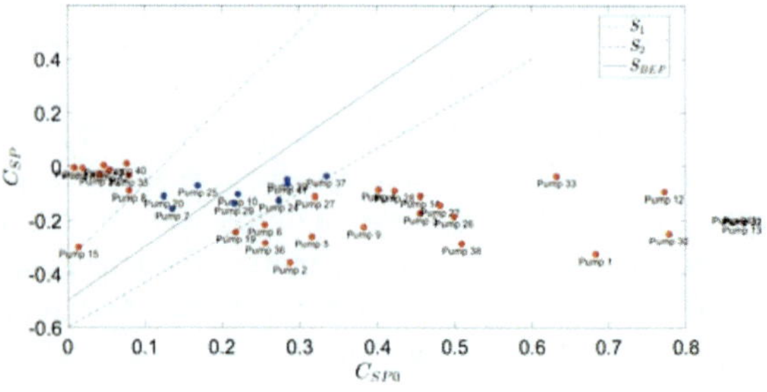


**Fig. 3** Results of the proposed algorithm for a simple branch hydraulic system. Blue points are inside the POR and red points are outside

**Table 1** Dimensions of the system used for the results discussion

| System | Type | $L_s$[(a)] (m) | $L_d$ (m) | $D_s$ (m) | $D_d$ (m) | $\epsilon$ (mm) | $\sum K_s$ | $\sum K_d$ | $H_{P0}$ (m) |
|---|---|---|---|---|---|---|---|---|---|
| 2 | Short-length | 5 | 45 | 0.09 | 0.078 | 0.001524 | 1.45 | 1 | 5 |

[a]Subscripts *s* and *d* indicate suction and discharge systems, respectively

**Table 2** Coefficients of the head and efficiency characteristic curves for pumps

| Pump | $a$ (s$^2$/m$^5$) | $b$ (s/m$^2$) | $H_{P0}$ (m) | $A$ (s$^2$/m$^6$) | $B$ (s/m$^3$) |
|---|---|---|---|---|---|
| 6 | −0.0072 | 0.1020 | 10.70 | −0.1241 | 5.7284 |
| 7 | −0.0028 | 0.0634 | 10.47 | −0.0577 | 4.1619 |
| 8 | −0.0011 | 0.0439 | 9.50 | −0.0221 | 2.6156 |

of the pumps were selected the numbers 6, 7 and 8 which permit to make different analyzes. The coefficients of the head and efficiency curves for these pumps are shown in Table 2.

Figure 4 shows the characteristic curves for two pumps operating outside the POR (pumps 6 and 8) and one pump operating inside the POR (pump 7). Comparing Figs. 3 and 4, it is possible to see that pumps at or above $S_1$ operate with $Q_{\text{op}} \leq Q_1$ and pumps at or below $S_2$ with $Q_{\text{op}} \geq Q_2$. Of course, pumps between $S_1$ and $S_2$ operate with $Q_1 \leq Q_{\text{op}} \leq Q_2$.

From this analysis, it is possible to highlight some important aspects. First, the closer the points are to $S_{\text{BEP}}$, the closer they are to their BEP. This conclusion is valid not only for points inside the POR envelope but for any point. This is an advantage of the proposed methodology. On the other hand, two points at the same "distance" from $S_{\text{BEP}}$ do not necessarily have the same efficiency. The geometrical distance in Fig. 3 simply shows how far that pump is from *its own* value of $\eta_{max}$. Two pumps at zero distance from $S_{\text{BEP}}$, i.e., exactly over the $S_{\text{BEP}}$ line, will operate at their own $\eta_{max}$ which, in the general case, is completely different. For pumps 6 and 8, for example, it is 75% and 79% respectively.

Nonetheless, satisfying Eq. 17 or, equivalently, falling inside the POR envelope in Fig. 3, yields the mathematical relation between the pump coefficients ($a$, $b$, $H_{P0}$, $A$ and $B$) and the system coefficients ($\alpha$ and $H_{S0}$) that guarantees a high efficiency value for a number of available pumps previously chosen to operate in a given system. For pumps working inside the POR, no more action is required because this is the condition to be identified by the model. Therefore, the operating flow rate falls between the POR limits, and the pump efficiency is close to the maximum value. For example, pumps 6, 7 and 8 operate with 94%, 99% and 89% of its maximum efficiency.

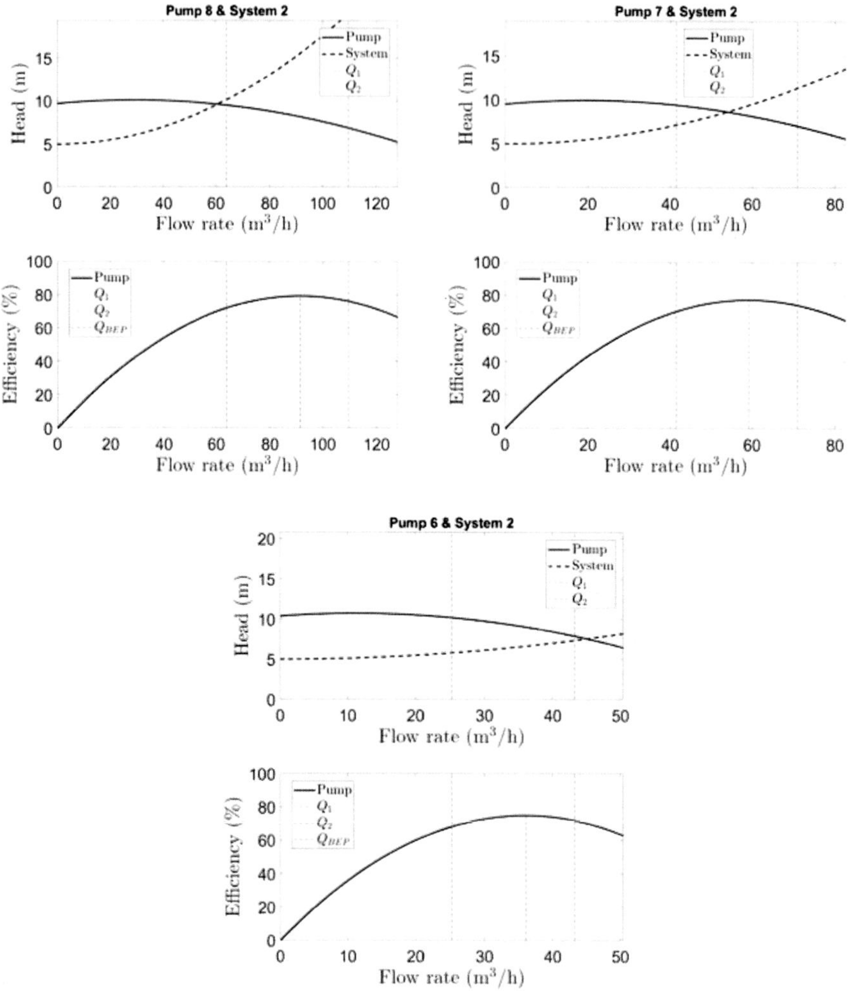


**Fig. 4** Characteristic curves

# 4 Conclusions

A mathematical model to identify if a pump operating in a given pumping system, was selected to operate efficiently is proposed. The model was validated by pairing seven different hydraulic systems with all the pumps on the database. Using system-pump non-dimensional coefficients, the model allows the designer to identify the position of the pumps inside the POR envelope, defined by lines $S_1$ and $S_2$. When a pump is above $S_1$, then $Q_{OP} \leq Q_1$. On the other hand, when a pump is below $S_2$, then $Q_{OP} \geq Q_2$. In both cases, the use of the pumps in the system is not recommended. Pumps operating inside the POR, fall between these limits, and the closer they are to the line $S_{BEP}$, the closer to their own BEP. In this case, it is possible to conclude that these pumps may be used in this system, i.e., they were efficiently selected. The results of this paper could be important for choosing the appropriate strategies for the efficient selection of centrifugal pumps.

## References


1. ETSU; CETIM; Reeves, DT, NESA, Tech. Univ. Darmstadt. Study on Improving the Energy Efficiency of Pumps. European Commission, UK (2001). http://www.jakob-albertsen.dk/komposit/Darmstadtrapport.pdf. Accessed on 20 April 2022
2. Copeland, C.; Carter, N.T. Energy-Water Nexus: The Water Sector's Energy Use R43200 (2017). https://crsreports.congress.gov/product/pdf/R/R43200. Accessed on 09/05/2022
3. HI(ANSI) HI 9.6.3 - 2017 Rotodynamic pumps - Guideline for Operating Regions
4. Siddique, M.H., Afzal, A., Samad, A.: Design optimization of the centrifugal pumps via low fidelity models. Math. Prob. Eng.. Article ID 3987594, 1–14, (2018). http://doi.org/10.1155/2018/3987594
5. Capurso, T., Bergamini, L., Torresi, M.: A new generation of centrifugal pumps for high conversion efficiency. Energy Convers. Manag. **256**, 115341 (2022). https://doi.org/10.1016/j.enconman.2022.115341
6. Abo Elyamin, G.R., Bassily, M.A., Khalil, K.Y., Gomaa, M.S.: Effect of impeller blades number on the performance of a centrifugal pump. Alexandria Eng. J. **58**(1), 39–48 (2019). https://doi.org/10.1016/j.aej.2019.02.004
7. Wu, C., Pu, K., Li, C., Wu, P., Huang, B., Wu, D.: Blade redesign based on secondary flow suppression to improve energy efficiency of a centrifugal pump. Energy. **246**, 123394 (2022). https://doi.org/10.1016/j.energy.2022.123394
8. Abdolahnejad, E., Moghimi, M., Derakhshan S.: Experimental and numerical investigation of slip factor reduction in centrifugal slurry pump. J. Braz. Soc. Mech. Sci. Eng. **43**(4), 179 (2021). https://doi.org/10.1007/s40430-021-02831-x
9. Li, X., Chen, B., Luo, X., Zhu, Z.: Effects of flow pattern on hydraulic performance and energy conversion characterisation in a centrifugal pump. Renew. Energy. **151**, 475–487 (2020). http://doi.org/10.1016/j.renene.2019.11.049
10. Liu, M., Tan, L., Cao. S.: Theoretical model of energy performance prediction and BEP determination for centrifugal pump as turbine Energy. **172**, 712–732 (2019). http://doi.org/10.1016/j.energy.2019.01.162
11. Zaman, D., Tiwari, M.K., Gupta, A.K., Sem. D.: Performance indicators-based energy sustainability in urban water distribution networks: A state-of-art review and conceptual framework. Sustain. Cities Soc. **72**, 103036 (2021). https://doi.org/10.1016/j.scs.2021.103036
12. Salomé, D., Meirelles, G., Melo, B., Bezerra, D.: Optimal pump selection for variable speed operation in water distribution network. Braz. J. Water Res. **25**, e49 (2020). https://doi.org/10.1590/2318-0331.252020200099
13. White, F.M.: Fluid Mechanics. McGraw-Hill Education, NewYork (2016)
14. Arun-Shankar, V.K., Subramaniam, U., Shanmugam, P., Hanigovszki, N.: A comprehensive review on energy efficiency enhancement initiatives in centrifugal pumping system. Appl. Energy. **181**, 495–513 (2016). https://doi.org/10.1016/j.apenergy.2016.08.070
15. Haaland, S.E.: Simple and explicit formulas for the friction factor in turbulent pipe flow. J. Fluids Eng. **105**, 89–90 (1983). https://doi.org/10.1115/1.3240948

# Reinforcement Learning with Probabilistic Boolean Networks in Smart Grid Models

**Pedro Juan Rivera Torres, Carlos Gershenson García, María Fernanda Sánchez Puig, Mario Franco, and Samir Kanaan Izquierdo**

**Abstract** In the field of Smart Grids, the necessity arises to continuously improve efficient and resistant, high-quality electrical power, concurrently handling all possible faults and failures. Reaching such goal requires high reliability in their components, appropriate maintenance, and a calculated occurrence of failures. Accurate operation of the system comprises those actions and contemporary methods to find, classify, and confine failures and faults, mimic and model processes with foreboding algorithms and analytics. This work showcases the use of an adaptive, self-organizing modeling structure, Probabilistic Boolean Networks, to forge ahead the knowledge of the dynamics of smart grid equipment, and to model and typify their conduct. We show that PBNs are indeed interchangeable with the basic Reinforcement Learning Cycle, where the model or agent has a correspondence with its surroundings and obtains a response as a reward signal. Dissimilar reward structures were constructed to represent favored actions. This knowledge can guide the PBN to avert fault conditions and failures.

---

P. J. R. Torres (✉) · C. G. García · M. F. S. Puig · M. Franco
Complexity Sciences Center (C3), National Autonomous University of Mexico, Ciudad de México, México
e-mail: pedro.rivera@c3.unam.mx

P. J. R. Torres · S. K. Izquierdo
Bioinformatics and Biomedical Signals Laboratory (B2SL), Facultat de Matemàtiques i Estadística, Centre de Recerca en Enginyeria Biomèdica (CREB), Universitat Politècnica de Catalunya, Barcelona, Spain

C. G. García
Research Institute of Systems and Applied Mathematics (IIMAS), National Autonomous University of Mexico, Ciudad de México, México

Lakeside Labs GmbH, Klagenfurt am Wörthersee, Austria

S. K. Izquierdo
Institut de Recerca Sant Joan de Déu, Esplugues de Llobregat, Barcelona, Spain

M. F. S. Puig
Faculty of Sciences, National Autonomous University of Mexico, Ciudad de México, México

 
O. Llanes-Santiago (ed.), *Proceedings of 19th Latin American Control Congress (LACC 2022)*, Studies in Systems, Decision and Control 464,
https://doi.org/10.1007/978-3-031-26361-3_19

## 1 Introduction

Electricity has altered humanity. The electrical power generation cycle (generation transmission and distribution) evolved from a commodity to a fundamental human need. It is produced in different manners, but its fundamental process has remained uniform: through the different (electromechanical, geothermal, solar, and other means) generation approaches, users receive it through the transmission and distribution network. They are circumscribed, directed transmission systems, controlled based on demand. Decentralization supplies and triggers demands, growing adept approaches in modeling and control of smart grids [1]. The application of Signal Processing techniques to this field has enabled the emergence of "Smart Devices" such as the Intelligent Power Router (IPR). This appliance [2] was influenced by IP Routing, and it has a certain amount of "intelligence", which permits line switching and load sharing. IPRs and similar devices allow the Electrical Power Distribution System (EPDS) to be dependable, resistent, malleable, and effective. With these appliances, choices can be actioned when perturbations are encountered, organizing with other apparatuses in their neighborhood to act upon to load changes, demands, fault conditions, and crises. The main elements of this appliance are presented here in Fig. 1.

EPDS with IPRs (EPDS-IPR) have autonomous service recovery, if an EPDS-IPR is arranged across the grid, programmed for exchanging data to control and rearrange the network with a particular rule set, when a perturbation unfolds, allowing survivability, and optimal system resource use. The design of EPDS-IPR is a complicated endeavor. IPRs have to be shaped with prearranged sequences and instructions on how to behave when a given group of conditions have happened. A separate task is to make them adaptive [3], and not pursue a pre-defined group of commands. A more agreeable scenario is the autonomy of the network and its self-reconfiguration in the occurrence of a perturbation. IPRs, when forming a grid, can be modeled as a Probabilistic Boolean Network (PBN), an adaptive setup that learns from its own long-run

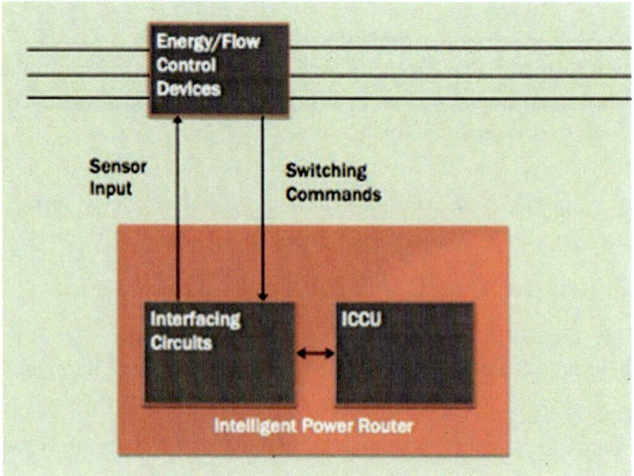


**Fig. 1** Elements of the intelligent power router

behavior and has abilities such as self-organization and resilience. Procedures for PBN-based modelling are showcased in [4–13], which validate the application of these methods to model processes and appliances in manufacturing and intelligent power with IPRs, empowering the simulation of different situations. They allow engineers to program these apparatuses properly and design resilient networks. We endow EPDS-IPRs with an intellect that permits survival in a large group of perturbation events, otherwise almost unpredictable. Methods that imitate living systems, also known as biomimetic [19], have been employed to examine and explain difficult problems in general, and notably in EPDS. PBNs (state-transition systems that fulfill the conditions for the Markov Property) are memoryless and do not rely on past system states). They were discovered by Shmulevich and Dougherty [13] by extension of Kauffman's N-K Networks concept [14, 15], combining the rule-based modeling quality of N-K networks while incorporating stochastic conduct. They are constructed with sets of constituent N-K networks (also known as Boolean Networks, or BN), that have probabilistic weights, in which every constituent N-K network is a selectable "context". Methods for incorporating PBNs to manufacturing engineering processes and systems were proposed in [9, 13], with further developments in [4–8, 10–12]. For this paper, the use of PBNs will be broadened to assess IPR trustworthiness, and for the examination of fault conditions leading to catastrophe. The main contributions herein presented are the exploration the capacity of PBN models to performing the standard Reinforcement Learning (RL) loop, exploring RL as a way to direct the EPDS-IPR's evolution, increasing reliability and resilience, looking to achieve a level of self-control and automation of the EPDS-IPR using RL.

## 2 Materials and Methods

Failure modes for a sole IPR in an EPDS-IPR have been characterized. The Probabilistic Symbolic Model Checker (PRISM) [17], was employed in validating the IPR and EPDS-IPR's formal computational correctness with Probabilistic Computational Tree Logic (PCTL). Our IPR models were constructed in PRISM using three modules: the IPR's operating environment, a module for the IPR's PBN, and finally its reward structure. The PBN's nodes states are in the following module, employs the nodes' state values in the first module and applies values derived from the Boolean predicting functions to migrate its adjacent state. Using the initial values and the IPR's fault conditions or failure modes, the state of its variables is modified, thus yielding the actual state of the device. The accuracy/reliability appraisal of the IPR's subcomponents are taken from [2].

Initially, PBN that represents the IPR is constructed, where each modeled subcomponent can be two different states; 0 (IPR is ON and 1 for OFF (established in [2])). For a particular node, the functions (Boolean) that assess the state of the device in time $t + 1$, called Predictor Functions, are executed and applied, for the current state of the nodes in $t$. As the second step, we build a matrix for all nodes, with their corresponding predictor/s. Only relevant nodes are considered for the calculation of

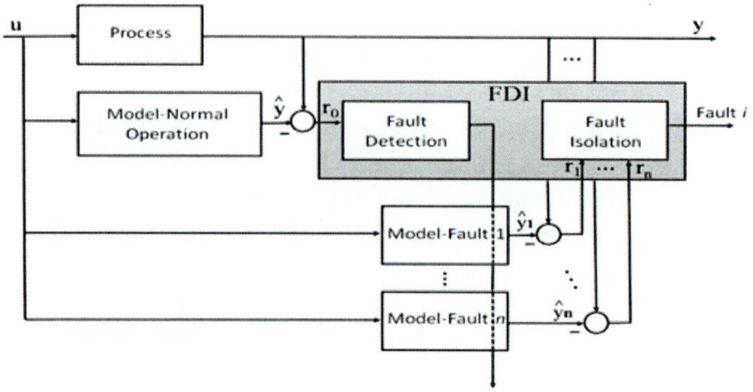


**Fig. 2** FDI method [18]

the predictors; and those that do not influence the state current node directly state are discarded for the analysis. The equations for the Boolean predictor functions of the observed node are derived from the connections established between relevant observed nodes, which define the state of the nodes. For each of them, there is a group or set of predictors (one or more), obtained through evaluation of the associations between observed and relevant nodes. The occurrence of states of every relevant node are examined, and then an assessment is formed about its future state in $t + 1$, due to the state of the EPDS's relevant nodes in time $t$. This proposed approach complies with the Fault Detection and Isolation (FDI) strategy that is described in [16], which we present in Fig. 2, with one model used for the description of the normal operation of a process and second for describing each of the failure modes.

PBNs are complex systems with attributes such as self-organization, and they organize into states called attractor states [16]. These are groups of repeating/cycling states that (in this work) relate to the failure modes exhibited by the IPR and EPDS. The presented models are similar in construction and semantics to those in [8]. Through the characterization of failure modes of the IPR (its faults and failures), our model can characterize the condition or state of their components/nodes to assess the correlation to the device's fault conditions. The reliability analysis in [2] was taken into consideration. Simulations were executed for the IPRs by modeling their relevant components, considering their Mean Time Between Failures (MTBF). They are able detect and isolate single and multiple EPDS-IPR element's failures.

## 3 Classification

Firstly, we with the classification of the IPR components' states, and following the method showcased in [16], the description of the system's failure modes initiates, beginning with its normal operation and finalizing with the modeling of the subsequent types of failure modes the EPDS-IPR. Subcomponents are perceived as per the description on page 4 (0 for correct operation, 1 for incorrect operation). The device's state is therefore a combination of states of its subcomponents. Breakers are redundant, given that every configuration of the IPR breaks into a series system, and their reliability as good as the one of its most unreliable component. Thus, the only manner of increasing the IPR's reliability is to increase the reliability of the most uncertain component, which translates to providing a pleonastic path to the breaker. The IPR has 16 states (Router, Software, Breaker1 Breaker2), from normal operation (0 0 0 0) to catastrophic failure (1 1 1 1). Some, like as the unresponsiveness or failure of one breaker, are identical. Thus, after a merge, we reduce it to 12 unique states. The following table classifies the IPR's failure mode by divining them into Categories, with the Types, Description, and the states for every fault condition in the IPR.

Relevant nodes for the IPR's PBN are described in [5]. Their state determines the failure mode in which the IPR is currently on, according to Table 1. In Cat. 1 the IPR operates correctly and operates the breakers on an AS but could also operate them unnecessarily. For Cat. 2, we describe the correct operation mode of the device. Cat. 3 is a catastrophic failure event. Finally in Cat. 4, the fault condition where the IPR does not operate in an AS and can switch breakers without the need to act when there is no AS is described. Table 2 showcases the predictors for all IPR component, for each of their configurations.

The prognosis of faults is allowed; the ones that are not causing catastrophic failure, but rather degrade into situations where the IPR operates even though it does not execute the assigned task to specifications. PRISM's Property verification was employed for the determination of the maximum occurrence probability of the fault conditions that may transform into to a fault or a failure.

**Table 1** IPR Failure mode classes [2]

| Category | 1 | 2 | 3 | 4 |
|---|---|---|---|---|
| Type | Fault | Normal Operation | Failure | Fault |
| Description | On a switching event (AS, active signal), the IPR operates as intended. On Inactive Signal (IS), IPR does not operate as intended | On AS, the IPR operates as intended. On IS, the IPR operates as intended | On AS, the IPR does not operate as intended. On IS, the IPR does not operate as intended | On AS, the IPR does not operate as intended. On IS, the IPR operates as intended |
| States | S9 | S0 | S3, S4, S10 | S1-2, S5-8, S11 |

**Table 2** Predictors and their corresponding Selection Probabilities for the IPR PBN

| Table | Predictor | Selection probability $c_J^{(i)}$ |
|---|---|---|
| $x_1$, software | $x_1(t+1) = x_1(t)$ | 1 |
| $x_2$, router | $x_2(t+1) = x_1(t) \wedge x_2(t)$ | 0.9611 |
| | $x_2(t+1) = x_1(t) \vee x_2(t)$ | 0.0389 |
| $x_3$, main breaker | $x_3(t+1) = x_1(t) \wedge x_3(t)$ | 0.9611 |
| | $x_3(t+1) = x_1(t) \vee x_3(t)$ | 0.0389 |
| $x_4$, secondary breaker | $x_4(t+1) = x_1(t) \wedge x_4(t)$ | 0.9611 |
| | $x_4(t+1) = x_1(t) \vee x_4(t)$ | 0.0389 |

## 4 Results and Discussion

Using PRISM [17] to quantitatively verify the model, and experiments were done with a PBN model of the IPR. The device's principal components were modeled, the relationship between these expressed as Boolean predictors functions. The set of PBN's nodes yields as output the total state of the apparatus. Time has been expressed in hours (h), and a numerical reward of '1' has been given to the favored state; that in which every subcomponent is operating normally. This constitutes the agent's feedback signal, given the actions it performs on the environment. The principal aim of the RL agent is to continue in the normal operation mode. Reward-based property experiments were performed as an assessment of the model's ability to fit the general RL loop. The RL agent's combined actions in its environment were evaluated, and also the agent's actions were examined the individually. The PRISM model was built with an MDP [20, 24], with its corresponding modules and predictors. The model's actions equate to the different classifications presented in Table 1 per the reliability analysis performed. The only missing aspect is assigning a reward for each action to be reinforced, which is the agent's environmental feedback. In PRISM, rewards structure can be employed in the model's construction to give These must currently be positive in PRISM, as the software does not assign costs to states or their sets. Multiple reward structures may be created for any model. With these, we can assess the outcome of the actions for our agent model. Reward-based experiments that give feedback about the maximum cumulative reward for unit of time consumed by an action or by a state were performed. The effects of the rewards were studied independently, given that even though we value the allowances of model checking, we cannot evaluate the effect of designating costs and compensations/rewards at the same time currently with this model checker. The current benefits of using PRISM currently outweigh its shortcomings. The first test performed was done to assess the maximum expected reward *Rmax* for the agent cooperating in the environment and performing any action available on its set. The PRISM reward structure gives a numerical reward of '5' for normal operation, a '1' to any fault condition or failure mode, and '0' to a catastrophic failure of the device (the larger reward belongs to the action to be reinforced more). We verified this PRISM PCTL property: *"Rmax = ? [C < = time]", with 'Rmax'* as the maximum reward property operator, 'C' for Cumulative Reward). In the following figure we present the findings of experiments evaluating the maximum cumulative reward of the RL agent in its environment.

The final reward value assessed is 42946. The RL agent executes actions and receives a signal as feedback from the environment as a reward for each action, which results in the linear plot from Fig. 3. As a reference Fig. 4 (left) presents a plot of the maximum probability of occurrence of the normal operation, that is approximated by a sigmoid curve.

The performed behavior has the biggest cumulative reward signal. States in this classification have a high occurrence probability. The rewards the rest of the actions

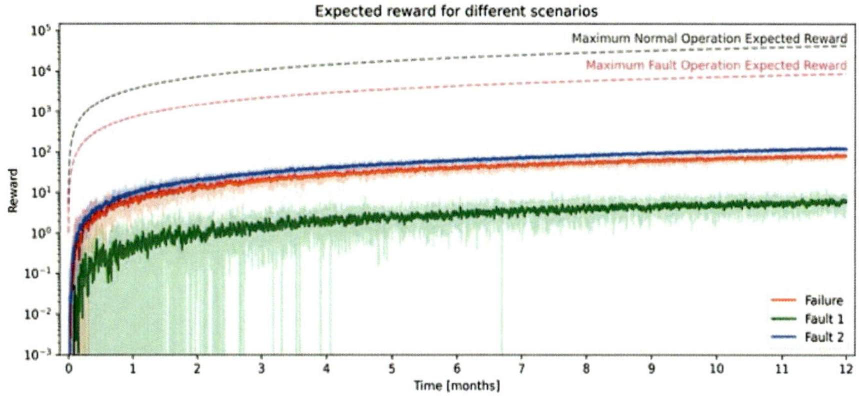


**Fig. 3** Expected Reward for different scenarios of the RL agent interacting with the environment

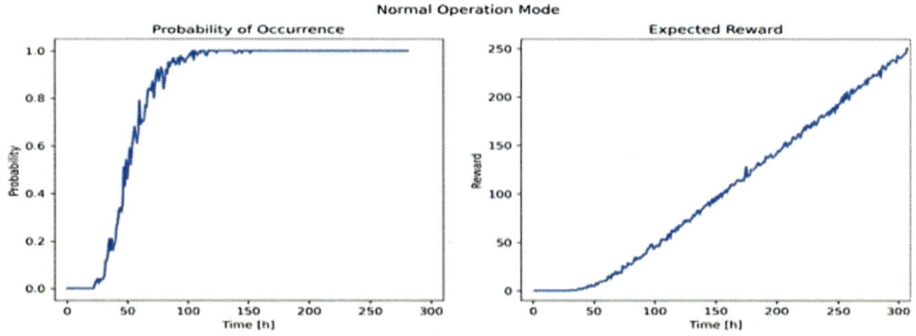


**Fig. 4** Results of the Maximum probability of occurrence experiment of the Normal Operation condition

have smaller cumulative rewards, which means they have lower probability. Rewards are low initially in its early hours of operation, which corresponds to "infant mortality" stage, and as the IPR achieves its steady state, the expected reward rate will increase almost in a linear fashion. The rest of the assessments were performed with similar PRISM properties. Figure 3 includes a graph for the Maximum Cumulative Reward for the Failure (in red) of the IPR in a year of operation. Here, the value for the reward is 73 h for one year in which the IPR was in Cat.3. Also included in Fig. 3 we have the results of an experiment of maximum cumulative reward for the Fault 1 (green) classification of the device in a year. The final maximum reward was 6 h in which the IPR was in this category. Finally, the figure includes the final value of the maximum reward for the Fault 2 classification of the IPR device in a full year (resulting in 118 h). The outcomes of these experiments are exportable as files that can serve to analyze the IPR model's behavior with analytical or machine learning tools. In the models, learning occurs as adaptation; the system adapts to survive perturbations. In complex systems, self-organization is achieved as a transition into steady states, the long-term conduct of the system. This feature of self-organization is the most essential form of learning (defining learning as a form of adaptation, and self-organization as an adaptive mechanism). Evolution of these systems may be externally directed, through the mechanism of intervention [8] so that unfavored states are averted. If instead of rewarding positively the normal operation mode of the device, we gave a reward to one of its failure modes, the resulting assessment would produce a maximum reward. These are correlated with the IPR's appraised reliability [2]. Thus, this result relates to the occurrence of the Cat. 1 fault condition. The validation through PRISM experiments of PBN modelled systems that can perform the standard RL cycle is an relevant step towards automatic system control through Machine Learning, with intrinsic control of the system.

## 5 Conclusions and Future Work

The Intelligent Power Router, a smart-grid device, was studied, and based on its reliability we used a modeling method that employs PBNs to make logical models with complex behavior. They have the capacity to self-organize into constituent Kauffman N-K networks with attractor cycles that can give us information about the IPRs long-term behavior. In our work, the behavior is equal to the states that can be assumed by the IPR's subcomponents and therefore, to the different fault conditions of the device. Probabilistic Boolean Networks were used to mimic IPRs with Reinforcement Learning, studying their evolution in time and their capacity to learn how to avoid undesirable states autonomously, increasing resilience. These were proposed as fundamental blocks with a new analysis method to solve problems Smart grid modeling, using RL that were validated by model checking.

Future work includes the use other RL techniques, like DQL, and answering fundamental questions, like if PBNs can be used as an alternative to other learning methods Artificial Neural Networks (ANN). We would need an equivalency to ANN neurons and PBNs. The authors would like to acknowledge UNAM's postdoctoral fellowship program for the support of this project.

## References


1. El Assri, N., Chabaa, S., Lmesri, K., Jallal, M. A., Zeroual, A.: Modeling techniques for decentralized energy systems applied in smart grids. In: E3S Web of Conferences, vol. 297. EDP Sciences (2021)
2. Irizarry-Rivera, A., Rodríguez-Martínez, M, Vélez, B. Vélez-Reyes, M., Ramírez-Orquín, A.R., O'Neill-Carrillo, E., Cedeño, J.R.: Intelligent power routers: distributed coordination for electric energy processing networks. In: Momoh, J., Mili, L. (eds.), Operation and Control of Electric Energy Processing Systems. Wiley, Hoboken, NJ, pp. 47–85 (2010)
3. Gershenson, C.: Facing Complexity: Prediction versus adaptation. In Massip, A., Bastardas, A. (eds.), Complexity Perspectives on Language, Communications and Society, pp. 3–14, Springer, Berlin (2013)
4. Rivera Torres, P.J., Kanaan Izquierdo, S.: Contributions to Reinforcement Learning through Probabilistic Boolean Networks, a poster presented in the Conference on Complex Systems, Held in Thessaloniki, Greece, October (2020)
5. Rivera Torres, P.J., Llanes Santiago, O.: Fault detection and isolation in smart grid devices using probabilistic boolean networks. In: Llanes Santiago, O., Cruz Corona, C., Silva Neto, A.J., Verdegay, J.L. (eds.), Computational Intelligence in Emerging Technologies for Engineering Applications. Studies in Computational Intelligence, vol. 872. Springer-Nature, Switzerland (2020)
6. Rivera Torres, P.J., Silva Neto, A.J., Llanes Santiago, O.: Multiple fault diagnosis in manufacturing processes and machines using probabilistic boolean networks. In: Martínez Álvarez, F., Troncoso Lora, A., Sáez Muñoz, J., Quintián, H., Corchado, E. (eds.), Advances in Intelligent Systems and Computing, SOCO 2019, vol. 950, pp. 355–365. Springer (2019)
7. Rivera-Torres, P.J., Serrano Mercado, E.I., Llanes Santiago, O., Anido Rifon, L.: Modeling preventive maintenance of manufacturing processes with probabilistic boolean networks with interventions. J. Intell. Manuf. **29**(8), 1941–1952 (2018a). Springer
8. Rivera-Torres, P.J., Serrano Mercado, E.I., Anido Rifon, L.: Probabilistic boolean networks and model checking as an approach for DFMEA for manufacturing systems. J. Intell. Manuf. **29**(6):1393–1413 (2018b). Springer
9. Rivera-Torres, P.J., Serrano Mercado, E.I., Anido Rifon, L.: Probabilistic boolean network modeling of an industrial machine. J. Intell. Manuf. **29**(4):875–890 (2018c). Springer
10. Rivera Torres, Pedro J., Llanes Santiago, O.: Model-based fault diagnosis of manufacturing processes and machines using probabilistic boolean networks. In: 19th Scientific Conference of Engineering and Architecture, ISP-CUJAE, La Habana, Cuba, November (2018d)
11. Rivera, T., Pedro, J.: Contribuciones al modelado teórico de sistemas de fabricación mediante Redes Booleanas Probabilísticas. Doctoral Dissertation, Universidade de Vigo, Spain, March (2017)
12. Rivera Torres, P.J., Serrano Mercado, E.I.: Probabilistic boolean network modeling as an aid for DFMEA in manufacturing systems. In: 18th Scientific Conference of Engineering and Architecture, ISP-CUJAE, La Habana, Cuba, November (2016)
13. Rivera-Torres, P.J., Seguel, J., Rodríguez-Martínez, M., Irizarry-Rivera, A.: Formal Methods for the Design and Analysis of Electrical Power Distribution Systems Endowed with Intelligent Power Routers, Submitted as Final Report for PO 4100307844—LMCO—2012, a grant funded by Lockheed-Martin (2012)

14. Kauffman, S.A.: Metabolic stability and epigenesis in randomly constructed genetic nets. J. Theor. Biol. **22**, 437–467 (1969)
15. Kauffman, S.A.: Homeostasis and differentiation in random genetic control networks. Nature **224**, 177–178 (1969)
16. Shmulevich, I., Dougherty, E.R.: Probabilistic Boolean Networks: Modeling and Control of Gene Regulatory Networks. SIAM, Philadelphia, PA, USA (2010)
17. Kwiatkowska, M.Z., Norman, G., Parker, D.: PRISM 4.0: verification of probabilistic real-time systems. In: Lecture Notes in Computer Science, vol. 6806, pp. 585–591 (2011)
18. Mendonça, L.F., Sousa, J.M., Sá da Costa, J.M.: An architecture for fault detection and isolation based on fuzzy methods. Expert Syst. Appl. **36**, 1092–1104 (2009)
19. Camps Echevarría L., Llanes-Santiago O., Silva Neto A.J. An approach for fault diagnosis based on bio-inspired strategies. In: IEEE Congress on Evolutionary Computation, Barcelona, España, pp. 1–7 (2010). https://doi.org/10.1109/CEC.2010.5586357
20. Grigoriev, A.S., Skorlygin, V.V., Grigoriev, S.A., Melnik, D.A., Filimonov, M.N.: A hybrid power plant based on renewables and electrochemical energy storage and generation systems for decentralized electricity supply of the northern territories. Int. J. Electrochem. Sci **13**, 1822–1830 (2018)
21. Liserre, M., et al.: Power routing: a new paradigm for maintenance scheduling. IEEE Ind. Electron. Mag. **14**(3), 33–45 (2020). https://doi.org/10.1109/MIE.2020.2975049
22. Orlov, Y.L., Baranova, A.V.: Bioinformatics of genome regulation and systems biology. Front. Genet. **11**, 625 (2020)
23. Zhang, D., Han, X., Deng, C.: Review on the research and practice of deep learning and reinforcement learning in smart grids. CSEE J. Power Energy Syst. **4**(3), 362–370 (2018). https://doi.org/10.17775/CSEEJPES.2018.00520
24. Zhang, H.T., Xu, F.Y., Zhou, L.: Artificial neural network for load forecasting in smart grid. In: 2010 International Conference on Machine Learning and Cybernetics, vol. 6, pp. 3200–3205. IEEE (2010)

# Selection of Model Structure Using PSO Method

**Carlos Alberto García Vázquez and Ana Isabel González Santos**

**Abstract** In this work, different variants of the particle swarm optimization (PSO) algorithm were applied to automate the stages of model structure selection, parameter estimation and model validation, within the system identification methodology. The application of the standard PSO algorithm and its modifications, such as a different way of calculating the inertial weight, the application of two termination criteria and the hybridization with genetic algorithms, were studied. All this was combined with the proposal of two variants of cost function that evaluated multiple validation criteria. Finally, combinations of algorithms and cost functions were analyzed to suggest those that solve the problem of obtaining a model that satisfies its user's quality expectations as fast as possible, from an input–output data set.

**Keywords** Particle swarm optimization · Model structure selection · Cost function

## 1 Introduction

In [3] a study of the system identification methodology was made, in which the automation of the stages of selection of the model structure, parameter estimation and validation of polynomial models, was proposed. In that work, the previous study of the system made feasible the use of the exhaustive search, testing all the possible models within certain established limits. However, this is not practical when the search space is large [4]. It is convenient in those cases to use an optimization method.

Although most of the researches consulted applied optimization related to parameter estimation, others, like [13], highlighted the importance of applying optimization to model structure selection, due to it is a time-consuming stage within system identification methodology. That research proposed the use of a genetic algorithm (GA) to select the structure of a neural network. However, there are other evolutionary algorithms to consider.

C. A. G. Vázquez (✉) · A. I. G. Santos
Universidad Tecnológica de La Habana José Antonio Echeverría (CUJAE), Havana, Cuba
e-mail: cagarcia@automatica.cujae.edu.cu

 
O. Llanes-Santiago (ed.), *Proceedings of 19th Latin American Control Congress (LACC 2022)*, Studies in Systems, Decision and Control 464,
https://doi.org/10.1007/978-3-031-26361-3_20

Ref. [6] made a survey in which GA, particle swarm optimization (PSO) and differential evolution (DE), were compared and summarized their main characteristics. It is highlighted the linear dependence on the solution time with respect to the population size as an advantage of PSO and DE over the GA. In PSO, the influence of the best solution on the population and the high tendency to premature convergence is notable. The study of the state of the art about PSO algorithm exposed in [17] showed a great interest of scientific literature about this subject and exposed the wide variety of PSO modifications, including hybridization with other algorithms.

Returning to system identification, Ref. [7] evaluated 7 variants of PSO applied to parameter estimation of an electric machine. Ref. [11] compared the performance of various algorithms based on populations, including PSO, to estimate the parameters of an IIR model. The exposed results showed that, for this particular problem, PSO is better than GA, but worse than the other evaluated algorithms. However, Ref. [2] applied successfully a modified PSO to estimate the model parameters of a piezoelectric actuator.

## 2 Automating the System Identification Methodology Using Optimization

The main idea of this research is to use an optimization method to find polynomial models with the required quality by selecting automatically the polynomial orders. This idea can be extended to other model structures, for automating the stages of model structure selection, parameter estimation and model validation, within the system identification methodology. To do this, we started from an input–output data set, and defined the range of orders of the polynomials in which the search for the model should be done, as well as the requirements that the model obtained must meet:

(1) A fit between the model output and the validation data greater than a minimum value.
(2) Uncorrelated residuals (the number of residuals to be accepted with evidence of correlation is specified).
(3) A maximum number of zeros and poles to be accepted outside the unit circle.
(4) A maximum number of model parameters.
(5) A range of values in which the gain must be found.

This idea implies that, according to the ranges of the polynomial orders, the search space where it is desired to find the model can be quite wide, thus the selection of the optimization method is important. On the other hand, every optimization method requires the evaluation of a cost function. Then, for to the problem to be solved by the optimization method, the cost function described by the diagram in Fig. 1 is proposed.

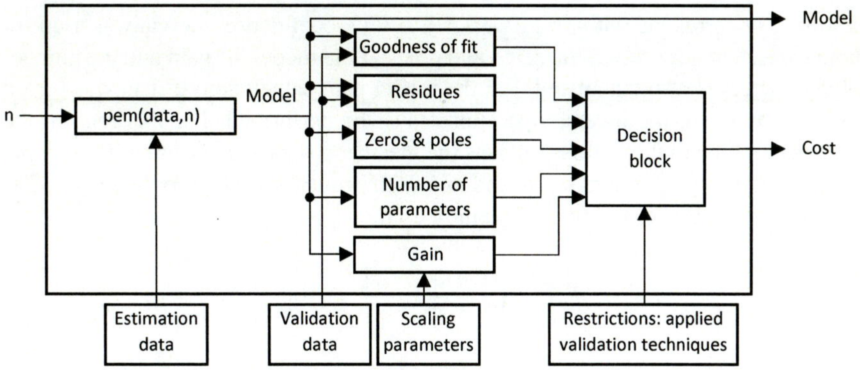


**Fig. 1** General design of the cost function

## 2.1 Cost Function

The cost function internally solves an optimization problem, by using the prediction error method (PEM) to estimate the model parameters, given a data set (estimation data) and the vector $n = [n_a n_b n_c n_d n_f n_k]$, which contains the order of the polynomials A, B, C, D and F and the time delay of the structure given by (1) and (2) [8, 10].

$$A\left(q^{-1}\right)y(k) = \frac{B\left(q^{-1}\right)}{F\left(q^{-1}\right)}u(k - n_k) + \frac{C\left(q^{-1}\right)}{D\left(q^{-1}\right)}e(k) \tag{1}$$

$$\begin{aligned} A\left(q^{-1}\right) &= 1 + a_1 q^{-1} + \cdots + a_{n_a} q^{-n_a} \\ B\left(q^{-1}\right) &= b_0 + b_1 q^{-1} + \cdots + b_{n_b} q^{-n_b} \\ C\left(q^{-1}\right) &= 1 + c_1 q^{-1} + \cdots + c_{n_c} q^{-n_c} \\ D\left(q^{-1}\right) &= 1 + d_1 q^{-1} + \cdots + d_{n_d} q^{-n_d} \\ F\left(q^{-1}\right) &= 1 + f_1 q^{-1} + \cdots + f_{n_f} q^{-n_f} \end{aligned} \tag{2}$$

After fitting a polynomial model, the goodness of fit $(Fit)$ between the model output and the system output, when both receive the same stimulus is calculated, using the validation data set (different from the estimation data set), according to Eq. (3). The autocorrelation coefficients of the residuals and the cross-correlation coefficients of the residuals with past inputs are calculated. A good model has the autocorrelation function of the residuals within a confidence interval and it has its residuals uncorrelated with past inputs [9]. A 99% confidence interval was used for both cases. The location of the zeros and poles of the model, its gain and the number of parameters is also calculated. For the model gain calculation, if data was scaled by some method, the inverse operations must be performed. For this reason, the parameters used in the scaling, such as the mean and standard deviation of the input and output data or their maximum and minimum values, must be supplied to the cost function, according to the scaling method used.

$$Fit = \left(1 - \frac{\|y - \hat{y}\|}{\|y - \overline{y}\|}\right) \cdot 100 \tag{3}$$

Finally, after performing the calculations previously mentioned, the results are fed to a decision block, which establishes the output value of the cost function. To calculate this value, two variants were considered:

(1) Rigid evaluation of the validation criteria taken into account.
(2) Output value penalization based on the validation criteria taken into account.

### 2.1.1 Decision Block with Rigid Evaluation of Validation Criteria

A minimum goodness of fit value is given, as well as maximum allowed numbers of autocorrelated residuals and residuals correlated with past values of the input, maximum allowed numbers of zeros and poles to be accepted outside the unit circle, a maximum number of parameters and a range for the model gain. If the calculation of the goodness of fit for the model output equals or exceeds the established minimum, and, at the same time, the amount of residuals with evidence of correlation does not exceed the established maximums, nor does the amount of zeros and poles, so as the number of parameters of the estimated model, and the model gain is within the allowed range, the output called "Cost" of the cost function will correspond to the result of $100 - Fit$. If any of these requirements is not met, that output will take the value 100. In both cases, the output called "Model" corresponds to the fitted model.

### 2.1.2 Decision Block with Output Penalization According to Validation Criteria

In this case, the following steps are followed to obtain the cost value:

Step 1: Penalization due to goodness of fit

$$\boldsymbol{If}\ Fit \geq Fit_{min}\ \boldsymbol{Then}\ Fit = Fit \cdot \frac{Fit}{100}$$

$$\boldsymbol{Else}\ Fit = 0$$

Step 2: Penalization due to residues

$$\boldsymbol{If}\ NResAcorr > NResAcorr_{max}\ \boldsymbol{Then}\ Fit = Fit \cdot 0.7^{NResAcorr - NResAcorr_{max}}$$

$$\boldsymbol{If}\ NResCcorr > NResCcorr_{max}\ \boldsymbol{Then}\ Fit = Fit \cdot 0.7^{NResCcorr - NResCcorr_{max}}$$

Step 3: Penalization due to zeros and poles

$$\boldsymbol{If}\ NPOut > NPOut_{max}\ \boldsymbol{Then}\ Fit = Fit \cdot 0.7^{NPOut - NPOut_{max}}$$

$$\boldsymbol{If}\ NZOut > NZOut_{max}\ \boldsymbol{Then}\ Fit = Fit \cdot 0.7^{NZOut - NZOut_{max}}$$

Step 4: Penalization due to number of parameters

$$Fit = Fit\left(1 - \frac{NParam}{NParam_{max}}\right)$$

Step 5: Gain penalization

$$\boldsymbol{If}\ Gain_{max} \neq \infty\ \boldsymbol{And}\ Gain_{min} \neq -\infty\ \boldsymbol{Then}$$

$$Gain_{med} = \frac{Gain_{max} + Gain_{min}}{2};\ Gain_{error} = \frac{Gain_{max} - Gain_{min}}{2}$$

$$\boldsymbol{If}\ Gain_{error} < |Gain - Gain_{med}| \leq 1.5Gain_{error}\ \boldsymbol{Then}$$

$$Fit = 0.8Fit$$

$$\boldsymbol{Else\ If}\ 1.5Gain_{error} < |Gain - Gain_{med}| \leq 2Gain_{error}\ \boldsymbol{Then}$$

$$Fit = 0.5Fit$$

$$\boldsymbol{Else}\ Fit = 0$$

where:

$NResAcorr$: Number of residuals with evidence of autocorrelation in the model

$NResAcorr_{max}$: Maximum allowed number of residuals with evidence of autocorrelation in the model

$NPOut$: Number of poles outside the unit circle in the model

$NPOut_{max}$: Maximum allowed number of poles outside the unit circle in the model

$NZOut$: Number of zeros outside the unit circle in the model

$NZOut_{max}$: Maximum allowed number of zeros outside the unit circle in a model

$NParam$: Number of model parameters
$NParam_{max}$: Maximum allowed number of model parameters
$Gain_{max}$: Maximum allowed model gain (if the system gain is known)
As in the previous case, the output called "Cost" of the cost function will correspond to the result of $100 - Fit$.

## 2.2 Optimization Method: PSO Algorithm

### 2.2.1 Standard PSO Algorithm

The PSO algorithm is inspired by the social behavior of animals, such as flocks of birds or schools of fish. The algorithm starts with a random population matrix, where the rows are called particles, moving on the cost surface with a speed [4]. The algorithm has $m$ particles and each one has $n$ elements, where $n$ is the dimension of the cost function. Each particle is associated with two vectors: the position of the particle $p$ and its speed $v$ [15]. The new position of the particles is calculated according to (4) and (5) at every iteration of the method [4, 15]. Algorithm 1 shows the sequence of steps of the standard PSO algorithm [4]

$$v_{m,n}^{new} = \omega v_{m,n}^{old} + c_1 r_1 \left(p_{m,n}^{local\,best} - p_{m,n}^{old}\right) + c_2 r_2 \left(p_{m,n}^{global\,best} - p_{m,n}^{old}\right) \quad (4)$$

$$p_{m,n}^{new} = p_{m,n}^{old} + v_{m,n}^{new} \quad (5)$$

where:
$v_{m,n}$: speed of the particle
$p_{m,n}$: particle variables (position)
$r_1, r_2$: random numbers in the range [0, 1]
$c_1, c_2$: learning factors (usually in the range [0, 2])
$p_{m,n}^{local\,best}$: position of the best local solution
$p_{m,n}^{global\,best}$: position of the best global solution
$\omega$: inertia (typically in the range [0, 1])

**Algorithm 1: Standard PSO**

**Step 1: initialize variables**

a) Set $m$ as the number of particles
b) Set $n$ as the dimension of the problem
c) Set the maximum number of iterations
d) Set the values of $c_1$ and $c_2$
e) Initialize particle positions, randomly distributed in the search space, and velocities

**Step 2: Evaluate the initial population**

a) Calculate the population cost (of each particle) using the cost function

**Step 3: Initialize the local minimum for each particle**

a) Establish the best cost of each particle as the costs resulting from the initial evaluation of each particle
b) Set the best position of each particle as the initial positions of the particles

**Step 4: Find the best particle in the initial population**

a) Set the global cost as the minimum value of the local minimum costs.
b) Set the position of the global minimum as the position of the particle corresponding to the minimum global cost

**Repeat as long as the number of iterations does not exceed the maximum number of iterations:**

**Step 5: Update speeds**

a) Set the maximum ($\omega_{max}$) and minimum ($\omega_{min}$) values of inertia
b) Calculate the value of inertia $\omega$
c) Assign values to $r_1$ and $r_2$
d) Calculate the new velocities according to equation (4).

**Step 6: Update the position of the particles**

a) Calculate the new positions according to equation (5)

**Step 7: Ensuring particles stay within the search space**

**Step 8: Evaluate the new population**

a) Calculate the population cost (of each particle) using the cost function
b) Update the minimum costs of each particle and the best positions of each particle
c) Update the global minimum cost
d) Update the best global position

### 2.2.2 PSO Applied to the Problem and Its Modifications

It is evident that the standard PSO cannot be applied as it is presented, since the vector of polynomial orders of polynomial models, in which the cost function is evaluated, is a vector of integer values. Then, algorithm 1 must be modified.

## *2.3 Discrete Optimization Problem*

The standard PSO algorithm is designed to solve continuous optimization problems, assuming by default that the algorithm measures the distance between solutions in the search space (particle space) based on the Euclidean distance and that, during continuous optimization, the cost associated with each particle changes in a small and reasonable range in the region adjacent to them. These premises are no longer assumed by default in discrete problems. The second premise guarantees that the algorithm can search effectively, while the first affects the second [1].

For the problem addressed in this work, the irregularity that can occur in the cost of adjacent points in the search space depends on the cost function. With the design of this function, previously exposed, it was tried to limit the radical changes in the cost from one point to another, at least in terms of its amplitude, in the case of the first variant of cost function. In the case of the second, a greater smoothness was also searched when moving in the search space. In this way, it is intended that the algorithm can work well based on the Euclidean distance, without the need to use other definitions of distance and the consequent increase in algorithm complexity.

Then, only one operation was added to algorithm 1: rounding the values of the position vector of each particle after it is calculated in step 6, part (a), taking them to the nearest integer. Regarding the initialization of the positions, prior to their first evaluation, it was also guaranteed that they were given as a function of integers. In step 1, part (e), a matrix $P_{m \times n}$ is created, in which each column is composed of random integers in the range $[a_j, b_j]$, with $j = 1, 2, \ldots, n$, which sets the limits of the search space in each dimension. For the estimation of polynomial models, $[a_1, b_1]$ corresponds to $[n_{a_{min}}, n_{a_{max}}]$, $[a_2, b_2]$ to $[n_{b_{min}}, n_{b_{max}}]$, $[a_3, b_3]$ to $[n_{c_{min}}, n_{c_{max}}]$, $[a_4, b_4]$ to $[n_{d_{min}}, n_{d_{max}}]$, $[a_5, b_5]$ to $[n_{f_{min}}, n_{f_{max}}]$ and $[a_6, b_6]$ to $[n_{k_{min}}, n_{k_{max}}]$. In this step the velocities are also initialized, creating a matrix $V_{m \times n}$ in which each column is composed of random numbers in the range $[-d_j, d_j]$ with $d_j$ given by (6) [15].

$$d_j = \frac{1}{2}(b_j - a_j) \tag{6}$$

## 2.4 Inertia Calculation

The relationship between local and global searches is maintained by adjusting the $\omega$ value. Therefore, an appropriate value of inertia can greatly improve the optimization ability of the algorithm and reduce the number of iterations [5]. Linear adaptation mechanisms for inertia are frequently used in the literature, such as the one shown in (7), proposed in [16], where $iter$ is the current number of iterations and $iter_{max}$ is the maximum number of iterations.

$$\omega = \omega_{max} - \frac{iter}{iter_{max}}(\omega_{max} - \omega_{min}) \tag{7}$$

However, according to [5], the linear decrease of $\omega$ is effective only when the algorithm searches close to the global optimum. Otherwise, the local search capacity is increased with the decrease of the inertial weight, causing the algorithm to fall into a local optimum. For this reason, that paper adopted a nonlinear dynamic inertial weight, which is updated according to (8). In this work, both updating mechanisms of $\omega$ are tested. For the one given by (7), $\omega_{max} = 1$ and $\omega_{min} = 0$, based on several tests. For the one given by (8), $\omega_{max} = 1$, $\omega_{min} = 0.4$, $C = 2$ and $p = 2$

$$\omega = 1 - \left[ \left( \omega_{max} \cdot e^{-\frac{iter-1}{iter_{max}}} \right) + \frac{1}{2C} \omega_{min} \cdot \log(iter) - \left| \frac{iter_{max} - iter}{p \cdot iter_{max}} \right| \right] \quad (8)$$

## 2.5 Termination Rules

Algorithm 1 terminates when all iterations have taken place. In order to reduce the time in which the result is obtained, the termination rules exposed in [15] were taken into account. The first rule is to compute the variance of the global minimum at each iteration, denoted as $\sigma^{(iter)}$ so that the algorithm ends when the condition given in (9) is met, where $iter_{last}$ is the last iteration where a new minimum was found.

$$\sigma^{(iter)} \leq \frac{\sigma^{(iter_{last})}}{2} \quad (9)$$

The second rule is based on similarity. Let $p^{(iter)}$ be a table with the best positions of each particle at iteration $iter$ and $E = 0$ at the beginning of the algorithm, at every iteration $iter > 1$, $g$ is calculated according to (10). Let $e$ be a small positive number, if $g < e$ then $E = E + 1$. The algorithm terminates when $E > E_{max}$, where $E_{max}$ is a small positive integer. In this work were used the values $E_{max} = 3$ and $e = 0.001$. Both rules were applied, so the algorithm ends the search if either rule is met.

$$g = \| p^{(iter)} - p^{(iter-1)} \| \quad (10)$$

## 2.6 Hybridization with Genetic Algorithms

For multimodal functions it is easy for the PSO algorithm to fall into local minima. The main cause of this is the lack of diversity in the population. This problem can be addressed by introducing the crossover operation [5]. The mutation operation can also be added, as is explained in [14]. In this work the following modification was made to algorithm 1: between steps (6) and (7) a mutation step and a crossover step were included. The applied mutation step consist of taking the updated position vector of each particle and randomly choosing an element of this vector to replace it with a random integer that is within the limits established for that element. The applied crossing step consist of taking each position vector resulting from the mutation step and replacing 50% of the elements in it with elements that occupy the same place in the vector of the best local position. Then, this operation is repeated, but using the vector of the best global position. For both cases, the position of the elements in the vectors to be replaced is chosen randomly at every iteration.

### 2.7 *Particle Removal*

Based on the PSO algorithm proposed in [14], in order to speed up the obtaining of results, this work applied particle elimination as an immediate step after step 8. The elimination begins after 20% of the maximum number of iterations have been carried out, while the number of particles exceeds 20% of the initial size of the population. Once both conditions are met simultaneously, at every iteration the 5% of the worst performing particles are removed from the population.

## 3 Experiment and Results

Based on the two ways of conceiving the cost function and the modifications to the standard PSO algorithm, an experiment is performed, in which the combinations of algorithms and cost functions are tested. The first variant of cost function was called FO1 and the second FO2. The modified standard PSO algorithm for solving the discrete problem was called IPSO, according to the nomenclature exposed in [12]. Next, eight IPSO versions were tested: IPSO1 (IPSO with linear decrease of $\omega$), IPSO2 (IPSO1 adding termination rules), IPSO3 (IPSO2 adding mutation and crossover), IPSO4 (IPSO3 adding particle removal), IPSO5 (IPSO with non-linear decrease of $\omega$), IPSO6 (IPSO5 adding termination rules), IPSO7 (IPSO6 adding mutation and crossover) and IPSO8 (IPSO7 adding particle removal).

For the experiment, a population of 20 particles was chosen, based on some tests previously done. It was chosen $iter_{max} = 100$, $c_1 = 1$ and $c_2 = 4 - c_1$, based on the algorithm proposed in [4]. Regarding the quality requirements of the models, $Fit_{min} = 70$, $NResAcorr_{max} = 0$, $NResCcorr_{max} = 0$, $NPOut_{max} = 0$, $NZOut_{max} = 0$, $NParam_{max} = 12$ were established. The data set dryer.dat from Daisy database was used. As there was no previous model gain estimation, this aspect was not taken into account in the validation. As limits of the orders of the polynomials, a range of values between 0 and 10 was established for each one. The MatLab® delayest command was used to estimate the value of the time delay, as well as the analysis of the cross-correlation graph between the output and the input. The results obtained by both methods were used to establish the range of the $n_k$ parameter. For each combination of algorithm with cost function, 30 runs were carried out. The results are shown in Tables 1 and 2, where *Mean* refers to the calculation of the mean value and *Std* to the calculation of the standard deviation.

In terms of model quality, the results of using FO2 are much better than the results of using FO1. FO2 allows a very stable behavior of fits. None of the 240 models provided by the experiment had evidence of autocorrelation of the residuals. For 3 of the algorithms there was no cross-correlation of the residuals with the past inputs and, for the other algorithms, the correlation detected was low. The restrictions imposed on the zeros and poles were fulfilled in all cases, while the number of parameters, in almost all the algorithms, was lower than the number of parameters of the models obtained using FO1. However, neither of the two objective functions guaranteed to meet the maximum number of parameters. Regarding the time in which the result is obtained, except for IPSO1, the results of FO1 are better, but this does not compensate its incapacity to guarantee obtaining models that pass the validation tests performed, especially the fit test and residual analysis.

Analyzing the algorithms used (in FO2, since it is not worth continuing the analysis with FO1), the fastest results were obtained with IPSO6, followed by IPSO8 and IPSO7, in that order, which have in common the combination of the non-linear decrease of $\omega$ and the application of termination rules. However, IPSO6 is the second function that violates the maximum parameter limit the most. On the other hand, IPSO7 obtained the third highest mean of the $Fit$ values, with a reasonable standard deviation from the mean value. In the rest of the validation tests it equals or exceeds IPSO6 and IPSO8, except in the residual analysis, where IPSO8 is better. Quite similar results can be noted between IPSO7 and IPSO8 in terms of the quality of the models and the time in which they are obtained. It should not be forgotten that, although IPSO4 was a few seconds slower, its performance in the validation tests was robust, successfully complying with all of them (it only slightly exceeds the maximum number of established parameters).

# 4 Conclusions

In this work it was shown that, by means of the PSO algorithm, it is possible to automate the stages of model structure selection, parameter estimation and validation of polynomial models. Given the presented problem of discrete nature, the effectiveness of the designed cost function with output penalization was tested, which manages to mitigate the problem of irregularity in the cost of adjacent points, and make the rounding method viable and successful to adapt the continuous PSO to the discrete problem, maintaining the simplicity of the algorithm in that sense. The advantage of using the two stopping rules was evidenced, although a detailed study has not been made for this problem. The influence of particle removal was not clear, so it is recommended to study it in more depth in the resolution of this type of problem. With the use of mutation and crossover, an improvement was observed in the compromise quality of the model Vs execution time of the algorithms with linear decrease of $\omega$. However, the algorithms with non-linear decrease of $\omega$ seemed to slow down a bit the

**Table 1** Results of algorithms with FO1

| | IPSO1 | IPSO2 | IPSO3 | IPSO4 | IPSO5 | IPSO6 | IPSO7 | IPSO8 |
|---|---|---|---|---|---|---|---|---|
| Mean fit | $-2{\cdot}10^{206}$ | $-10{\cdot}10^{56}$ | $-8{\cdot}10^{176}$ | $-5{\cdot}10^{203}$ | $-4{\cdot}10^{79}$ | $-4{\cdot}10^{178}$ | $-1{\cdot}10^{258}$ | $-1{\cdot}10^{165}$ |
| Std fit | $\infty$ | $5{\cdot}10^{57}$ | $\infty$ | $\infty$ | $2{\cdot}10^{80}$ | $\infty$ | $\infty$ | $\infty$ |
| Fit > 70% | 8 | 15 | 3 | 6 | 9 | 12 | 5 | 4 |
| Fit > 80% | 5 | 10 | 4 | 2 | 12 | 1 | 4 | 6 |
| Fit > 90% | 1 | 0 | 0 | 1 | 4 | 0 | 0 | 2 |
| MeanNResAcorr | 1.93 | 3.73 | 2.13 | 4.73 | 2.07 | 2.07 | 4 | 2.07 |
| StdNResAcorr | 1.64 | 9.15 | 2.34 | 10.5 | 2.02 | 2.33 | 9.27 | 1.87 |
| MeanNResCcorr | 4.77 | 3.03 | 4.13 | 4.63 | 3.23 | 5.47 | 4.17 | 4.77 |
| StdNResCcorr | 6.86 | 3.65 | 4.40 | 4.88 | 5.02 | 5.99 | 3.99 | 5.46 |
| MeanNPOut | 0.63 | 0.77 | 1.23 | 0.83 | 0.43 | 1.07 | 0.93 | 0.77 |
| StdNPOut | 0.85 | 0.97 | 1.14 | 0.95 | 0.50 | 0.83 | 0.94 | 0.86 |
| MeanNZOut | 0.77 | 0.63 | 0.73 | 0.93 | 0.53 | 0.73 | 0.90 | 0.80 |
| StdNZOut | 1.22 | 1.13 | 1.11 | 1.68 | 0.78 | 1.11 | 1.32 | 1.24 |
| MeanNParam | 25.73 | 25.87 | 26.30 | 24.57 | 25.20 | 22.70 | 23.53 | 24.00 |
| StdNParam | 6.55 | 6.30 | 7.48 | 7.10 | 7.38 | 5.88 | 5.95 | 6.23 |
| Mean time (sec) | 794.01 | 16.66 | 17.48 | 17.47 | 739.91 | 15.67 | 17.14 | 17.46 |
| Std time (sec) | 81.92 | 0.96 | 1.23 | 1.32 | 62.66 | 0.77 | 5.08 | 3.77 |

**Table 2** Results of algorithms with FO2

| | IPSO1 | IPSO2 | IPSO3 | IPSO4 | IPSO5 | IPSO6 | IPSO7 | IPSO8 |
|---|---|---|---|---|---|---|---|---|
| Mean fit | 90.24 | 89.77 | 90.21 | 90.08 | 89.75 | 89.37 | 90.15 | 89.79 |
| Std fit | 1.74 | 1.19 | 0.91 | 0.95 | 2.01 | 1.94 | 1.28 | 1.72 |
| %Fit > 80% | 30 | 30 | 30 | 30 | 30 | 30 | 30 | 30 |
| %Fit > 90% | 20 | 15 | 19 | 18 | 20 | 15 | 19 | 15 |
| Mean NResAcorr | 0 | 0 | 0 | 0 | 0 | 0 | 0 | 0 |
| Std NResAcorr | 0 | 0 | 0 | 0 | 0 | 0 | 0 | 0 |
| Mean NResCcorr | 0 | 0.03 | 0 | 0 | 0.20 | 0.23 | 0.10 | 0.03 |
| Std NResCcorr | 0 | 0.18 | 0 | 0 | 0.41 | 0.43 | 0.31 | 0.18 |
| Mean NPOut | 0 | 0 | 0 | 0 | 0 | 0 | 0 | 0 |
| Std NPOut | 0 | 0 | 0 | 0 | 0 | 0 | 0 | 0 |
| Mean NZOut | 0 | 0 | 0 | 0 | 0 | 0 | 0 | 0 |
| Std NZOut | 0 | 0 | 0 | 0 | 0 | 0 | 0 | 0 |
| Mean NParam | 15.27 | 17.20 | 15.50 | 14.23 | 24.93 | 24.67 | 18.27 | 18.50 |
| Std NParam | 4.72 | 5.56 | 4.34 | 3.55 | 7.07 | 5.28 | 4.71 | 4.61 |
| Mean time | 691.34 | 219.91 | 75.85 | 68.76 | 865.10 | 25.99 | 56.26 | 55.07 |
| Std time | 57.06 | 223.87 | 29.20 | 31.24 | 96.63 | 8.77 | 29.76 | 34.68 |

obtaining of the result, with an apparent improvement in the quality of the models, an issue that, for this type of problem, it is also recommended to study in more depth.

## References


1. Chen, Q., Peng, Y., Zhang, M., Yin, Q.: Application analysis on PSO algorithm in the discrete optimization problems. J. Phys: Conf. Ser. **2078**, 1–15 (2021)
2. Feng, Y., Li, Y.: System identification of micro piezoelectric actuators via rate-dependent Prandtl-Ishlinskii hysteresis model based on a modified PSO algorithm. IEEE Trans. Nanotechnol. **20** (2021)
3. García, C.A., González, A.I., Pérez, V.: Metodología y algoritmo de validación para identificar modelos de una unidad manejadora de aire. RIELAC **42**, 105–121 (2021)
4. Haupt, R.L., Haupt, S.E.: Practical Genetic Algorithms. Wiley, New Jersey (2004)
5. Jia, L., Zhao, X.: An improved particle swarm optimization (PSO) optimized integral separation PID and its application on central position control system. IEEE Sens (2019)
6. Kachitvichyanukul, V.: Comparison of three evolutionary algorithms: GA, PSO, and DE. Ind. Eng. Manag. Syst. **11**, 215–223 (2012)
7. Liu, Z., Wei, H., Li, X.H.: Global Identification of electrical and mechanical parameters in PMSM drive based on dynamic self-learning PSO. IEEE Trans. Power Electron. **33**, 10858–10871 (2018)
8. Ljung, L.: System Identification: Theory for the User, p. 07458. Upper Side River, NJ (1999)
9. Ljung, L.: System Identification Toolbox™ User's Guide. The MathWorks Inc., Natick (2015)
10. Roffel, B., Betlem, B.: Process Dynamics and Control. Modeling for Control and Prediction. The Atrium, Southern Gate, Chichester, West Sussex PO19 8SQ. Wiley, England

11. Singh, S., Ashok, A., Kumar4, M. & Rawat5, T.K.: Adaptive infinite impulse response system identification using teacher learner based optimization algorithm. Appl. Intell. (2018)
12. Strasser, S., Goodman, R., Sheppard, J., Butcher, S.: A new discrete particle swarm optimization algorithm. In: ACM GECCO. Denver, CO, USA (2016)
13. Talib, R., Nabil, N., Choi, W.: Optimization-based data-enabled modeling technique for HVAC systems components. Buildings [Online] **10** (2020)
14. Tan, J.S., Goh, S.L., Sura, S., Kendall, G. Sabar, N.R.: Hybrid particle swarm optimization with particle elimination for the high school timetabling problem. Evol. Intell. (2020)
15. Tsoulos, I.G., Tzallas, A., Karvounis, E.: Improving the PSO method for global optimization problems. Evol. Syst. (2020)
16. Shi, Y., Eberhart, R.C.: Parameter Selection in particle swarm optimization. In: Evolutionary Programming VII. Lecture Notes in Computer Science, vol. 1447, pp. 591–600 (1998)
17. Zhang, Y., Wang, S., Ji, G.: A comprehensive survey on particle swarm optimization algorithm and its applications. Math. Probl. Eng. **2015**, 1–38 (2015)

# Neuro-Adaptive PID Helicopter Controller Based on Atomic Functions

**O. F. Garcia–Castro, L. E. Ramos–Velasco, E. Escamilla–Hernández, R. Garcia–Rodriguez, M. A. Vega–Navarrete, C. R. Domínguez–Mayorga, and L. N. Oliva–Moreno**

**Abstract** A feature of wavelets is that signals can be analyzed in time-frequency, unlike Fourier theory, which only does so in frequency. For many years, the most used wavelet functions for control applications have been Haar, Morlet, Rasp, Polywog, and Shannon, among others. This paper presents new functions called Atomic Function (AF) as activation functions in a radial basis neural network (RBF). Although, they maintain the same properties, one of the principal advantages of the AF is that they are easier to implement. The control of the Quanser helicopter with two degrees of freedom is presented to prove the effectiveness of the AF. In the proposed approach,

O. F. Garcia–Castro · L. E. Ramos–Velasco (✉) · R. Garcia–Rodriguez · M. A. Vega–Navarrete · C. R. Domínguez–Mayorga
Postgraduate Program of Aerospace Engineering, Metropolitan Polytechnic University of Hidalgo, Boulevard Acceso a Tolcayuca 1009, Ex Hacienda San Javier, 43860 Tolcayuca, Hidalgo, Mexico
e-mail: lramos@upmh.edu.mx

O. F. Garcia–Castro
e-mail: 213220003@upmh.edu.mx

R. Garcia–Rodriguez
e-mail: rogarcia@upmh.edu.mx

M. A. Vega–Navarrete
e-mail: mvega@upmh.edu.mx

C. R. Domínguez–Mayorga
e-mail: cdominguez@upmh.edu.mx

E. Escamilla–Hernández
ESIME Culhuacan, National Polytechnic Institute, Avenida Santa Ana 1000, San Francisco Culhuacan, Culhuacan CTM V, 04260 Mexico City, Ciudad de México CDMX, Mexico
e-mail: eescamillah@ipn.mx

L. N. Oliva–Moreno
Tecnología e Innovación, UPIIH Hidalgo, National Polytechnic Institute, Carretera Pachuca - Actopan Kilómetro 1+500, Distrito de Educación, Salud, Ciencia, 42162 San Agustín Tlaxiaca, Hidalgo, Mexico
e-mail: loliva@ipn.mx



O. Llanes-Santiago (ed.), *Proceedings of 19th Latin American Control Congress (LACC 2022)*, Studies in Systems, Decision and Control 464,
https://doi.org/10.1007/978-3-031-26361-3_21

the RBF neural network is used for: *a)* the input-output identification of the system and *b)* auto-tuning PID controllers. Numerical simulations show the performance of the closed-loop system together with the AF under different conditions.

## 1 Introduction

For many years the approximation theory by neural networks is based on *traditional* activation functions as binary, linear, sigmoid or hyperbolic tangent to name a few. However, to overcome some drawbacks of the learning algorithms, as vanishing gradient problem, new activation functions has been proposed recently principally for deep learning as softmax, ReLu, swish and hard-swish.

In 1975, Rvachev [15, 16] introduced an atomic function theory. A function called *atomic* is a compactly supported solution of functional-differential equation with constant coefficients. Although the atomic function (AF) is used for digital processing signals [3, 9, 12]. The property of spline-like function approximation has drawn the attention recently for control applications. The spline theory is based on the fact that polynomials can approximate any function over a finite segment [4]. The AF belongs to a class of finite functions where its value is not zero in a local segment. This function represents a special linear differential equation with constant coefficients with shifted arguments [6]. The simplest and most important AF, called up-function, $up(x) = 1$, is defined as a Fourier series. One of the principal characteristics of this function is that its derivative can be expressed as a function itself shifted and stretched by a factor of 2. In the present work, it is employed this new $up(x)$ as activation functions in the radial basis (RBF) neural network to control a two-degree-of-freedom (2 DOF) helicopter.

The usefulness of proportional integral derivative (PID) controllers is based on their widespread application to most control systems. Such schemes are tuned at the job site employing manual or automatic tuning techniques [1]. The traditional PID controller consists of three coefficients: integral, proportional and derivative; each of which contains a fixed constant gain [2]. Although a virtue of PID controls is characterized by dispense with the mathematical model of the plant [1]. However, the constant gain is a disadvantage for many applications, such as in aircraft, where the system is subject to different perturbations. To overcome this many self-tuning methods have been proposed [5, 13, 18, 19]. Unlike previous approaches, in this paper self-tuning PID gains are proposed based on RBF neural network where the activation functions are atomic functions. At the same time, the RBF neural network is used to identify the dynamics of the model, which, together with the infinite impulse response (IIR) filters are part of the identification stage [14]. Thus, the RBF network's adaptation algorithm is used to simultaneously approximate the dynamical model and self-tunning PID. In order to verify the performance of the proposed approach, the control of the 2 DOF Quanser helicopter is proposed.

The rest of the paper is organized as follows: in Sect. 2 mathematical preliminaries on atomic functions are presented, while in Sect. 3 the PID-RBFAF-IIR algorithm is exposed directly on the 2 DOF helicopter. On the other hand, in Sect. 4 the results obtained from the numerical simulations in closed loop are shown. Finally, in the Sect. 5 conclusions are given.

## 2 Preliminaries: The Theory of Atomic Functions

In the present work, the mother of the AFs $up(x)$ is used, which presents a better performance in the harmonic analysis. Some of the advantages of this function over the classical algebraic trigonometric polynomials is that they only satisfy homogeneous linear differential equations with constant coefficients, and the AF's are compatible infinitely differentiable solutions of compact support defined in the real numbers $\mathbb{R}$ for equations with shifted argument, i.e. the AFs belong to the class of 'finite functions' equal to zero everywhere except a local segment of an $x$-axis (for example, $-1 < x < 1$) and satisfying a special type of linear functional-differential equations with constant coefficients and 'shifted' arguments [17]

$$Lf(x) = \lambda \sum_{\gamma=1}^{M} c(\gamma) f(ax - b(\gamma)),\ |a| > 1 \tag{1}$$

where $L$ is a linear differential operator with constant coefficients, i.e. $Lf(x) = f^{(n)}(x) + a_1 f^{(n-1)}(x) + a_2 f^{(n-2)}(x) + \ldots + a_{n-1} f^{'}(x) + a_n y(x)$. If $a = 1$ and $b(k) \equiv 0$ $(k = 1, M)$ equation (1) becomes an ordinary differential equation. The simplest and the most important is the function $up(x)$ and represented via Fourier series by the formula [11, 17]

$$up(x) = \frac{1}{2\pi} \int_{-\infty}^{\infty} e^{jwx} \prod_{\gamma=1}^{\infty} \frac{sin(w2^{-\gamma})}{w2^{-\gamma}} dw \tag{2}$$

this equation can be used as a formal definition of AF.

### 2.1 The "Mother" Atomic Function $up(x)$ and Its Main Properties

From the aforementioned definitions one can obtain immediately main properties of the function $up(x)$:

1. Function $up(x)$ is even, i.e., $up(x) = up(-x)$.
2. Its maximum value is $up(0) = 1$. Its integral equals unity, i.e., $\int_{-1}^{1} up(x)dx = 1$.

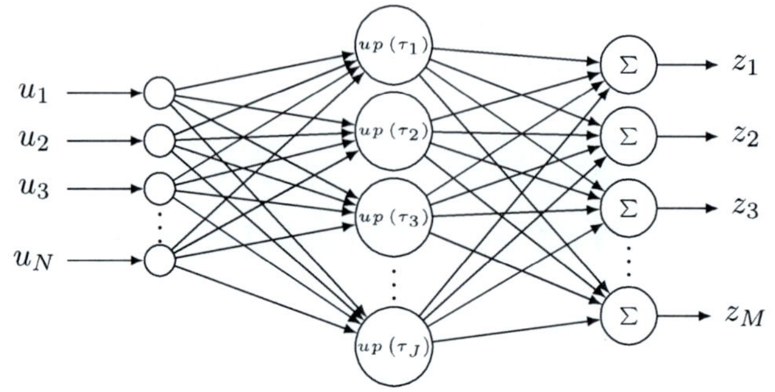


**Fig. 1** RBF Neural diagram with an AF as activation function with $\tau_i = (k - q_i)/s_i$, where $q_i$ and $s_i$ are the AF parameters: translations and dilation, respectively

3. The first derivative of the function $up(x)$ has a simple expression: $up'(x) = 2\ up(2x+1) - 2\ up(2x-1)$. This property states that, its derivative can be expressed via the function itself shifted and stretched by the factor of 2.

Since the RBF is in discrete time, it is necessary to know the atomic function $up(x)$ in discrete time since it is used as the activation function in the wavenet (see Fig. 1). Thus, the discrete representation of (2) and its derivative are given by:

$$up(\tau_i) = \sum_{l=0}^{L} \cos(\pi l \tau_i) \prod_{h=1}^{H} \begin{cases} 1 & , \ \pi l/2^h = 0 \\ \dfrac{2^h \sin(\pi l/2^h)}{\pi l} & , \ \pi l/2^h \neq 0 \end{cases} \tag{3}$$

$$up'(\tau_i) = 2up\,(2\tau_i + 1) - 2up\,(2\tau_i - 1) \tag{4}$$

where $L$, $H$ are the desired approximation levels and they are given by designer, in our case we give $L = H = 20$.

## 3 PID-RBFAF-IIR Control Scheme for the Quanser Helicopter

The PID-RBFAF-IIR control scheme is shown in Fig. 2. It is made up of four blocks: *(i)* Plant, which consists of a 2 DOF helicopter, whose dynamics are represented by linear differential equations; *(ii)* Identification (RBFAF-IIR), in this block the input-output identification of the helicopter is made; while in *(iii)* Auto-tuning are carried out where the gains of the PID controllers are adjusted online; finally, we have the *(iv)* Discrete PID controllers. Below is a detailed description of each of the blocks.

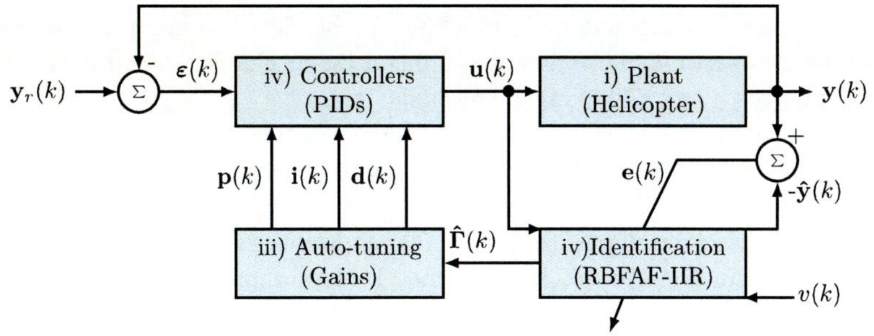


**Fig. 2** Block diagram of the PID-RBFAF-IIR scheme for the 2 DOF Quanser helicopter

## *(i) Plant (Helicopter)*

The 2 DOF helicopter and its free-body diagram are shown in Fig. 3, the dynamics that describes the behavior of the helicopter is given by two coupled transfer functions [10]:

$$H_{\theta\theta}(s) = \frac{y_\theta(s)}{V_\theta(s)} = \frac{2.361}{s^2 + 9.26s}, \qquad H_{\phi\theta}(s) = \frac{y_\phi(s)}{V_\theta(s)} = \frac{0.2402}{s^2 + 3.487} \tag{5}$$

$$H_{\theta\phi}(s) = \frac{y_\theta(s)}{V_\phi(s)} = \frac{0.07871}{s^2 + 9.26s}, \qquad H_{\phi\phi}(s) = \frac{y_\phi(s)}{V_\phi(s)} = \frac{0.7895}{s^2 + 3.487} \tag{6}$$

where $y_\theta$ and $y_\phi$ are the model outputs for each axis of motion, In addition, Fig. 3 also shows the coupling of the model.

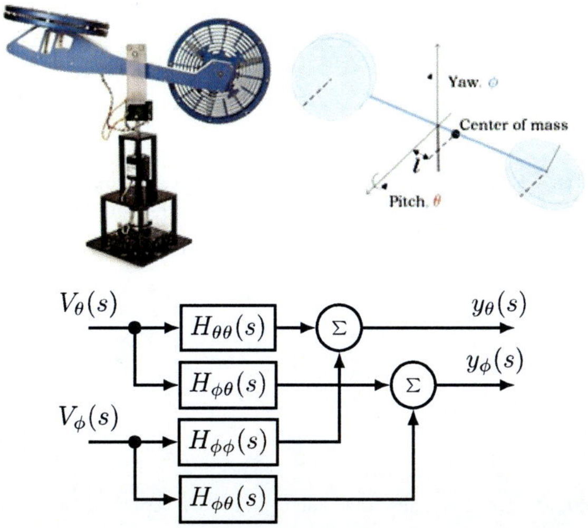


**Fig. 3** Model of the 2 DOF Quanser helicopter, the free body diagram [10] and the coupling diagram based on transfer functions

Additionally, the helicopter has a block where the electrical interpretation of the control signal is performed $\mathbf{u} = [u_\theta,\ u_\phi]^{\mathrm{T}}$, that is the mechanical-electrical relationship that is given by the following expressions

$$V_\theta = 0.4899u_\theta - 0.7387u_\phi, \quad V_\phi = 0.0793u_\phi - 0.7387u_\theta. \tag{7}$$

where $u_\theta$ and $u_\phi$ are the signals from each PID controller, while $V_\theta$ ($\pm 24$ V) and $V_\phi$ ($\pm 15$ V) are the voltages for pitch motor and yaw motor, respectively.

### *(ii) RBFAF-IIR Identification*

In this block, the input-output dynamics of the system are identified and its structure is composed of a RBF neural network [8] whose activation functions are AF, as shown in Fig. 4.

Similarly, the Fig. 5 shows the internal configuration of the IIR filter blocks whose outputs $\hat{y}_\theta$ and $\hat{y}_\phi$ are the estimated positions of the helicopter. It should be noted that these filters "prune" the neurons that have little contribution in the identification process and with it an input-output structure is obtained [7].

The neural network and IIR filters parameters in their matrix form are: the control signal $\mathbf{u} = [u_\theta,\ u_\phi]^{\mathrm{T}}$, the translation and dilation parameters are $\mathbf{q} = [q_1,\ q_2,\ \ldots,\ q_J]^{\mathrm{T}}$ and $\mathbf{s} = [s_1,\ s_2,\ \ldots,\ s_J]^{\mathrm{T}}$, respectively; the hidden layer are the AFs, renamed as $\psi = [\psi_1,\ \psi_2,\ \ldots,\ \psi_J]^{\mathrm{T}} = [up(\tau_1),\ up(\tau_2),\ \cdots,\ up(\tau_J)]^{\mathrm{T}}$, the

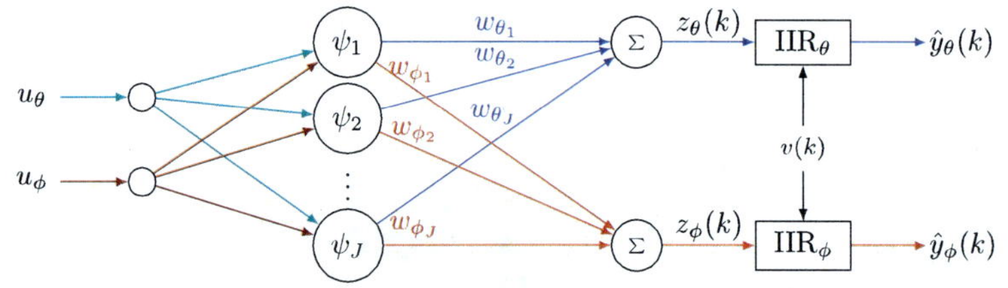


**Fig. 4** RBFAF neural network diagram with IIR filters

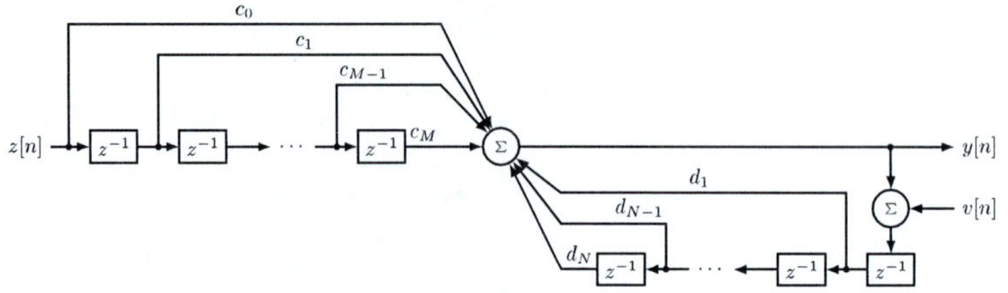


**Fig. 5** IIR filter structure where $M$ and $N$ represent the number of feedback and feed-forward coefficients, respectively

neural network guideline $\hat{\Gamma} = [\hat{\Gamma}_\theta,\ \hat{\Gamma}_\phi]^{\mathrm{T}}$, the neural network output $\mathbf{z} = [z_\theta,\ z_\phi]^{\mathrm{T}}$, the estimated position $\hat{\mathbf{y}} = [\hat{y}_\theta,\ \hat{y}_\phi]^{\mathrm{T}}$; and the synaptic weights matrices $\mathbf{W} \in \mathbb{R}^{2\times J}$, the filters coefficients $\mathbf{C} \in \mathbb{R}^{2\times M}$ and $\mathbf{D} \in \mathbb{R}^{2\times N}$ are:

$$\mathbf{W} = \begin{bmatrix} w_{\theta 1},\ w_{\theta 2},\ \cdots,\ w_{\theta J} \\ w_{\phi 1},\ w_{\phi 2},\ \cdots,\ w_{\phi J} \end{bmatrix},\quad \mathbf{C} = \begin{bmatrix} c_{\theta 0},\ c_{\theta 1},\ \ldots,\ c_{\theta M} \\ c_{\phi 0},\ c_{\phi 1},\ \ldots,\ c_{\phi M} \end{bmatrix},\quad \mathbf{D} = \begin{bmatrix} d_{\theta 1},\ d_{\theta 2},\ \ldots,\ d_{\theta N} \\ d_{\phi 1},\ d_{\phi 2},\ \ldots,\ d_{\phi N} \end{bmatrix}. \tag{8}$$

The neural network output $\mathbf{z}(k)$ is given by $\mathbf{z}(k) = \mathbf{u}^{\mathrm{T}}(k)\mathbf{W}(k)\psi^{\mathrm{T}}(k)$, to get the estimated position $\hat{\mathbf{y}}(k)$, the output $\mathbf{z}(k)$ passes through the IIR filters, which is

$$\hat{\mathbf{y}}(k) = \hat{\Gamma}(k) + \mathbf{D}\hat{\mathbf{Y}}(k)\mathbf{v}(k) \tag{9}$$

where $\mathbf{v}(k) = [v(k),\ v(k)]^{\mathrm{T}}$ is the persistent excitation signal of the filters and

$$\hat{\Gamma}(k) = \mathbf{C}\mathbf{z}(k),\ \hat{\mathbf{Y}}(k) = \begin{bmatrix} \hat{y}_\theta(k-1)\ \hat{y}_\theta(k-2) \cdots \hat{y}_\theta(k-N) \\ \hat{y}_\phi(k-1)\ \hat{y}_\phi(k-2) \cdots \hat{y}_\phi(k-N) \end{bmatrix}. \tag{10}$$

The optimization of the described parameters is carried out using the back-propagation algorithm where the cost function $\mathbf{E}(k)$ is defined as

$$\mathbf{E}(k) = \frac{1}{2}\sum_t [\mathbf{e}_t(k)]^2 \tag{11}$$

for which estimation error is required $\mathbf{e}_t(k) = \mathbf{y}_t(k) - \hat{\mathbf{y}}_t(k)$, where $\mathbf{y}_t$ indicates the system output and $\hat{\mathbf{y}}_t$ is the estimated output by the neural network, $t = \{\theta, \phi\}$. To minimize the cost function $\mathbf{E}(k)$ we use the partial differential equation with respect to each RBFAF-IIR parameter as shown below

$$\frac{\partial \mathbf{E}}{\partial \mathbf{W}(k)} = \mathbf{u}(k)\mathbf{e}^{\mathrm{T}}(k)\mathbf{C}(k)\psi_a^{\mathrm{T}}(\tau) = U\mathbf{I}_e(\mathbf{C}\psi_a)^{\mathrm{T}} \tag{12}$$

$$\frac{\partial \mathbf{E}}{\partial \mathbf{q}(k)} = -\mathbf{e}(k)\mathbf{u}(k)\mathbf{C}^{\mathrm{T}}(k)\psi_b = U(\mathbf{I}_f(\mathbf{C}\psi_b)^{\mathrm{T}}) \tag{13}$$

$$\frac{\partial \mathbf{E}}{\partial \mathbf{s}(k)} = \tau\frac{\partial \mathbf{E}}{\partial \mathbf{q}(k)} = \frac{\partial \mathbf{E}}{\partial \mathbf{q}(k)} \odot \tau \tag{14}$$

$$\frac{\partial \mathbf{E}}{\partial \mathbf{C}(k)} = -\mathbf{u}(k)\mathbf{e}^{\mathrm{T}}(k)\mathbf{z}(k-M) = U\mathbf{I}_e\mathbf{z}^{\mathrm{T}} \tag{15}$$

$$\frac{\partial \mathbf{E}}{\partial \mathbf{D}(k)} = -\mathbf{v}(k)\mathbf{e}^{\mathrm{T}}(k)\hat{\mathbf{y}}(k-N) = v\mathbf{I}_e(\hat{\mathbf{y}}\psi_a)^{\mathrm{T}} \tag{16}$$

where $U = u_\theta(k) + u_\phi(k)$, $\mathbf{I}_e = \mathrm{diag}\{e_\theta(k),\ e_\phi(k)\}$, $\mathbf{I}_f = [e_\theta(k),\ e_\phi(k)]^{\mathrm{T}}$, $\tau = [\tau_1,\ \tau_2,\ \cdots,\ \tau_J]$ and

$$\psi_a = \begin{bmatrix} \psi_1(\tau_1) & \psi_2(\tau_2) & \cdots & \psi_J(\tau_J) \\ \psi_1(\tau_1-1) & \psi_2(\tau_2-1) & \cdots & \psi_J(\tau_J-1) \\ \vdots & \cdots & \ddots & \vdots \\ \psi_1(\tau_1-M) & \psi_2(\tau_2-M) & \cdots & \psi_J(\tau_J-M) \end{bmatrix}, \quad \psi_b = \begin{bmatrix} \frac{\partial \psi_1(\tau_1)}{\partial s_1} & \cdots & \frac{\partial \psi_J(\tau_J)}{\partial s_J} \\ \frac{\partial \psi_1(\tau_1-1)}{\partial s_1} & \cdots & \frac{\partial \psi_J(\tau_J-1)}{\partial s_J} \\ \vdots & \ddots & \vdots \\ \frac{\partial \psi_1(\tau_1-M)}{\partial s_1} & \cdots & \frac{\partial \psi_J(\tau_J-M)}{\partial s_J} \end{bmatrix}. \quad (17)$$

The parameters update complies with the following rule

$$\lambda(k+1) = \lambda(k) + \mu_\lambda \Delta\lambda(k), \quad \Delta\lambda(k) = -\frac{\partial \mathbf{E}}{\partial \lambda(k)} \quad (18)$$

where $\lambda$ can be **C**, **D**, **W**, **q** or **s**, with its corresponding learning rate $\mu_\lambda$.

For programming purposes, the third representation of each of the Eqs. (12)–(16) was used, where $\odot$ represents the Hadamard operator.[1]

## *(iii) Auto Tuning of Proportional, Integral and Derivative Gains*

This block is responsible for adjusting the gains of the controller $\mathbf{p}(k) = [p_\theta, \ p_\phi]^T$, $\mathbf{i}(k) = [i_\theta, \ i_\phi]^T$ y $\mathbf{d}(k) = [d_\theta, \ d_\phi]^T$, which is done as follows, respectively [7]

$$\mathbf{p}(k) = \mathbf{p}(k-1) + \mu_p \mathbf{e}(k)\hat{\Gamma}(k)[\varepsilon(k) - \varepsilon(k-1)] \quad (19)$$

$$\mathbf{i}(k) = \mathbf{i}(k-1) + \mu_i \mathbf{e}(k)\hat{\Gamma}(k)\varepsilon(k) \quad (20)$$

$$\mathbf{d}(k) = \mathbf{d}(k-1) + \mu_d \mathbf{e}(k)\hat{\Gamma}(k)[\varepsilon(k) - 2\varepsilon(k-1) + \varepsilon(k-2)] \quad (21)$$

where $\mu$ is the learning rate, $\hat{\Gamma}(k)$ is the directive of the neural network given by (10), the signal $\varepsilon(k)$ is the tracking error which is defined as $\varepsilon(k) = \mathbf{y}_r(k) - \mathbf{y}(t)$ with $\mathbf{y}_r(k)$ as the reference signal and $\mathbf{y}(k)$ as the output signal.

## *(iv) PID Controllers*

In this block, the control signal is generated $\mathbf{u}(k) = [u_\theta, \ u_\phi]^T$ which causes the system output to vary $\mathbf{y}(k)$ in order to approximate the reference signal $\mathbf{y}_r(k)$. This stage only needs the gains values $\mathbf{p}(k)$, $\mathbf{i}(k)$ and $\mathbf{d}(k)$ as well as the tracking error $\varepsilon(k)$. Then the control signal $\mathbf{u}(k+1)$ is given by

$$\mathbf{u}(k+1) = \mathbf{u}(k) + \mathbf{p}(k)[\varepsilon(k) - \varepsilon(k-1)] + \mathbf{i}(k)[\varepsilon(k)] + \mathbf{d}(k)[\varepsilon(k) - 2\varepsilon(k-1) + \varepsilon(k-2)]. \quad (22)$$

[1] Let the vectors $a, b, c \in \mathbb{R}^{n\times 1}$ where $c_x = a_x \cdot b_x$ with $x \in \{1, 2, ..., n\}$.

## 4 Numerical Simulation Results

The LabVIEW 2013 platform is used For the programming of the PID-RBFAF-IIR controller, and this is applied on the linear model of the Quanser helicopter to control the positions in pitch and yaw axes. A pre-training was made to obtain the initial values of the network parameters shown in Table 1.

From the above, a closed-loop simulation is performed using the AF function to track a trajectory $\mathbf{y}_r(k)$, this is generated for each axis of motion applying the following [7]:

$$y_{r_\theta} = \begin{cases} 28\sin(\pi k/250), & 0 \le k < 2.5 \\ 30, & 2.5 \le k < 5 \\ -30, & 5 \le k < 7.5 \\ 12\sin(\pi k/250) + 4\sin(\pi k/320) + 8\sin(\pi k/100), & 7.5 \le k < 10 \end{cases} \quad (23)$$

and

$$y_{r_\phi} = \begin{cases} 120\sin(2\pi k/1000), & 0 \le k < 5 \\ 100\sin(2\pi k/1000) + 20\sin(2\pi k/100), & 5 \le k < 10 \end{cases} \quad (24)$$

**Table 1** Values obtained and proposed for PID-RBFAF-IIR for the atomic function

| Parameter | Value |
|---|---|
| Neurons, $J$ | 3 |
| Delays, $M$ | 3 |
| Advances, $N$ | 2 |
| Parameter | Value |
| $\mu_{\mathbf{W}}$ | $3.80\times10^{-8}$ |
| $\mu_{\mathbf{q}}$ | $3.80\times10^{-8}$ |
| $\mu_{\mathbf{s}}$ | $3.80\times10^{-8}$ |
| $\mu_{\mathbf{C}}$ | $5.15\times10^{-8}$ |
| $\mu_{\mathbf{D}}$ | $5.90\times10^{-5}$ |
| $v(k)$ | $3.36\times10^{-2}$ |
| Parameter | Value |
| $\mu_{\mathbf{p}}$ | [1.00, 0.90] |
| $\mu_{\mathbf{i}}$ | [0.80, 0.60] |
| $\mu_{\mathbf{d}}$ | [0.75, 0.80] |
| $\mathbf{p}(k)$ | $[2.90 \times 10^{-2}, 9.04 \times 10^{-3}]$ |
| $\mathbf{i}(k)$ | $[1.58 \times 10^{-2}, 1.10 \times 10^{-2}]$ |
| $\mathbf{d}(k)$ | $[1.45 \times 10^{-2}, 5.50 \times 10^{-3}]$ |

The response obtained from the plant and the tracking error are shown in Fig. 6. This shows that the PID-RBFAF-IIR is able to identify the dynamics of the linear model and limit the tracking error.

## 4.1 Auto-Tuning PID Gains

From the simulation show in Fig. 6, the graph in Fig. 7 is the generated control signal $\mathbf{u}(k)$. In addition, the gains $\mathbf{p}(k)$, $\mathbf{i}(k)$ y $\mathbf{d}(k)$ show a constant change to achieve the desired trajectory, this behavior can be seen in Fig. 8.

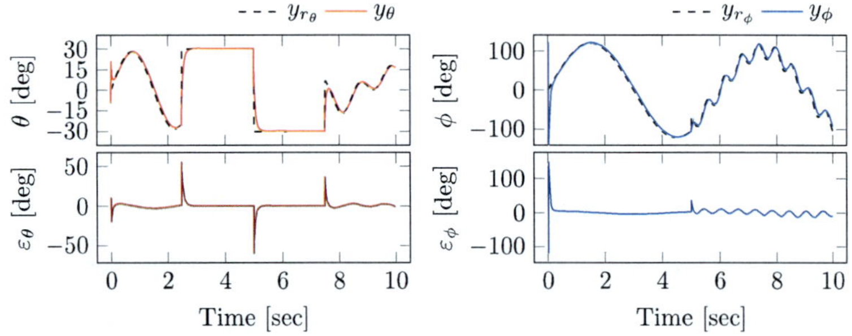


**Fig. 6** Linear model identification in closed loop control and tracking errors, where $y_{r_\theta}$ and $y_{r_\phi}$ are the reference values for the pitch and yaw axes, respectively

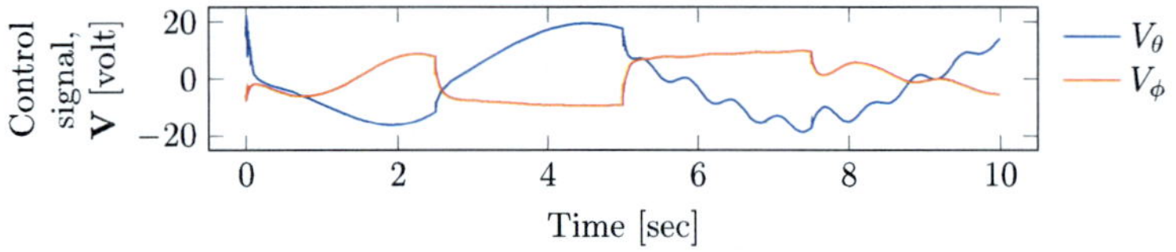


**Fig. 7** Behavior of the control signal $\mathbf{V}(k) = [V_\theta(k),\ V_\phi(k)]^T$ for the pitch $\theta$ and yaw $\phi$ axes

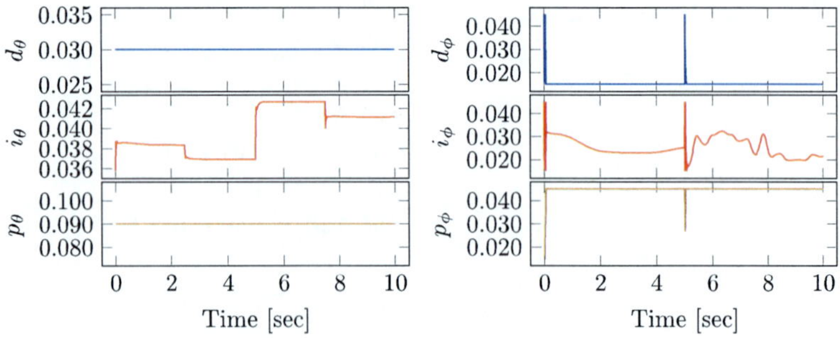


**Fig. 8** Behavior of the PID controller gains for the pitch $\theta$ and yaw $\phi$ axes, respectively

## 5 Conclusions

To the best of our knowledge, this is the first time that the AF have been applied in a control scheme. Based on the results shown in the Sect. 4, it is possible to conclude that the proposed PID-RBFAF-IIR control scheme based in Atomic Functions for the linear two-degree-of-freedom helicopter model is feasible due to its good closed-loop performance.

## References

1. Åström, K.: PID Controllers. Instrument Society of America, Research Triangle Park, N.C. (1995)
2. Åström, K.: Advanced PID Control. Instrument Society of America, Research Triangle Park, N.C. (1996)
3. Brysina, I., Makarichev, V.: Generalized atomic wavelets. Radioelectronic and Computer Systems, pp. 23–31 (2018)
4. Budunova, K.A., Kravchenko, V.: On a new method for approximation of squares of atomic functions $h_a(x)$ by nonnegative rational fractions. J. Commun. Technol. Electron. **66**, 1252–1265 (2021)
5. Emamia, S.A., Paolo, C., Banazadeh, A.: Neural network-based flight control systems: present and future. Annu. Rev. Control. **59**, 97–137 (2018)
6. Eremenko, S.: Atomic machine learning. J. Neurocomput. (3) (2018)
7. Garcia-Castro, O.F.: Controladores PID wavenet aplicados a un modelo de helicóptero de 2 GDL. Bachelor's thesis, Universidad Politécnica Metropolitana de Hidalgo, México (2020)
8. Haykin, S.: Neural networks, a comprehensive foundation. Pearson, EUA (1999)
9. Hernandez-Matamoros, A., Fujita, H., Escamilla-Hernandez, E., Perez-Meana, H., Nakano-Miyatake, M.: Recognition of ECG signals using wavelet based on atomic functions. Biocybernet. Biomed. Eng. **40**(2), 803–814 (2020)
10. Inc., Q.: User manual 2 DOF helicopter experiment, set up and configuration. Markham, Ontario (2012)
11. Kravchenko, V.F., Kravchenko, O.V., Konolalov, Y.Y., Budunova, K.: Atomic functions theory: history and modern Results: pioneer of atomic functions theory. In: IEEE Ukrainian Microwave Week, pp. 619–623 (2020)
12. Kravchenko, V.F., Peres-Meana, H.M., Ponomaryov, V.I.: Adaptive Digital Processing of Multidimensional Signals with Applications. Fizmatlit, Moscow (2009)
13. Ouyang, Y., Dong, L., Xue, L., Sun, C.: Adaptive control based on neural networks for an uncertain 2-DOF helicopter system with input deadzone and output constraints. IEEE/CAA J. Autom. Sin. **6**(3), 807–815 (2020)
14. Ramos-Velasco, L.E., Domínguez-Ramírez, O., Parra-Vega, V.: Wavenet fuzzy PID controller for nonlinear MIMO systems: experimental validation on a high-end haptic robotic interface. Applied Soft Computing, pp. 199–205 (2016)
15. Rvachev, V.A.: About some classes of infinitely differentiable functions. Methods Dyn. Syst. Anal. **7**(3–5) (1983)
16. Rvachev, V.A.: Compactly supported solutions of functional-differential equations and their applications. Russ. Math. Surv. **45**(1), 87–120 (1990)
17. Rvachev, V.L.: Theory of r-functions and their applications. Naukova Dumka, Kiev (1982)

18. Zhao, Z., He, W., Yang, J., Li, Z.: Adaptive neural network control of an uncertain 2-DOF helicopter system with input backlash and output constraints. Neural Comput. Appl. (2022)
19. Zou, T., Wu, H., Sun, W., Zhao, Z., Ouyang, Y., L., Dong, L., Sun, C.: Adaptive neural network sliding mode control of a nonlinear two-degrees-of-freedom helicopter system. Asian J. Control 199–205 (2022)

# Predictor Based Control Scheme for High-Order Delayed Systems with One Unstable Pole

**Alán Ricardo Pineda Higuera, Luis Alberto Barragan Bonilla, María Margarita Vargas Mondragón, Basilio del Muro Cuéllar, Juan Francisco Márquez Rubio, and Gonzalo Isaac Duchén Sánchez**

**Abstract** In this paper, a modification of the Smith Predictor that allows the control of high-order systems with an unstable pole and time delay is presented. For the design of the observer-predictor scheme, an injection scheme consisting of two design parameters ($g_1$ and $g_2$), which allows estimating the signal of interest before it is delayed is used. Likewise, the necessary and sufficient stability conditions are obtained. These stability conditions assure the existence of the proposed strategy. This control scheme is evaluated via numerical simulations.

## 1 Introduction

Time delays appear in the modelling of dynamic systems. These time delays produce problems for the analysis and design of control strategies. Some examples of delayed systems are presented in hydraulic actuators [3], thermoforming processes [4] and teleoperated systems [8]. A common and simple approach to dealing with these problems is to ignore the delay term when it is small enough relative to the system dynamics. The problem arises when the delay time is relatively large and it can not be ignored for the analysis and design of control strategies. As a result, when the control loop is closed, the delay term appears in the characteristic equation of the transfer function, which leads to an infinite number of roots.

To tackle the stability problem of time-delayed systems, several proposals for controlling such class of systems have been reported in the literature. One of the proposed solutions is the use of traditional controllers, such as Proportional (P), Integral (I), Derivative (D) controllers, and their combinations, the Proportional-Integral (PI), Proportional-Derivative (PD), and Proportional-Integral-Derivative (PID) controllers, the latter being the most used controller in the industry [1, 9, 11].

A. R. Pineda Higuera (✉) · L. A. Barragan Bonilla · M. M. Vargas Mondragón ·
B. del Muro Cuéllar · J. F. Márquez Rubio · G. I. Duchén Sánchez
Instituto Politécnico Nacional, Escuela Superior de Ingeniería Mecánica y Eléctrica (ESIME) Unidad Culhuacan, Av. Santa Ana No. 1000, Coyoacan, Mexico
e-mail: alanpinedah@gmail.com

 
O. Llanes-Santiago (ed.), *Proceedings of 19th Latin American Control Congress (LACC 2022)*, Studies in Systems, Decision and Control 464,
https://doi.org/10.1007/978-3-031-26361-3_22

An ideal way to control delayed systems is to make use of the internal signal before it is delayed. With this intermediate signal, it is possible to apply any control strategy considering the delay free system. Generally, these intermediate signals are inaccessible due to the fact that the time delay is immersed within the system. For this reason, devices have been developed that allow estimating these signals. These devices are known as predictors. The first of them, the well known Smith Predictor [12], allows estimating the signal of interest before it is delayed. However, its main limit lies in the fact that it only works for stable systems. Therefore, several modifications have been developed that allow estimating the signal or signals of interest before they are affected by the time delay for unstable plants [2, 5, 6, 10].

In this paper, a modification of the Smith Predictor that allows the control of high order systems with time delay and one unstable pole is proposed. Likewise, the demonstration for the estimation of the system signals is presented, and the necessary and sufficient conditions are obtained to apply the proposed control scheme based on an observer-predictor. The work is organized in the following form; Sect. 2 establishes the class of systems considered in this work and the problem statement. In Sect. 3 two preliminary results are presented, the first preliminary result provides stability conditions through the use of a PD type controller. The second preliminary result corresponds to an injection scheme, this result will be used later. In Sect. 4, the observer-predictor scheme proposed in this paper is presented. In Sect. 5 the simulation results are presented. Finally in Sect. 6 general conclusions are presented.

## 2 Problem Statement

The class of systems considered in this paper is given by,

$$\frac{Y(s)}{U(s)} = G(s)e^{-\tau s} = \frac{\beta}{(s-a)\prod_{i=1}^{n}(s+b_i)}e^{-\tau s}, \tag{1}$$

where $U(s)$ and $Y(s)$ are input and output signals, respectively, $\tau \geq 0$ is the time delay assumed to be known, $a > 0$ is the position of the unstable pole, $b_1, b_2, ..., b_n > 0$ are values associated to the positions of stable poles, and $\beta > 0$ is the system proportional gain.

To show the stability problem associated to time-delay systems, consider a proportional control strategy based on an output feedback loop of the form,

$$U(s) = [R(s) - Y(s)]k, \tag{2}$$

which produces a closed-loop given by,

$$\frac{Y(s)}{R(s)} = \frac{kG(s)\,e^{-\tau s}}{1 + kG(s)e^{-\tau s}}. \tag{3}$$

where the characteristic equation is given by,

$$1 + kG(s)e^{-\tau s} = 0. \tag{4}$$

The roots of Eq. (4) are all the values of the complex variable $s = \sigma + j\omega$ such that (4) is satisfied. Using the Euler identity, Eq. (4) can be rewritten as,

$$1 + kG(s)(cos(\omega\tau) - jsin(\omega\tau)) = 0. \tag{5}$$

Due to the sine and cosine functions being involved in Eq. (5), the equation has an infinite number of solutions. Therefore, the term $e^{-\tau s}$, located in the characteristic Eq. (4) of the transfer function (3) hinderss the stability analysis and classic control approaches present considerable limitations for the design of control strategies for this class of systems.

This paper proposes a control scheme based on an observer-predictor for unstable high order systems with time delay given by (1). In addition, the necessary and sufficient conditions assuring the existence of this proposed strategy are presented.

## 3 Preliminary Results

In this section, two preliminary results are stated. These results will be used later to obtain stability conditions for the observer-predictor based control presented in this paper.

### 3.1 Stability Condition for a PD Controller

**Lemma 1** ([7]) *Consider the system given by* (1), *with a PD-type controller given by* $C_{PD}(s) = k_p(1 + k_d s)$, *and the control law* $U(s) = [R(s) - Y(s)]C_{PD}(s)$. *Then, there exist* $k_p$ *and* $k_d$ *gains such that the closed-loop system* $Y(s)/R(s)$ *is stable if and only if,* $\tau < \frac{1}{a} - \sum_{i=1}^{n} \frac{1}{b_i} + \sqrt{\frac{1}{a^2} + \sum_{i=1}^{n} \frac{1}{b_i^2}}$.

Note that the closed-loop transfer function of the system associated with Lemma 1 is,

$$\frac{Y(s)}{R(s)} = \frac{\beta(k_p + k_p k_d s)e^{-\tau s}}{(s-a)\prod_{i=1}^{n}(s+b_i) + (\beta k_p + \beta k_p k_d s)e^{-\tau s}}, \tag{6}$$

where the characteristic equation is,

$$(s-a)\prod_{i=1}^{n}(s+b_i)+(\beta k_p+\beta k_p k_d s)e^{-\tau s}=0. \tag{7}$$

This preliminary result will be used later.

## 3.2 Injection Scheme

The injection scheme is presented in Fig. 1, which allows the stabilization of the class of systems given by (1). This injection scheme will be used in the main result of this work. This section presents the stability conditions under which the injection scheme allows the stabilization of this class of systems.

**Lemma 2** *Consider the class of systems given by* (1) *and the output injection shown on Fig. 1, where* $g_1$ *and* $g_2$ *are free design parameters. Then, there exist* $g_1$ *and* $g_2$ *gains such that the closed-loop system* $Y(s)/U(s)$ *is stable if and only if* $\tau < \frac{1}{a}-\sum_{i=1}^{n}\frac{1}{b_i}+\sqrt{\frac{1}{a^2}+\sum_{i=1}^{n}\frac{1}{b_i^2}}$.

***Proof*** Consider the injection scheme shown in Fig. 1. Where the corresponding transfer function $Y(s)/U(s)$ is,

$$\frac{Y(s)}{U(s)}=\frac{\beta e^{-\tau s}}{(s-a)\prod_{i=1}^{n}(s+b_i)+(-g_1 a+g_2\beta+g_1 s)e^{-\tau s}}, \tag{8}$$

where characteristic equation is given by,

$$(s-a)\prod_{i=1}^{n}(s+b_i)+(-g_1 a+g_2\beta+g_1 s)e^{-\tau s}=0. \tag{9}$$

Observe the characteristic Eqs. (7) and (9) are similar. Therefore, it can be concluded that the stability condition for the Eq. (7) given in Lemma 1, is the same

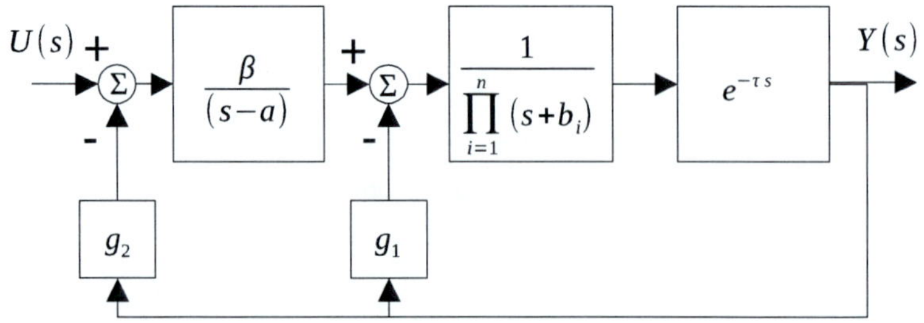


**Fig. 1** Injection scheme

as the condition for Eq. (9). Likewise, observe that for Eq. (7), $k_p$ and $k_d$ are the free design parameters, while for Eq. (9), the free design parameters are $g_1$ and $g_2$, where the relationship between both pairs of parameters can be obtained as $-g_1 a + g_2\beta = \beta k_p$, and $g_1 = \beta k_p k_d$. Then, there exist $g_1$ and $g_2$ gains such that the system described by (1) with the injection scheme illustrated in Fig. 1 is stable if and only if $\tau < \frac{1}{a} - \sum_{i=1}^{n}\frac{1}{b_i} + \sqrt{\frac{1}{a^2} + \sum_{i=1}^{n}\frac{1}{b_i^2}}$. □

Once the condition of the Lemma 2 is satisfied, it is possible to obtain the parameters $g_1$ and $g_2$. Now, a methodology for obtaining these parameters is proposed.

1. The first step is to select a value for $\alpha$ from the interval,

$$\tau - \frac{1}{a} + \sum_{i=1}^{n}\frac{1}{b_i} < \alpha < \sqrt{\frac{1}{a^2} + \sum_{i=1}^{n}\frac{1}{b_i^2}}. \tag{10}$$

2. With the selected $\alpha$ value, it is possible to observe that the following function,

$$f(\omega) = -\tau\omega + atan\left(\frac{\omega}{a}\right) - \sum_{i=1}^{n} atan\left(\frac{\omega}{b_i}\right) + atan(\alpha\omega), \tag{11}$$

has two solutions for $f(\omega) = 0$. The first one is with $\omega = 0$, and the second one with a value $\omega_c > 0$.

3. With the values $\alpha$ and $\omega_c$ from steps 1 and 2, it is possible to obtain the intervals $g_1$ and $g_2$ with,

$$\beta\alpha a\prod_{i=1}^{n} b_i < g_1 < \beta\alpha a\prod_{i=1}^{n} b_i f_2(\omega_c), \tag{12}$$

$$a\frac{g_1}{\beta} + a\prod_{i=1}^{n} b_i < g_2 < a\frac{g_1}{\beta} + a\prod_{i=1}^{n} b_i f_2(\omega_c), \tag{13}$$

with,

$$f_2(\omega_c) = \sqrt{\frac{\left(1 + \left(\frac{\omega_c}{a}\right)^2\right)\prod_{i=1}^{n}\left(1 + \left(\frac{\omega_c}{b_i}\right)^2\right)}{1 + \alpha^2\omega_c^2}}. \tag{14}$$

With the $g_1$ and $g_2$ values given by Eqs. (12) and (13) respectively, it is possible to implement the injection scheme presented in Fig. 1.

# 4 Estimation strategy

In this section, the proposed observer-predictor scheme is shown in Fig. 2, along with the proof of the adequate estimation. Consider a space-state representation for system (1) given by,

$$\begin{bmatrix} \dot{z}(t) \\ \dot{X}(t) \end{bmatrix} = \begin{bmatrix} a & 0 \\ B_1 & A_1 \end{bmatrix} \begin{bmatrix} z(t) \\ X(t) \end{bmatrix} + \begin{bmatrix} \beta \\ 0 \end{bmatrix} u(t), \tag{15}$$

$$y(t+\tau) = \begin{bmatrix} 0 \; C_1 \end{bmatrix} \begin{bmatrix} z(t) \\ X(t) \end{bmatrix}, \tag{16}$$

with,

$$A_1 = \begin{bmatrix} -b_1 & 1 & \dots & 0 \\ \vdots & \vdots & \ddots & \vdots \\ 0 & 0 & b_{n-1} & 1 \\ 0 & 0 & 0 & b_n \end{bmatrix}, B_1 = \begin{bmatrix} 0 \; 0 \; \dots \; 1 \end{bmatrix}^T, \text{ and, } C_1 = \begin{bmatrix} 1 \; \dots \; 0 \; 0 \end{bmatrix}.$$

**Theorem 1** *Consider the system given by* (15)–(16) *with the observer-predictor scheme shown in Fig. 2. Then, there exist* $g_1$ *and* $g_2$ *such that* $\lim_{t\to\infty}[\hat{\omega}(t) - \omega(t)] = 0$, *if and only if* $\tau < 1/a - \sum_{i=1}^{n} 1/b_i + \sqrt{1/a^2 + \sum_{i=1}^{n} 1/b_i^2}$,

with,

$$\omega(t) = \begin{bmatrix} z(t) \\ X(t) \end{bmatrix}, \tag{17}$$

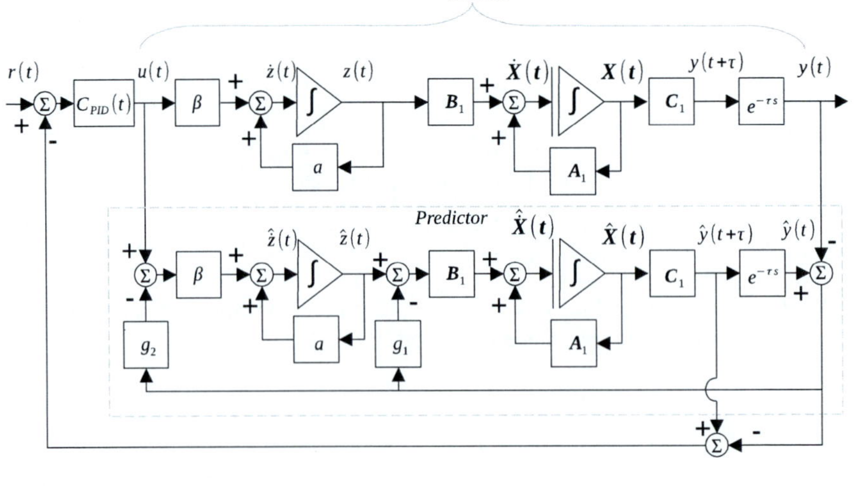


**Fig. 2** Proposed observer scheme

and,

$$\hat{\omega}(t) = \begin{bmatrix} \hat{z}(t) \\ \hat{X}(t) \end{bmatrix}. \tag{18}$$

***Proof*** For the proof of convergence in the observer-predictor scheme shown in Fig. 2, it is not necessary to take into account the controller $C_{PID}(s)$ as it is independent from the estimation strategy.

The prediction error is defined as $e_\omega(t) = \hat{\omega}(t) - \omega(t)$, and the output error as $e_y(t) = \hat{y}(t) - y(t)$. Thus, it is possible to describe the behavior of the error dynamic as,

$$\begin{bmatrix} \dot{e}_z(t) \\ \dot{e}_x(t) \\ e_y(t+\tau) \end{bmatrix} = \begin{bmatrix} a & 0 & -g_2\beta \\ B_1 & A_1 & -g_1 B_1 \\ 0 & C_1 & 0 \end{bmatrix} \begin{bmatrix} e_z(t) \\ e_x(t) \\ e_y(t) \end{bmatrix}. \tag{19}$$

Consider the space-state realization of system (8) corresponding to the injection scheme illustrated in Fig. 1, which can be expressed as,

$$\begin{bmatrix} \dot{z}(t) \\ \dot{X}(t) \\ y(t+\tau) \end{bmatrix} = \begin{bmatrix} a & 0 & -g_2\beta \\ B_1 & A_1 & -g_1 B_1 \\ 0 & C_1 & 0 \end{bmatrix} \begin{bmatrix} z(t) \\ X(t) \\ y(t) \end{bmatrix} + \begin{bmatrix} \beta \\ 0 \\ 0 \end{bmatrix} u(t). \tag{20}$$

Observe that in the error dynamic (19), there is not an input signal, therefore, in order to fulfill that $e_\omega(t) \to 0$ when $t \to \infty$, it is required that the error dynamic of the system (19) is stable. Now, it is important to note that the error dynamic (19) is identical to the dynamic in system (20) (corresponding to the injection shown in Fig. 1). Thus, the stability condition, $\tau < \frac{1}{a} - \sum_{i=1}^{n} \frac{1}{b_i} + \sqrt{\frac{1}{a^2} + \sum_{i=1}^{n} \frac{1}{b_i^2}}$, corresponding to system (20) is then extended to the error dynamic of system (19), which ensures the convergence of the $\omega(t)$ and $\hat{\omega}(t)$ signals. □

**Remark 1** Once the condition of the Theorem 1 is satisfied, corresponding to the estimation of the $\hat{\omega}(t)$ signal, it is possible to take the estimated signal, $\hat{y}(t+\tau)$, to design a $C_{PID}(s)$ controller considering system (1) without delay. In this paper, a PID-type controller is proposed as given by,

$$C_{PID}(s) = k_p + \frac{k_i}{s} + k_d s, \tag{21}$$

which can be designed using a root locus analysis.

## 5 Simulation Results

**Example 1**

Consider the third-order system containing one unstable pole, two stable poles, and a time delay given by,

$$\frac{Y(s)}{U(s)} = \frac{1}{(s-1)(s+3)(s+4)}e^{-1.2s}, \tag{22}$$

According to the proposed observer-predictor based control presented in this paper, the stability condition of the proposed control strategy is given by Theorem 1. For this example, observe the maximum delay allowed is $\tau < 1.5$. Due to the delay of the plant being $\tau = 1.2$, it is possible to apply the proposed control strategy. Then, it is possible to apply the methodology described in Sect. 3 to obtain the $g_1$ and $g_2$ parameters of proposed observer-predictor scheme.

1. An $\alpha$ value is selected from the interval presented in the Eq. (10). In this case, the interval is given by $0.783 < \alpha < 1.083$, and a value of $\alpha = 0.96$. is selected.
2. With $\alpha = 0.96$, it is possible to observe that function (11) has two solutions for $f(\omega) = 0$. One with $\omega = 0$, and the second one with $\omega_c = 0.5891$. This is shown in Fig. 3.
3. With the previously selected parameters, $\alpha$ and $\omega_c$, it's possible to obtain the stabilizing $g_1$ and $g_2$ intervals using Eqs. (12) and (13). The intervals for this example are $11.52 < g_1 < 11.98$ and $23.75 < g_2 < 24.2393$. For the simulation, the values $g_1 = 11.75$ and $g_2 = 23.99$ are selected. Note that Fig. 4 shows the stabilizing region for $g_1$ and $g_2$ considering the entire interval of $\alpha$.

For the design of the control $C_{PID}(s)$ given in (21) the following parameters are used, $k_p = 225$, $k_i = 75$ and $k_d = 150$.

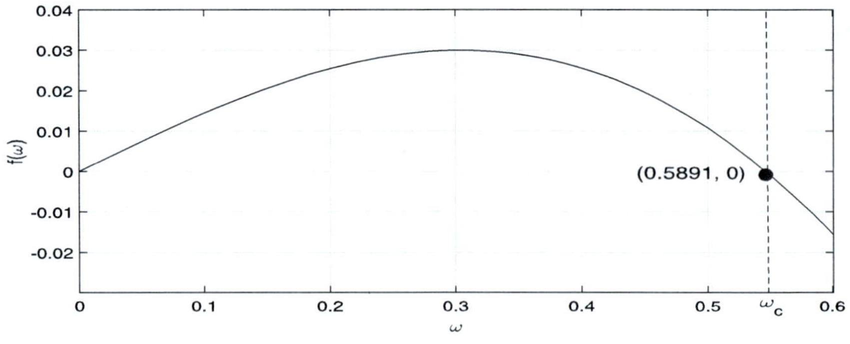


**Fig. 3** Function $f(w)$ for Example 1

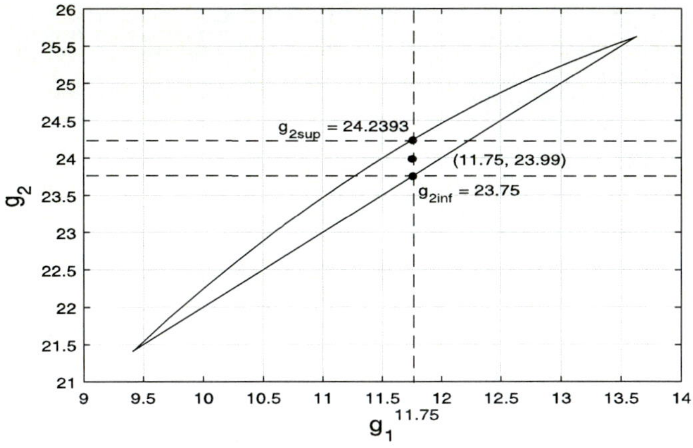


**Fig. 4** Stability region for parameters $g_1$ and $g_2$

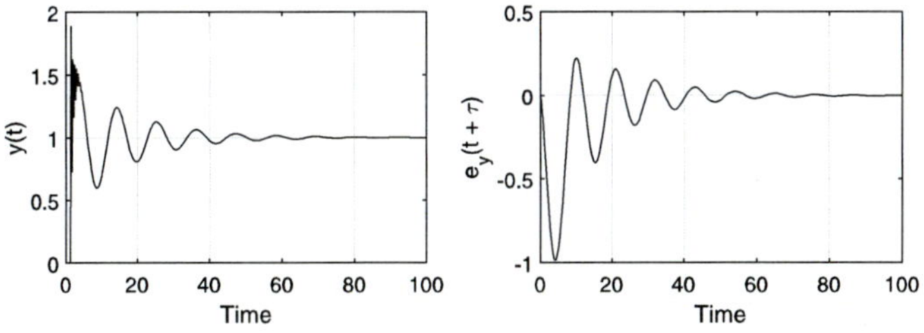


**Fig. 5** Output response and error convergence of the system from Example 1

In Fig. 5, it is possible to observe both that the output response of the system is stable in stationary state given an $R(s)$ step input with amplitude 1 and the convergence of the signal of interest. The initial conditions for the plant states of 0.1 are also considered.

**Example 2**

Consider the third-order system previously examined in [7], which contains one unstable pole, two stable poles, and a time delay given by,

$$\frac{Y(s)}{U(s)} = \frac{0.2}{(s-0.2)(s+0.5)(s+2)} e^{-0.5s}, \tag{23}$$

Using the methodology presented in [7], it is possible to propose an adequate $C_{PID2}(s) = k_p(1 + k_i/s + k_d s)$ controller for (23) with the parameters $k_p = 2$, $k_i = 0.05$, and $k_d = 0.5$. By using an equivalent $C_{PID}(s)$ controller in the pro-

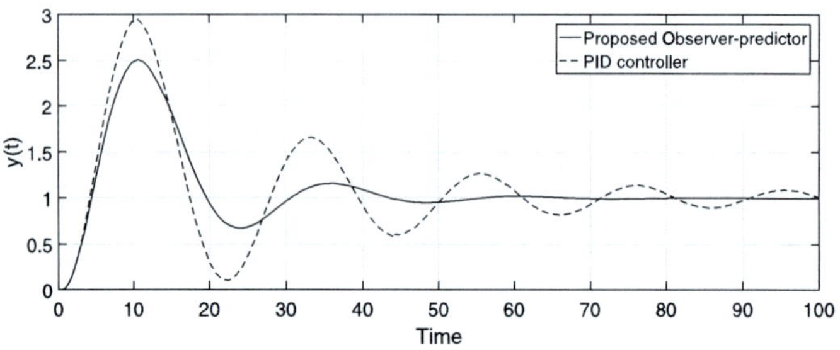


**Fig. 6** Output comparison between the proposed control scheme and the stabilization by simple controllers described in [7]

posed observer-predictor scheme considering $k_p = 2$, $k_i = 0.1$, $k_d = 1$, $g_1 = 1$, and $g_2 = 1$, it is possible to observe the improved response to a step input when using the proposed scheme. Said comparison is shown in Fig. 6.

## 6 Conclusions

Systems with time delay present a significant challenge when analysing and designing control strategies. In this paper, a control scheme based on an observer-predictor is presented. The proposed control scheme allows the control of a particular class of linear system under the delay condition of $\tau < 1/a - \sum_{i=1}^{n} 1/b_i + \sqrt{1/a^2 + \sum_{i=1}^{n} 1/b_i^2}$. For the design of the observer-predictor scheme, an injection scheme with two design parameters ($g_1$ and $g_2$) is used, which allows adequately estimating the signal of interest; this allows the implementation of any control strategy for the class of systems approached in this work without considering the delay time. The stability conditions are improved when implementing P/PI controllers with respect to previously reported works [7].

## References

1. Arrieta, O., Vilanova, R., Visioli, A.: Proportional-integral-derivative tuning for servo/regulation control operation for unstable and integrating processes. Ind. Eng. Chem. Res. **50**(6), 3327–3334 (2011)
2. Cruz-Díaz, C., del Muro-Cuéllar, B., Duchén-Sánchez, G., Márquez-Rubio, J.F., Velasco-Villa, M.: Observer-based pid control strategy for the stabilization of delayed high order systems with up to three unstable poles. Mathematics **10**(9) (2022). https://doi.org/10.3390/math10091399
3. Deng, W., Yao, J., Wang, Y., Yang, X., Chen, J.: Output feedback backstepping control of hydraulic actuators with valve dynamics compensation. Mech. Syst. Signal Process. **158**, 107769 (2021). https://doi.org/10.1016/j.ymssp.2021.107769

4. Kali, Y., Ayala, M., Rodas, J., Saad, M., Doval-Gandoy, J., Gregor, R., Benjelloun, K.: Time delay estimation based discrete-time super-twisting current control for a six-phase induction motor. IEEE Trans. Power Electron. **35**(11), 12570–12580 (2020). https://doi.org/10.1109/TPEL.2020.2995773
5. Karan, S., Dey, C.: Modified smith predictor-based p-pd control for pure integrating delay dominated processes. Can. J. Chem. Eng. (2021)
6. Kumar, M.P., Narayana, K.V.L.: Multi control scheme with modified smith predictor for unstable first order plus time delay system. Ain Shams Eng. J. **9**(4), 2859–2869 (2018)
7. Lee, S.C., Wang, Q.G., Xiang, C.: Stabilization of all-pole unstable delay processes by simple controllers. J. Process Control **20**(2), 235–239 (2010)
8. Park, J., Lee, T., Liu, Y., Chen, J.: Dynamic Systems with Time Delays: Stability and Control (2019). https://doi.org/10.1007/978-981-13-9254-2
9. Pawar, S.N., Chile, R.H., Patre, B.M.: Predictive extended state observer-based robust control for uncertain linear systems with experimental validation. Trans. Inst. Meas. Control **43**(12), 2615–2627 (2021)
10. Ren, W., Luo, Y., He, Q., Zhou, X., Deng, C., Mao, Y., Ren, G.: Stabilization control of electro-optical tracking system with fiber-optic gyroscope based on modified smith predictor control scheme. IEEE Sensors J. **18**(19), 8172–8178 (2018)
11. Silva, G.J., Datta, A., Bhattacharyya, S.P.: PID controllers for time-delay systems (2005)
12. Smith, O.J.M.: Close control of loops with dead time, p. 53. Eng. Prog., Vol, Chem (1957)

# Observer Based Control of a Scalar Neutral Delayed System

**J. F. Marquez Rubio, A. Urquiza Castro, R. J. Vazquez Guerra, B. del Muro Cuéllar, and E. Aranda Bricaire**

**Abstract** The general problem of control for neutral delayed systems (not just scalar case) has been widely analyzed by considering solutions based on time domain approach. As a consequence, sufficient conditions have been obtained following the mentioned approach. Instead, in this work the stability analysis is based on the frequency domain approach. It is proposed an observer based control for a particular neutral delayed system, in the scalar case. Convergence of the observer is solved using Nyquist stability criteria. The proposal is evaluated by means a numerical example.

## 1 Introduction

Time delay appears frequently in dynamic systems, these systems can be classified in physical, chemical, biological, economic phenomena, among others, having a high impact on the industrial processes; time delays are caused by various circumstances, such as the transport of material, energy or information. Processes with time delays are often a source of instability and performance degradation when it is desired to control them. In recent years, various solutions have been presented to deal with the problem of stabilizing time delay systems. One of them is based on prediction or state observer strategies, which consists of counteracting the time delay and its effects.

J. F. Marquez Rubio · A. Urquiza Castro · B. del Muro Cuéllar
Instituto Politécnico Nacional ESIME Culhuacán, Av. Santa Ana No. 1000, Coyoacán 04440, CDMX, México

R. J. Vazquez Guerra (✉)
Instituto Politécnico Nacional CECyT No. 7 Cuauhtémoc, Ermita Iztapalapa No. 3241, Iztapalapa 09570, CDMX, México
e-mail: rjjvg@yahoo.com.mx

E. Aranda Bricaire
CINVESTAV México, Ciudad de México, México
e-mail: earanda@cinvestav.mx



O. Llanes-Santiago (ed.), *Proceedings of 19th Latin American Control Congress (LACC 2022)*, Studies in Systems, Decision and Control 464,
https://doi.org/10.1007/978-3-031-26361-3_23

One of the most popular prediction strategies that can be found in the literature is the Smith predictor (PS), this strategy provides an estimate of future output by means of an observation scheme [10]. The main limitation of this strategy is that it can only be applied to stable systems. In order to solve this problem, several modifications of the original structure proposed by Smith can be found in the literature, that are intended to be used in unstable systems [1, 2, 6, 7, 12].

On the other hand, a particular class of delayed systems are time-delayed neutral systems, which are characterized by systems involving the derivative of the delayed state, so the stability in this case is a more complex issue because the rate of change of the state depends not only on the delayed state, but also on the rate of change of the past state. In recent years, special attention has been given to the problem of stability analysis of linear time-delayed neutral systems. For example, the work presented in [3] obtains the necessary conditions for the stability of a time-delayed neutral system by using a new stability criterion for a scalar case. The peculiarity of these conditions is that they depend exclusively on the Lyapunov delay matrix. Another work that deals with the case of neutral systems with time delay is presented in [5], where the sufficient condition for the stabilisation of the delay-free system is obtained. However, these methods are based on state feedback, so they cannot be applied to systems where not all states are accessible.

On the other hand, the issue of observer-predictor based control for time-delay neutral systems has also received little attention. A work presented in the literature that addresses this class of systems is given in [9], which uses the Lyapunov stability method combined with matrix inequality techniques; an observer-predictor based controller is implemented in the system, which makes the closed-loop system asymptotically stable. Another related work is published in [4]. Its approach is to make use of a linear inequality matrix (LMI) to design an observer-based feedback control scheme. This last works presents general solutions providing just necessary stability conditions due to LMI involved in the analysis.

In this paper, it is considered the class of neutral delayed systems, specifically the scalar case. The proposed control strategy is based on observer design, where necessary and sufficient conditions are provided. Moreover, such stability conditions are related to the plant parameters, which represents an advantage under the works previously presented in literature.

This paper is organized as follows. In Sect. 2 the class of systems and the ideal proposed control strategy are presented. Section 3 presents the main results of this work. In Sect. 4, simulation results are presented. Finally, in Sect. 5 some conclusions are provided.

## 2 Class of System and Ideal Proposed Control Method

Consider a scalar neutral delayed system given by,

$$\dot{x}(t) + d\dot{x}(t-\tau) = ax(t) - bx(t-\tau) + u(t) \tag{1}$$

$$y(t) = cx(t) \tag{2}$$

where $\tau$ is the time-delay, $a > 0$, $u(t)$ and $y(t)$ are the input and output signals. Block diagram of the system (1)–(2) can be seen in Fig. 1. It is assumed that the measurement of the internal variables of the system is not available. However, we could proposed an ideal control strategy where as the measurement of signals $\omega_1$ and $\omega_2$ (which are functions of the state $x(t)$) were available, of the form,

$$u(t) = d\dot{x}(t-\tau) + bx(t-\tau) - kx(t-\tau) \tag{3}$$

or

$$u(t) = \omega_1(t-\tau) + \omega_2(t-\tau) - kx(t-\tau) \tag{4}$$

The proposed control strategy (4) is illustrated in Fig. 1. Notice that the terms $\omega_1(t-\tau_1)$ and $\omega_2(t-\tau)$ in (4) eliminate the effects of the $d\dot{x}(t-\tau)$ and $bx(t-\tau)$ terms of the original system. This proposed control allows to simplify the design of the control of the whole system. This simplification takes place when (4) is substituted in (1), giving the following closed-loop,

$$\dot{x}(t) = ax(t) - kx(t-\tau) \tag{5}$$

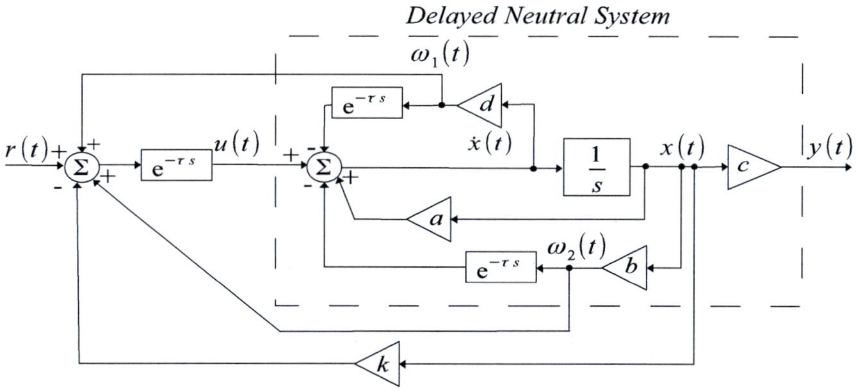


**Fig. 1** Ideal strategy

Notice that the resultant closed-loop system (5) corresponds to a first order unstable system controlled by a proportional action $k$. It is well known that system (5) is stable if and only if $\tau < \frac{1}{a}$, and the stabilizing parameter $k$ is given by,

$$a < k < a\sqrt{1+\left(\frac{\omega_c}{a}\right)^2} \tag{6}$$

with the first positive $\omega_c$ satisfying $atan(\frac{\omega}{a}) = \tau\omega$. Unfortunately, under the assumption previously made, the measurement of signals $\omega_1$ and $\omega_2$ is not available from the system. Therefore, in this work an observer strategy is designed to estimate the signals $\omega_1$ and $\omega_2$, which implies the adequate estimation of the state $x(t)$.

# 3 Main Results

## *3.1 Delayed Proportional Control*

In this section, an easy way to solve the stabilization problem of a proposed delayed-proportional control for the scalar neutral delayed system given by (1)–(2) is presented. Thus, the associated transfer function to the system (1)–(2) is given by,

$$\frac{Y(s)}{U(s)} = \frac{c}{(s-a) + be^{-\tau s} + sde^{-\tau s}} \tag{7}$$

Now, consider a proposed delayed-proportional control for the system (7) with a unitary output feedback of the form,

$$U(s) = \bar{k}e^{-\tau s}(R(s) - Y(s)) \tag{8}$$

where $R(s)$ is the new input reference. In this way, the closed loop transfer function is,

$$\frac{Y(s)}{R(s)} = \frac{\bar{k}ce^{-\tau s}}{(s-a) + (b + sd + \bar{k}c)e^{-\tau s}} \tag{9}$$

The closed-loop characteristic equation of (9) is,

$$(s-a) + (ds + b + \bar{k}c)e^{-\tau s} = 0 \tag{10}$$

Also, notice that a state space representation of the closed-loop system (9) can be obtained by substituting $u(t) = \bar{k}(r(t-\tau) - cx(t-\tau))$ in (1), obtaining,

$$\dot{x}(t) + d\dot{x}(t-\tau) = ax(t) - (b + \bar{k}c)x(t-\tau) + \bar{k}r(t-\tau) \tag{11}$$

On the other hand, consider an auxiliary system given by,

$$Q(s) = \frac{\bar{Y}(s)}{\bar{U}(s)} = \frac{(ds + b + \bar{k}c)e^{-\tau s}}{(s - a)}, \tag{12}$$

a unitary output feedback of the form $\bar{U}(s) = \bar{R}(s) - \bar{Y}(s)$. Thus, it is easy to see that the characteristic closed-loop equation of $\frac{\bar{Y}(s)}{\bar{R}(s)}$ is the same as (10). Therefore, the stability properties of the characteristic equation (10) can be analyzed by using the stability Nyquist criteria by using the auxiliary system (12). The stability Nyquist establishes that the characteristic equation of $\frac{\bar{Y}(s)}{\bar{R}(s)}$ (or equivalently the characteristic equation (10)) is stable if and only if the trajectory on the Nyquist diagram of (12) encircles once the point $-1 + 0j$ in counter clockwise direction. In this way, a practical way to stabilize the characteristic equation (10) consists in test some values of the parameter $\bar{k} > 0$ and determinate if there exists a value of the parameter $\bar{k}$ such that the Nyquist diagram of (12) encircles once the point $-1 + 0j$ in counter clockwise direction. This can be done easily by plotting the Nyquist diagram of (12) for different values of $\bar{k}$, notice that the parameters $a$, $b$, $c$ and $d$ are taken directly from the plant.

## 3.2 Proposed Observer

Consider de augmented scalar neutral delayed system given by (1) as,

$$\dot{x}(t) + d\dot{x}(t - \tau) = ax(t) - bx(t - \tau) + \bar{u}(t - \tau) \tag{13}$$

$$y(t) = cx(t) \tag{14}$$

Notice that the difference between system (1)–(2) and (13)–(14) is the delay-time at the input in (13). Then, consider a proposed observer for the augmented system (13)–(14) shown in Fig. 2, where $\hat{\dot{x}}(t)$ is the estimation of $\dot{x}(t)$, $\hat{x}(t)$ is the estimation of $x(t)$ ,and therefore $\hat{\omega}_i(t)$ is the estimation of $\omega_i(t)$, for $i = 1, 2$; $\bar{k}$ is a free-parameter design.

The main aim of the proposed observer in Fig. 2 is to estimate the variables $\omega_1(t)$, $\omega_2(t)$ and $x(t)$. However, since the variables $\omega_1(t)$, $\omega_2(t)$ are functions of $x(t)$, it is enough to ensure the convergence on the error in the state $x(t)$ defined as $e_x(t) = x(t) - \hat{x}(t)$, i.e., it is desirable to get $lim_{t\to\infty}(x(t) - \hat{x}(t)) = 0$ as $t \to \infty$. Therefore, consider the dynamic error as $\dot{e}_x = \dot{x}(t) - \hat{\dot{x}}(t)$ or,

$$\dot{e}_x(t) + d\dot{e}_x(t - \tau) = ae_x(t) - (b + \bar{k}c)e_x(t - \tau) \tag{15}$$

Notice that the dynamic error (15) has the same stability property as (11) by considering that the input $\bar{k}r(t - \tau)$ in (11) does not affect the stability property. Therefore,

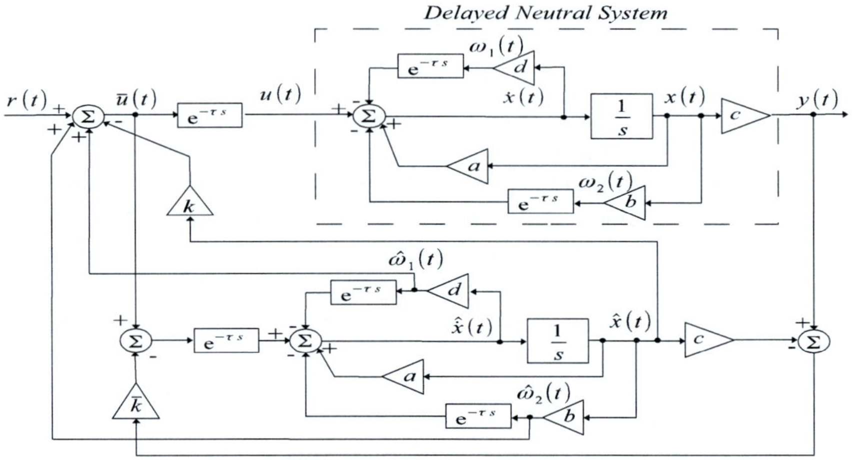


**Fig. 2** Observer strategy for neutral delayed systems

$lim_{t\to\infty}(x(t) - \hat{x}(t)) = 0$ as $t \to \infty$ if and only if there exists a value of the parameter $\bar{k}$ such that the Nyquist diagram of (12), encircles once the point $-1 + 0j$ in counter clock-wise direction.

To find the stabilizing parameter $\bar{k}$, it should be plotted the Nyquist diagram of (12) for different values of $\bar{k} > 0$. Therefore, the values of $\bar{k}$ such that the Nyquist diagram of (12), encircles once the point $-1 + 0j$ in counter clockwise direction can be used to the proposed observer in Fig. 2.

## 3.3 Proposed Observer Based Control

In this section, the ideas proposed in Sect. 2 are recovered to be used by considering the proposed observer in Sect. 3.2. In Sect. 3.2 an adequate observer was designed, and the convergence of the estimated variables was ensured. Now, in this Section, it is implementing the proposed control law (4) by considering the estimated variables taken from the observer, giving as a result,

$$u(t) = \hat{\omega}_1(t - \tau) + \hat{\omega}_2(t - \tau) - k\hat{x}(t - \tau) \tag{16}$$

where $\tau < \frac{1}{a}$ should be satisfied and the parameter $k$ is obtained by using (6). The complete proposed observer based control is shown in Fig. 2.

## 4 Simulation Results

In this section, the performance of the proposed observer is evaluated through an academic example to illustrate the effectiveness of the results here exposed.

Let us consider a scalar neutral delayed system given by,

$$\dot{x}(t) + 0.01\dot{x}(t - 0.5) = x(t) - x(t - 0.5) + u(t) \tag{17}$$

$$y(t) = x(t) \tag{18}$$

First, the observer design is regarded. Therefore, $lim_{t\to\infty}(x(t) - \hat{x}(t)) = 0$ as $t \to \infty$ if and only if there exists a value of the parameter $\bar{k}$ such that the Nyquist diagram of (12) (with $a = 1, b = 1, c = 1, d = 0.01$ and $\tau = 0.5$), encircles once the point $-1 + 0j$ in counter clockwise direction. In this way, Fig. 3 shows the Nyquist

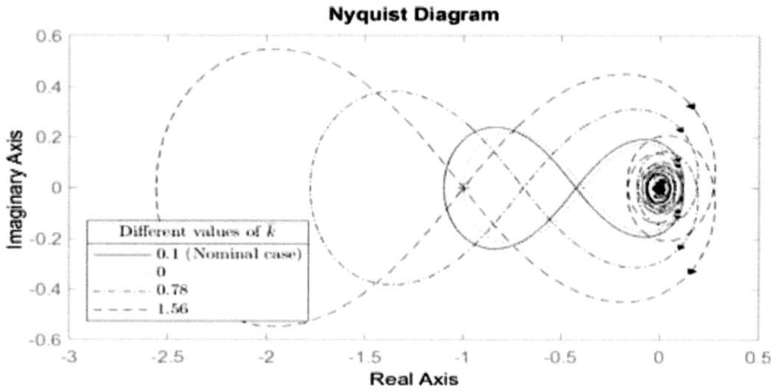


**Fig. 3** Nyquist diagram for different values of $\bar{k}$

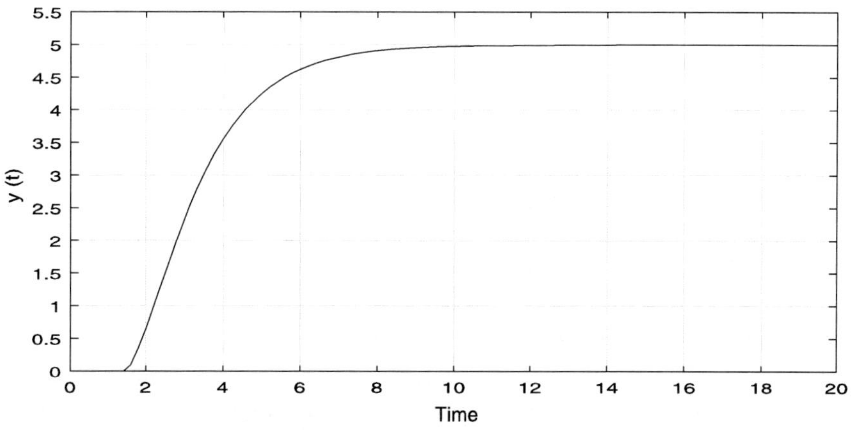


**Fig. 4** Output response in nominal case

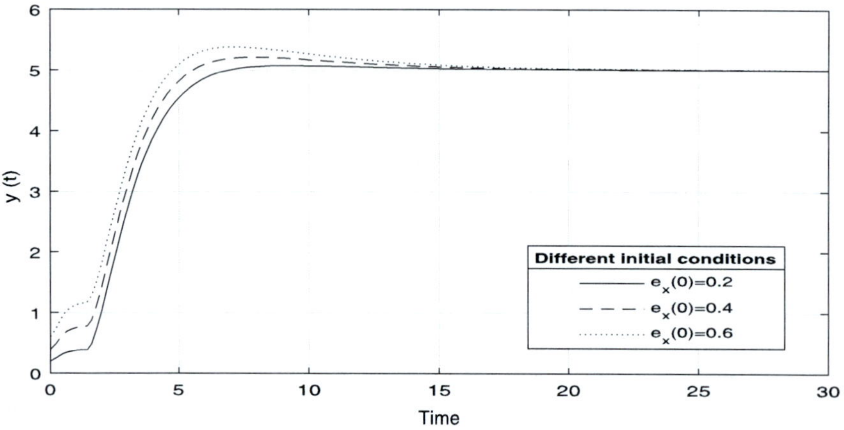


**Fig. 5** Output response with different initial conditions

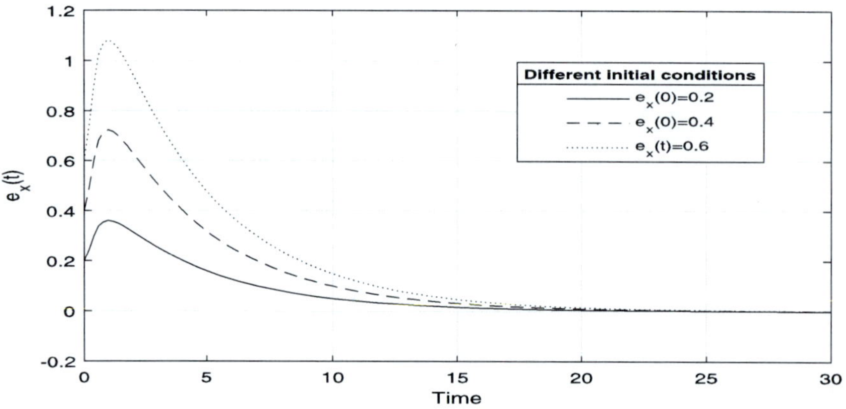


**Fig. 6** Output estimation error, $e_x(t) = x(t) - \hat{x}(t)$ for different initial conditions

diagrams of (12) for different values of $\bar{k} > 0$. Note that the achievable interval of $\bar{k}$ is $0 < \bar{k} < 1.56$. For the simulation experiment, $\bar{k} = 0.1$ is selected for the proposed observer, Fig. 3 shows the Nyquist diagram for this value of $\bar{k}$.

As a second step, the design of the proposed proportional control is presented. To this end, the stabilizing values of $k$ can be obtained by expression (6), giving as a result $1 < k < 2.53$. For the simulation experiment, the stabilizing parameter $k$ used to the control strategy is $k = 1.2$.

In order to illustrate the behavior of the proposed control strategy, some simulations are performed; Fig. 4 shows the closed-loop behavior of the controlled system in the nominal case, i.e. initial conditions equal to zero. Figure 5 shows the closed-loop behavior of the controlled system, when initial conditions different from zero are considered in the state of the plant. Figure 6 shows the output estimation error, $e_x(t) = x(t) - \hat{x}(t)$, with different initial conditions.

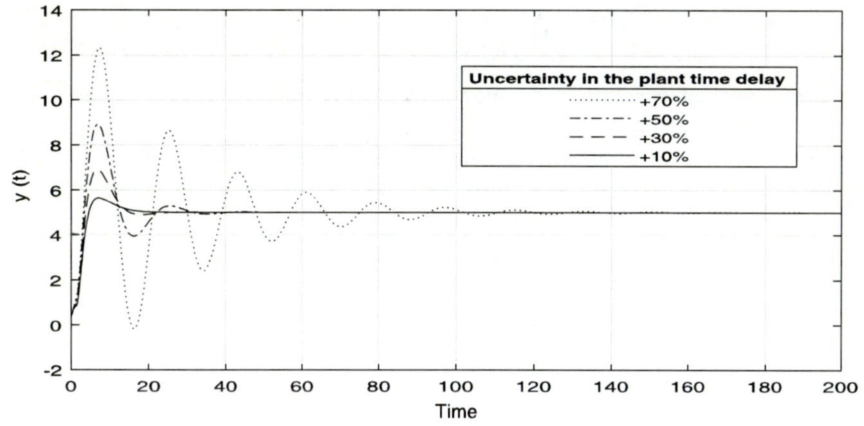


**Fig. 7** Output response considering time delay uncertainty and initial condition of $e_x(0) = 0.4$

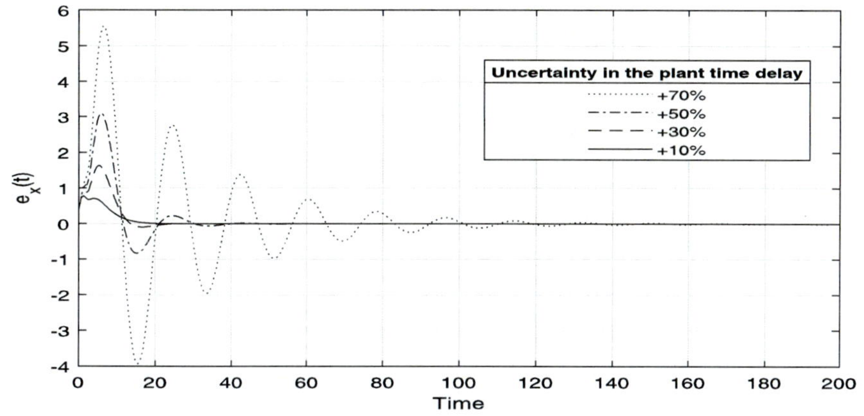


**Fig. 8** Output estimation considering time delay uncertainty and initial condition of $e_x(0) = 0.4$

The output response performance is presented in Fig. 7 when initial condition in the plant is set at $e_x(0) = 0.4$ and different uncertainties in the time delay $\tau$ is present in the plant. The corresponding output estimation error to the last experiment is presented in Fig. 8.

## 5 Conclusions

This work presents a solution of control for a neutral delayed systems using a frequency domain approach. The general idea of the control is decreasing/eliminating the effects of the some internal variables of the system. An observer-estimator is designed for estimate the internal variables of the system. Once that such effects are eliminated, a simple proportional controller allows to stabilize the original simplified plant. Necessary and sufficient conditions are derived for the observer design as well as the control for the plant based on the estimated variables, which are in general more desirable conditions than just sufficient conditions.

## References


1. Astrom, K., Hang, C., Lim, B.: A new smith predictor for controlling a process with an integrator and long dead-time. IEEE Trans. Autom. Control **39**(2), 343–345 (1994). https://doi.org/10.1109/9.272329
2. Buenfil, R., D.M.B.: Predictor hírido para sistemas lineales con retardo de tiempo. AMCA, 111–116 (2013)
3. Gomez, M.A., Egorov, A.V., Mondie, S.: Necessary stability conditions for neutral type systems with a single delay. IEEE Trans. Autom. Control **62**(9), 4691–4697 (2017). https://doi.org/10.1109/tac.2016.2625738
4. Lien, C.H.: Guaranteed cost observer-based controls for a class of uncertain neutral time-delay systems. J. Optim. Theory Appl. **126**(1), 137–156 (2005). https://doi.org/10.1007/s10957-005-2665-2
5. Ma, W.B., Adachi, N., Amemiya, T.: Delay-independent stabilization of uncertain linear systems of neutral type. J. Optim. Theory Appl. **84**(2), 393–405 (1995). https://doi.org/10.1007/bf02192121
6. Matausek, M., Micic, A.: A modified smith predictor for controlling a process with an integrator and long dead-time. IEEE Trans. Autom. Control **41**(8), 1199–1203 (1996). https://doi.org/10.1109/9.533684
7. Márquez-Rubio, J., del Muro-Cuéllar, B., Velasco-Villa, M., Álvarez Ramírez, J.: Control basado en un esquema observador para sistemas de primer orden con retardo. Revista mexicana de ingeniería química **9**(1), 43–52 (2010)
8. Niculescu, S.I.: Delay Effects on Stability. Springer, London (2001). https://doi.org/10.1007/1-84628-553-4
9. Park, J.H.: On the design of observer-based controller of linear neutral delay-differential systems. Appl. Math. Comput. **150**(1), 195–202 (2004). https://doi.org/10.1016/s0096-3003(03)00209-1
10. Smith, O.J.M.: Closer control of loops with deadtime. Chem. Eng. (1957)
11. Amemiya, Takashi: Delay-independent stabilization of linear systems. Int. J. Control **37**(5), 1071–1079 (1983). https://doi.org/10.1080/00207178308933029
12. Torrico, B.C., Normey-Rico, J.E.: 2DOF discrete dead-time compensators for stable and integrative processes with dead-time. J. Process Control **15**(3), 341–352 (2005). https://doi.org/10.1016/j.jprocont.2004.05.006

# Low Cost Systems and Biomedical Applications

# ESP32 Based Low-Power and Low-Cost Wireless Sensor Network

**Alejandro Perdomo-Campos, Iliover Vega-González, and Jorge Ramírez-Beltrán**

**Abstract** The development of increasingly efficient sensor networks and remote terminal units (RTUs) is a highly relevant task in applications that require remote acquisition of information for processing, remote control of certain processes and fault diagnosis. In this context, the search for efficient ways to transmit information becomes necessary. In some applications there are several nodes of the same type acquiring data simultaneously in remote places, for which the development of a remote reading mechanism based on wireless communication technologies is required. In others, it is necessary to trigger actions through commands issued remotely when some event is detected. In this work, it is proposed a simple, low power and low-cost solution for the development of a LoRa sensor network consisting of several nodes and a single-channel gateway based on the ESP32 System-on-Module (SoM) for reading measurement nodes in water supply systems. The solution designed takes advantage of ESP32 potential for wireless communications, facilities of the chip SX1276 and the features of LoRa as communication technology.

## 1 Introduction

In the development of remote terminal units (RTUs), one of the fundamentals to take into account is the way data is transmitted. The concept of RTU can include all those nodes of a system that execute remote reading actions or implement remote control mechanisms. This way, we call RTU to any edge device of a network that is used to acquire and transmit data wirelessly (remote reading) or to trigger actions, events

A. Perdomo-Campos (✉) · I. Vega-González · J. Ramírez-Beltrán
Technological University of Havana, "José Antonio Echeverría", 114 Street E/Ciclovía and Rotonda, Marianao, Havana, Cuba
e-mail: aperdomoc@tele.cujae.edu.cu

I. Vega-González
e-mail: ciivega@cime.cujae.edu.cu

J. Ramírez-Beltrán
e-mail: jorgeramirezcihcujae@gmail.com



O. Llanes-Santiago (ed.), *Proceedings of 19th Latin American Control Congress (LACC 2022)*, Studies in Systems, Decision and Control 464,
https://doi.org/10.1007/978-3-031-26361-3_24

or processes on other associated devices through commands sent from some central node of the network (remote control).

In the development of RTUs for data acquisition, data loggers and smart meters play a fundamental role. Their design criteria must take into account as a distinctive element the way data is storaged or read. In a large number of applications, devices are designed to operate in remote and difficult access places, powered by batteries. In these cases, the storage of data recorded in local devices memories implies the need to move to their location for reading. In order to make easier and speed up the reading process, remote reading systems can be implemented to allow information to be obtained remotely. In this sense, wireless communication technologies play a fundamental role.

Taking into account that most of the applications that include remote reading systems for remote data acquisition conceive the existence of sensor nodes as edge devices, gateways as information concentrators and remote servers as final destination, they are in most cases related to the Internet of Things (IoT). There are many wireless communication technologies and protocols used in IoT applications, according to their range and power consumption characteristics. Overviews and comparisons of wireless technologies used for IoT are provided in [1, 2]. In [3] two general categories are defined: LPLANs (Low Power Local Area Networks) with a range of less than 1000 m (IEEE 802.15.4, IEEE P802.1ah, Bluetooth/LE, etc.) and LPWANs (Low Power Wide Area Networks) with a range greater than 1000 m (LoRa, Sigfox, DASH7, etc.).

Among all these technologies, stand out those that are special for certain applications due to their long-range possibilities and low power consumption; that is the case of technologies such as LoRa, that belongs to the group of the aforementioned LPWANs.

A review of LoRa technology is provided in [3]. It is a low-power, long-range and low transmission rate wireless communication technology designed for applications where devices need to be powered by batteries for long time periods. Its use is optimal particularly when it is necessary to transmit small volumes of data over long distances with high energy efficiency.

Based on the study of LoRa technology, the implementation of a LoRa sensor network is proposed for the monitoring of water supply systems. The network designed bases its packet forwarding method on a single-channel gateway implemented with low-cost resources. Multichannel gateways are more expensive on the market, and their implementation by own means may be a complex problem [4]. In [5] technical and economic advantages offered by an own-designed single-channel gateway solution based on the chip SX1276 are experimentally verified.

In this work it is proposed a solution with RTUs and LoRa gateway based on ESP32, whose main goal is their simple design that make devices to be low-cost and low-power consumers. It was applied in the Technological University of Havana "José Antonio Echeverría" (CUJAE) water supply system.

## 2 LoRa and ESP32

LoRa technology uses Semtech's proprietary Chirp Spread Spectrum (CSS) modulation with Forward Error Correction (FEC) channel coding. This modulation used by LoRa makes it immune to the Doppler Effect and suppresses the need for high-precision crystal oscillators embedded in the systems to synchronize signal transmissions. This way the manufacturing cost of both transmitters and receivers can be reduced.

LoRa specification includes a well-defined frame structure, which is used by modulators/demodulators manufactured by Semtech to operate. The frame structure is composed of a preamble, an optional header, a 255-byte payload and an optional CRC at the end of the frame.

Sometimes the term LoRa is used when referring to the LoRaWAN standard, a link protocol based on LoRa physical layer that has become the most popular way to communicate edge devices with gateways through LoRa, due to its condition of free standard. However, the terms LoRa and LoRaWAN should not be confused. Implementations for LoRaWAN-based applications are described in [6, 7].

LoRaWAN is an ALOHA-type MAC link protocol developed by the LoRa Alliance to standardize the use of the LoRa physical layer between devices. It uses several fields for carrying control information: control flags, commands defined in the standard itself, addresses assigned to devices and complementary data used by the protocol to implement its communication algorithms [3]. In certain applications, due to their simplicity, the use of so many control fields may become unnecessary, especially if data payload to be transmitted per frame does not exceed the order of a few bytes.

The proposed solution does not include the use of LoRaWAN; in exchange, it uses LoRa physical layer assuming an own implemented protocol. There is evidence of other LoRa networks implementations without the use of LoRaWAN looking precisely for flexibility in the management of the technology [8], even for leaks detection and location in water distribution systems applications [9].

Not using LoRaWAN in the proposed solution will reduce power consumption of network nodes, by reducing the volume of data to be transmitted in each packet and, this way, the time delay in packet transmission. In addition, due to the ALOHA medium access strategy used by LoRaWAN, strategies for synchronization between nodes must be implemented, which also has an impact on device consumption [10]. Instead of this, the proposed solution implements a medium access strategy by Carrier Sense Multiple Access (CSMA) through Channel Activity Detection (CAD) mechanisms available to the LoRa modems used, which eliminates the need for synchronization between nodes.

According to the goals pursued with the solution proposed, ESP32 comes as the most suitable platform to use. It has great facilities for the implementation of low power consumption strategies along with high processing capability and embedded peripherals for wireless communication. This allows to use ESP32 in the design of edge devices as well as middle-end devices in a wireless sensor network. Other platforms do not possess all these features combined, some of which can be briefly mentioned.

Arduino, in spite of being maybe the most popular platform in the development community, has great limitations in peripherals as well as in power consumption. Another popular alternative are single board computers (SBC), which have great computational power, but their considerable power consumption makes them inappropriate for the application. Other microcontrollers-based and embedded platforms have similar limitations in terms of power consumption or processing capability. MSP430 seems to be the most attractive competitor in this sense; however, its lack of peripherals dedicated to wireless communication makes ESP32 the appropriate candidate. Moreover, many commercial ESP32 development boards include LoRa chips and antennas in a single board, which makes easier the development of LoRa solutions.

## 3 Design of the Sensor Network

The sensor network proposed consists of several nodes whose primary objective is the monitoring of water meter devices located at different points of the hydraulic network. This way it will be possible to asses water consumption in the water supply system.

Some basic conditions must be met in order to ensure the success of the solution designed. RTUs must be located within a ratio of 5 km from the gateway to make LoRa communication possible with commercial antennas attached to development boards. At the same time, gateway must be located next to a Wi-Fi Access Point connected to the network where the server is to be hosted. Water meter devices where RTUs are meant to work on, must possess an electrical output or a mechanical device that allows the generation of a signal through the coupling of a sensor element to measure water consumption. Regarding future work, water pipes must also have at these points entrances needed for pressure measurement as well.

### *3.1 Proposed Network Architecture*

The proposed network architecture is shown in Fig. 1. Sensor nodes will operate physically located next to the water meters to obtain data. These will periodically obtain water consumption in fixed time intervals, and will transmit data through LoRa to a single-channel gateway. The latter will be in charge of decoding the information and sending it through a TCP/IP network via Wi-Fi to an SQL database in a web server.

Unlike the original conception of LoRaWAN, in this case, gateway is not a simple intermediary. It must receive packets coming from several nodes, carry out the handshaking exchange with them to ensure the success of communication and process

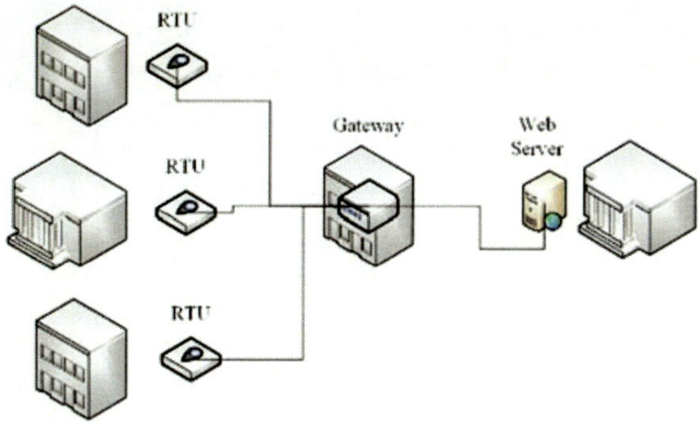


**Fig. 1** Proposed architecture for LoRa sensor network in a water supply system

packets data to be sent to the server. At this point, an own communication protocol algorithm (explained in detail in later sections) between the nodes and gateway is proposed.

Gateway used is single-channel, assuming the limitations this implies. However, this does not affect the operation of the network as long as the nodes transmit at different times, ensuring that transmissions do not concur or implementing recovery mechanisms in the event that this occurs.

## *3.2 Hardware and Operating Principle of the RTU*

ESP32 based designed RTUs base their operation on a conception of very low power consumption, exploiting sleep modes of the ESP32 and the associated resources, particularly the ULP coprocessor able to maintain its operation when the main processor is in deep sleep mode, which allows monitoring external events at a very low energy cost.

RTUs receive an electrical pulse train output provided by modern water meters. In the testing scenario, these were equipped with a Cyble device, which provided them with the pulse train output. With the use of Cyble, pulses were generated through the activation of a potential-free contact due to inductance variations in a resonant circuit, induced by the passing of water through the meter. This way, the Cyble generates a pulse for each unit of volume that passes through the pipe, and ESP32 must count the number of pulses received to determine the total volume of water that passed through the pipe in a time window.

A real time clock DS3231 is used to keep the time base of the events. Its alarms are used to delimit time windows in which consumption will be measured. Time windows of one minute are delimited. Therefore, every minute, DS3231 alarm will activate, indicating the active time window has ended. At this time ESP32 will read a pulse counter implemented by software and obtain the average flow rate per minute of measurement, which can be translated into water consumption, this being the data to be transmitted by LoRa to the gateway.

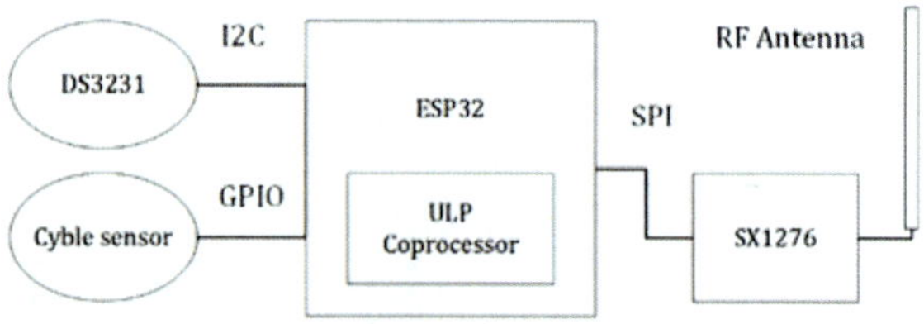


**Fig. 2** Functional block diagram of a terminal unit

To minimize power consumption in RTUs, ESP32 program algorithm sets the initial conditions and goes into deep sleep mode. From that point, the counting of pulses will be carried out by the ULP coprocessor through polls to the pin that must be connected to the Cyble sensor output. The coprocessor will also attend, by polling too, DS3231 alarm pin, and in case of detecting a low-level transition, it will wake up the main processor from deep sleep mode so that it can read the value of the pulse count variable, transmit it by LoRa to the gateway and return to deep sleep; ULP coprocessor, after waking up the main core, will communicate with DS3231 by $I^2C$ to reset the alarm flag and open the new time window of one minute.

The pins used for DS3231 alarm connection and the Cyble output must be RTC GPIOs, so that they remain active in deep sleep mode. These must be attended by polling due to the lack of interrupt mechanisms in the ULP coprocessor. Software debounce strategies were implemented. $I^2C$ communication the coprocessor must establish with DS3231 to disable the alarm flag must be implemented by bitbanging, since the coprocessor does not have functional hardware that allows $I^2C$ to be implemented through the corresponding peripheral.

For LoRa communication, Semtech's SX1276 integrated circuit is proposed. LoRa communication can be implemented by writing via SPI into the corresponding registers to configure the transmitter appropriately. Figure 2 illustrates the structure of the terminal node in a functional block diagram.

According to RTUs hardware needs, the use of ESP32 Heltec Wi-Fi LoRa 32 V2 commercial development kit is proposed as a first approach to test the model. This kit integrates the ESP32 and the SX1276 LoRa chip in the same development board, leaving only the Cyble sensor and DS3231 RTC as external components in the hardware design.

### *3.3 Hardware and Operating Principle of the Single-Channel Gateway*

For the design of a single-channel gateway that receives data coming from sensor nodes and forwards them through Wi-Fi to an external server, a LoRa communication system is required, as in the nodes, but in this case, it also demands hardware for Wi-Fi transmission. In this context, ESP32 becomes even more attractive to perform the task, given its embedded peripherals for Wi-Fi communication.

The use of the same ESP32 Heltec Wi-Fi LoRa 32 V2 development kit, which has all necessary hardware resources, is proposed to implement the gateway. This turns gateway into a low-cost device based on commercial components.

ESP32 software must be properly designed so that it configures SX1276 to act as a packet receiver device and implements the exchange with transmitters. The algorithmic conception of the ESP32 program is based on continuously monitoring SX1276 packet reception flag and, when it activates, reading the packet via SPI by accessing to the corresponding registers, validate the packet and being in the presence of a valid one, package its payload into an https packet to send via Wi-Fi over the network to the web server.

Unlike nodes, gateway operation will not be limited in consumption, as it must be located in a safe place where it can receive power directly from the electrical network.

### 3.4 Design of a Communication Protocol

To achieve communication between nodes and the gateway with error recovery strategies, a communication protocol was designed. The protocol needs every node in the network to be assigned a unique identifier (ID). The gateway stores in memory the IDs of all network nodes, having its own ID.

Gateway is constantly checking SX1276 packet reception flags to act when the arrival of a packet occurs. When a node has information to transmit, it activates its channel activity detection mechanism (CAD) and when it stops detecting channel activity, it sends data in a LoRa packet, instantly configuring itself as a receiver for a certain time to wait for a confirmation of receipt that must be sent by the gateway. The payload structure of packets sent by nodes is described below, also shown in Fig. 3:

Byte 0: Node ID
Byte 1: Control Key ($0 \times 67$)
Byte 2: Control Key ($0 \times 73$)
Byte 3: Control Key ($0 \times 72$)
Bytes 4–5: Payload (16-bit value of pulse count variable).

Node ID is used by the gateway to identify the transmitting source whereas bytes contained in the control key are used as a data validation mechanism by software in addition to the intrinsic SX1276 FEC hardware mechanism.

| Node ID | Control Key (0x67) | Control Key (0x73) | Control Key (0x72) | Total of pulses (H) | Total of pulses (L) |
|---|---|---|---|---|---|

**Fig. 3** Payload structure for LoRa packets transmitted by data-logger nodes

**Fig. 4** Structure of the ACK packet payload transmitted by the gateway

| Gateway ID | Control Key (0x67) | Control Key (0x73) | Control Key (0x72) | ACK Value (0x01) |
|---|---|---|---|---|

When the gateway receives a packet and checks if it is valid, it extracts data contained in its payload and sends a 5-byte payload LoRa ACK packet to the node. Then it sends received data via Wi-Fi to a SQL database on the server using https as the application layer protocol. The structure of ACK packets is described below and shown in Fig. 4:

Byte 0: Gateway ID
Byte 1: Control key (0 × 67)
Byte 2: Control key (0 × 73)
Byte 3: Control key (0 × 72)
Byte 4: ACK value (0 × 01).

Designed algorithm includes delays at various moments of the communication process in order to guarantee synchronism, as well as a failure recovery strategy based on repeating packets transmission a finite number of times while ACK packets are not received.

## 4 Experimentation

Once the LoRa gateway and RTUs were ready to use, a virtual LAMP server was installed in one of the university network servers. A series of tests were carried out inside the university campus to verify the effectiveness of the communication protocol designed, as well as the reach and security of transmissions. The tests yielded satisfactory results. Figure 5 shows a fragment of a table with data extracted from the database and displayed in an html file in the browser of a PC that accesses the web server through the network.

These records correspond to a test in which the system operation was simulated by transmitting a constant value from a node at random instants of time with the aim of testing the communication channel and the protocol designed. By receiving the known value transmitted it was ensured that data sent from an RTU was correctly received by the gateway and redirected to the web server with no errors.

ESP32 Heltec Wi-Fi LoRa 32 V2 price is currently about 35,00 USD, whereas DS3231 price is about 8,00 USD. This makes every RTU to have an estimated price of 43,00 USD, and only 35,00 USD for the gateway, which represents a low cost as commercial gateways price is about 1000,00 USD.

RTUs make potential use of ESP32 deep sleep operation mode, by waking up just once a minute. The wake-up process imposes a delay of about 200 ms before entering the main program. Time delay of the routine for a successful LoRa transmission, measured using one of ESP32 64-bit high resolution hardware timers, is about 150 ms

**Fig. 5** Fragment of table with constant values sent from a node to the server

| id | nodeID | value | reading_time |
|---|---|---|---|
| 45 | 1 | 45 | 2022-03-08 17:56:23 |
| 44 | 1 | 45 | 2022-03-08 17:50:32 |
| 43 | 1 | 45 | 2022-03-08 17:50:24 |
| 42 | 1 | 45 | 2022-03-08 17:44:47 |
| 41 | 1 | 45 | 2022-03-08 17:44:33 |
| 40 | 1 | 45 | 2022-03-08 17:43:34 |
| 39 | 1 | 45 | 2022-03-08 17:43:24 |
| 38 | 1 | 45 | 2022-03-08 17:41:45 |
| 37 | 1 | 45 | 2022-03-08 17:41:36 |

before ESP32 returns into deep sleep. Time out of deep sleep mode may then be calculated to 350 ms, a result that shows dynamic consumption is insignificant in a one-minute time window.

Current consumption of ESP32 Heltec Wi-Fi LoRa 32 V2 board was measured during a transmission. To reduce ESP32 power consumption, CPU frequency was reduced to 80 MHz and single core mode was used. With the original board, peaks of 77 mA were registered in this operation mode, whereas in deep sleep mode a static consumption of 19.65 mA was registered. ESP32 consumes only about 150 uA in deep sleep mode when using the ULP coprocessor, and SX1276 remains also in sleep mode when not transmitting, with a consumption of 0.2 uA. The rest of power is consumed by the board components, as ESP32 Heltec Wi-Fi LoRa 32 V2 is not a board optimized for low power operation. However, static consumption of the board can be reduced to 2 mA by physically removing the unused components. Under these conditions, lifetime of a 20 Ah battery may be estimated to be 257 days.

Other boards with low power design could be used instead, but the absence of a LoRa chip embedded in them forces to add LoRa as an external module. However, this alternative will be evaluated in future work. The use of FireBeetle ESP32 kit along with an SX1276 independent module seems to be the best possible choice as FireBeetle is a board optimized for low power consumption, reaching only 0.2 mA in deep sleep mode. With this hardware components, power consumption of RTUs would be drastically reduced. Both FireBeetle boards and SX1276 modules prices are about 10,00 USD, so the principle of being low-cost would maintain.

## 5 Conclusions

With the implementation of the proposed sensor network, a low-cost and low-consumption wireless communication system is available. The implemented single-channel gateway is capable of serving as many nodes as needed as long as temporary independence is ensured in transmissions. It is very useful having a device capable of solving a problem whose solution would have been expensive to acquire commercially. In addition, most commercial gateways operate over LoRaWAN, which also contrasts with the objectives that were sought.

The development of this solution allowed to evaluate the performance of LoRa communication without being subject to complex protocols. It was verified that low power consumption alternatives may be found in ESP32 development boards.

Having a real time centralized database of a variable on a web server provides all the advantages to obtain patterns of the variable being measured and predict its curve behavior, which can be remarkable for making technical and/or administrative decisions. It will also contribute to faster fault diagnosis and a more effective reduction of the effects that these can cause in water services, which translates into a higher quality of service and a reduction in losses of this valuable resource.

As future work, it will be added to RTUs the possibility of pressure measurement for wireless transmission, pressure being the most important magnitude to monitor in order to perfection leaks detection mechanisms. The use of different hardware platforms that provide lower consumption alternatives will also be considered.

## References

1. Nolan, K.E., Guibene, W., Kelly, M.Y.: An evaluation of low power wide area network technologies for the Internet of Things. In: 2016 International Wireless Communications and Mobile Computing Conference (IWCMC), pp. 439–444 (2016)
2. Ikpehai, A., Adebisi, B., Rabie, K.M., Anoh, K., Ande, R.E., Hammoudeh, M., Gacanin, H., Mbanaso, U.M.: Low-power wide area network technologies for Internet-of-Things: a comparative review. IEEE Internet Things J 1 (2018)
3. Augustin, A., Yi, J., Clausen, T., Townsley, W.: A study of LoRa: long range & low power networks for the Internet of Things. Sensors **16**, 1466 (2016). https://doi.org/10.3390/s16091466
4. Zhou, Q., Zheng, K., Hou, L., Xing, J., Xu, R.: Design and implementation of open LoRa for IoT. IEEE Access **7**, 100649–100657 (2019)
5. Jovalekic, N., Drndarevic, V., Darby, I., Zennaro, M., Pietrosemoli, E., Ricciato, F.: LoRa transceiver with improved characteristics. IEEE Wirel. Commun. Lett. 1 (2018)
6. Paredes-Parra, J.M., García-Sánchez, A.J., Mateo-Aroca, A., Molina-Garcia, A.: An alternative Internet-of-Things solution based on LoRa for PV power plants: data monitoring and management. Energies **12**, 881 (2019)
7. Lin, S., Ying, Z., Zheng, K.: Design and implementation of location and activity monitoring system based on LoRa (2019). ArXiv190201947Cs
8. Liu, S., Xia, C., Zhao, Z.: A low-power real-time air quality monitoring system using LPWAN based on LoRa. In: 2016 13th IEEE International Conference on Solid-State and Integrated Circuit Technology (ICSICT), pp. 379–381 (2016)

9. Cattani, M., Boano, C.A., Steffelbauer, D., Kaltenbacher, S., Günther, M., Römer, K., Fuchs-Hanusch, D., Horn, M.: Adige: an efficient smart water network based on long-range wireless technology. In: Proceedings of the 3rd International Workshop on Cyber-Physical Systems for Smart Water Networks. ACM, New York, NY, USA, pp. 3–6 (2017)
10. Rizzi, M., Depari, A., Ferrari, P., Flammini, A., Rinaldi, S., Sisinni, E.: Synchronization uncertainty versus power efficiency in LoRaWAN networks. IEEE Trans Instrum Meas **68**, 1101–1111 (2019). https://doi.org/10.1109/TIM.2018.2859639

# Low-Cost Portable System for the Estimation of Air Quality

**Erik Hernandez-Rodriguez, David Kairúz-Cabrera, Alain Martinez, Rosa Amalia González-Rivero, and Olivier Schalm**

**Abstract** Air pollution is currently considered one of the main problems facing mankind. Traditional systems that measure air quality variables are expensive and their access is mostly limited to wealthy regions. Low- and middle-income countries often cannot access this equipment because they do not have the resources to afford it. Alternative methodologies involving the use of low-cost sensors in air quality monitoring systems have therefore emerged in response to both academic and citizen concerns. The main issue regarding the use of low-cost sensors is the quality of the data. This paper presents the design of a portable low-cost concept prototype of a monitoring system, capable of estimating air quality from data with a minimum assured quality for decision-making. The system passed a series of low-cost calibration processes in order to assure the quality of the measurements provided by the selected sensors. An attempt has been made to achieve a product that is as cheap as possible but capable of delivering reliable measurements. The collected data during operation is shown in a 52North IoT platform with geospatial reference. The concentrations measured in the routes through the city of Santa Clara show concentrations close to the risk for health.

E. Hernandez-Rodriguez · D. Kairúz-Cabrera · A. Martinez (✉) · R. A. González-Rivero
Universidad Central "Marta Abreu" de Las Villas, Santa Clara, Cuba
e-mail: amguardia@uclv.edu.cu

E. Hernandez-Rodriguez
e-mail: ehrodriguez@uclv.cu

D. Kairúz-Cabrera
e-mail: dkairuz@uclv.cu

R. A. González-Rivero
e-mail: rogrivero@uclv.cu

O. Schalm
Antwerp Maritime Academy, Antwerp, Belgium
e-mail: olivier.schalm@hzs.be



O. Llanes-Santiago (ed.), *Proceedings of 19th Latin American Control Congress (LACC 2022)*, Studies in Systems, Decision and Control 464,
https://doi.org/10.1007/978-3-031-26361-3_25

## 1 Introduction

Air pollution is contemplated as an international issue with several dimensions, mainly in health and environmental aspects [15]. Low and middle incomes countries report most of the deaths related with low air quality in the world [16]. These nations often lack the necessary resources to measure the pollutants present in the air with an adequate surface coverage. The 2030 Agenda for Sustainable Development includes air pollution as the fifth risk factor (globally recognized as significant) associated with the risk of death from non-communicable diseases [19]. In Cuba, studies have been carried out that demonstrate the relationship between pollution and the impact on human health [4].

Systems dedicated to air quality monitoring and surveillance consists mainly on expensive mobile or fixed stations [3]. These structures require quality control checks, frequent calibration and qualified professionals to be operated. Such methods offer high quality data, however, sometimes such level of accuracy are not required for decision making [2]. That is the reason why companies like AQMesh, Envirowatch, or Libelium have created systems that allow the surveillance of air quality with lower costs than traditional stations, but still not reachable for some low and middle income countries [1, 20]. The scientific community has increased interest in low-cost air pollution technologies and as result, many academic research projects and citizen science projects have been conducted using low-cost sensors (LCS) [7, 13, 29].

Advances in LCS and the availability of electronic components have made it possible to build low-cost measurement systems in different parts of the world. These alternative methodologies answer claims for both scientific questions and end-user applications providing a wealth of information [26]. The remaining issue concerning low cost sensing procedures is the quality of the data when comparing with reference devices [14, 23].

Engineering design is a problem-solving approach and a systematic and iterative process. It includes steps like the improvement of existing designs under constraints such as size, weight, communications possibility, costs, and other limitations [12]. The right design for portable devices ensures real civilian engagement and reliable data collection for geospatial analysis of individual exposure [22].

Literature regarding the design and use-cases of air pollution surveillance systems using LCS and open-source hardware and software consists of several sensors interfaced with one or more computing elements [9–11, 30]. It is reflected that most of the deployed systems use Raspberry Pi, Arduino, and NodeMCU as a computing elements, however, advances in technologies and integration provide new possibilities. Specifically designed boards appear as a solution to many engineering problems and have been used also in air quality monitoring systems [17, 27]. These devices include several components in a single development board making it easier to get portability and compactness. Studies also claim that major primary air pollutants include gaseous pollutants: sulfur dioxide ($SO_2$), nitrogen dioxide ($NO_2$), carbon monoxide (CO), ozone ($O_3$), and volatile organic compounds ($VOCs$) and particulate matter ($PM$) [28].

According to the aforementioned, this paper presents the design of a portable low-cost prototype capable of making reliable measurements of some of the major primary air pollutants. The system passed a series of low-cost calibration processes and the collected data is shown in an IoT platform. The objective is to develop a concept prototype of a low-cost monitoring system capable of estimating air quality from data with a minimum assured quality for decision-making.

## 2 Materials and Methods

This design is a direct result of a long term research carried out in the Universidad Central "Marta Abreu" de Las Villas with the purpose of measuring air quality. Three systems have been builded using low-cost elements, one fixed (HZS-GARP-AQ-03 [24] and two mobile (HZS-GARP-AQ-01-02) [18, 25]. The new prototype is based on the experiences of the previous designs. The structure has improved features such as size, number of measured variables, and power consumption.

### 2.1 *Hardware Description*

The system, named as HZS-GARP-AQ-04, is based on the LilyGo TTGO T-Beam T22 V1.1 development board. This board incorporates several functionalities in a single package that were previously distributed in different separated layouts. This device contains: GPS NEO-M8N, storage via PSRAM, LORA communication (868 MHz), as well as an ESP32 module that provides processing and Wi-Fi/Bluetooth communication. In this way, the tasks accomplished in previous designs by the Arduino Mega, ESP01, GPS NEO-6M/NEO-8M, and a data logging shield are now integrated into an small form factor new system, reducing volume, energy consumption and increase reliability. The resolution of the integrated analog-to-digital converter (ADC) is 12 bits, this surpass the need for extra 16 bits ADC, in order to measure analog voltages from gas sensors accurately (normally 1 mV by ppb). Figure 1 shows a general scheme of how the main hardware elements of the system interact and a picture of it. A customized shield is included on top of the TTGO T-Beam, to interconnect the different sensors.

Table 1 presents the sensors integrated into the system, they have been selected after performance versus price evaluation across several tests during research. NOVA SDS011 sensor is used to measure particulate material [26]. AM-2315 sensor gives information about ambient temperature and relative humidity outside of the system container while a BME-280 provides the same information plus atmospheric pressure inside to compensate for the effect of these variables on the gas sensors measurements (details of how compensation is done can be found at [24]). The Analog Front End (AFE) provided by Alphasense accommodates 4 electrode A-series sensors for measuring of the levels of $NO_2$, $O_3$, $SO_2$, and CO. Contrary to previous designs,

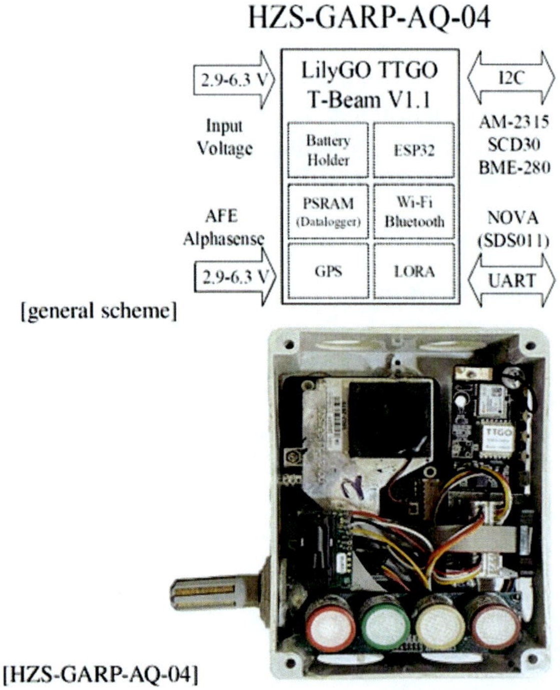


**Fig. 1** Structure of the HZS-GARP-AQ-04

**Table 1** Components of the system

| Component | Price (USD) | Energy | Characteristics |
|---|---|---|---|
| AM2315: (RH/T) | 30 | 500 μA | 0–100%/−40–80 °C |
| Sensirion SCD30: (CO2/RH/T) | 60 | 19 mA | 400–10000 ppm |
| NOVA SDS011: (PM2.5,PM10) | 28 | 80 mA | 0.0–999.9 $\mu g/m^3$ |
| OX-A431: (O3) | 51 | 650 μA | 0–20 ppm |
| SO2-A4: (SO2) | 51 | 650 μA | 0–50 ppm |
| NO2-A43F: (NO2) | 51 | 650 μA | 0–20 ppm |
| CO-AX: (CO) | 51 | 650 μA | 0–2000 ppm |
| AFE Air quality Gas Sensor | 98 | 2.5 mA | 4 sensors capability |
| BME280: (RH/T/AP) | 5 | 3.6 μA | −40–85 °C/300–1100 hPa |
| LILYGO TTGO T-Beam V1.1 | 37 | 180 mA | ESP32, GPS, LORA |
| 1x Samsung ICR18650-26 | 5.5 | – | Capacity: 3200 mAh |
| Polystyrene box | 6 | – | 176×126×57 mm |
| Total (12) | 473.5 | | |

Sensirion SCD30 replaces COZIR-A for the measurement of $CO_2$ concentration. According to the datasheets provided by the manufacturers, both sensors have similar characteristics, however, the SCD30 has a lower price (1/3 of COZIR-A). The tests carried out in the laboratory support this (Fig. 3).

Main power is provided by one Samsung 18650 lithium element. The TTGO T-beam module has a battery holder for this type of lithium element on its back that facilitates portability. The integrated AX192 power management unit on the board is responsible for supplying the appropriate voltages to each element, as well as monitoring and charging the battery. The TTGO T-Beam microcontroller and the rest of the digital sensors used go into low-power mode at times when no readings are being obtained. The analog gas sensors are directly connected to the battery through the AFE. They are not disconnected from the battery if the sampling frequency is less than 10 min. This is due to the transient instability that is generated in it for approximately 5 min after being energized.

Table 1 shows a list of prices for each element on the system, as well as its main characteristics and maximum energy consumption. The total price is less than 500 USD, taking into account the production cost of the custom shield and not including design engineering costs. This starting price is about 4 times lower than commercial LCS-based devices with similar features currently on the market.

Additionally, the system requires a group of routers distributed throughout the area to be explored. In this case, the open-source dual channels LoRa Gateway LG02-868 from the firm Dragino was selected. The indoor antennas were changed by 4.5dbi outdoor omni-directional fiberglass antennas. Indoor equipment was selected to be able to place it within public entities with wired internet connection

## 2.2 *Software Description*

The programming was carried out in the integrated development environment (IDE) version 1.8.19 provided by Arduino. Figure 2 shows the workflow used, it can be seen that it is similar to previous versions and to the one described in the specialized bibliography on the subject. The software initializes the hardware elements and checks the presence of the sensors. Sample times and variables related to both the analog and digital GPIOs are configured according to user specifications and then perform a data check. This process involves comparing the measurements with the maximum and minimum limits previously established, which makes the reading valid or not and leads to the storage of the information.

The device is connected to a 52North IoT platform using LoRa connection from the TTGO T-beam to the LG02-868 Gateways. Information is stored on internal PSRAM during operation as a backup in the case data network cannot be reached. The platform allows to access material of interest and observe geospatial and time distribution of air pollutants. Is possible to filter this information according to region, time, and pollutants as well as see graphics with air quality performance. Readings taken by the system can be displayed directly on a cell phone using the Serial Bluetooth Terminal

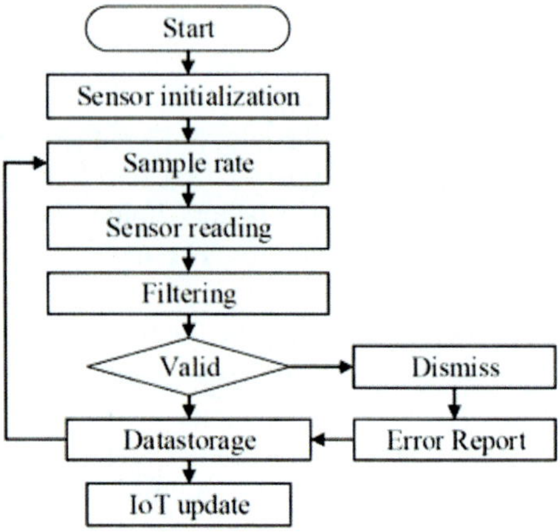


**Fig. 2** Flow diagram

application. This is done in order to obtain fast local feedback of the measurements and to check the correct operation of the device. This communication could be used to change the sampling periods without the need to reprogram the system. To save energy the connection is made via Bluetooth Low Energy (BLE) and only transmits over this channel when a device is connected.

## *2.3 Calibration*

All calibration experiments are performed in a basic equipped laboratory. The calibration of the temperature, relative humidity, and $SO_2$ was described in [18] and a similar process is applied to the rest of the sensors on this version.

The results, for the calibration process of $CO_2$ sensors: COZIR-A, and Sensirion are shown in Fig. 3. For the calibration process, both sensors were placed inside a hermetic plastic container with two fans to achieve the homogenization of the system, after that a three-stage procedure was carried out. The first step is to measure

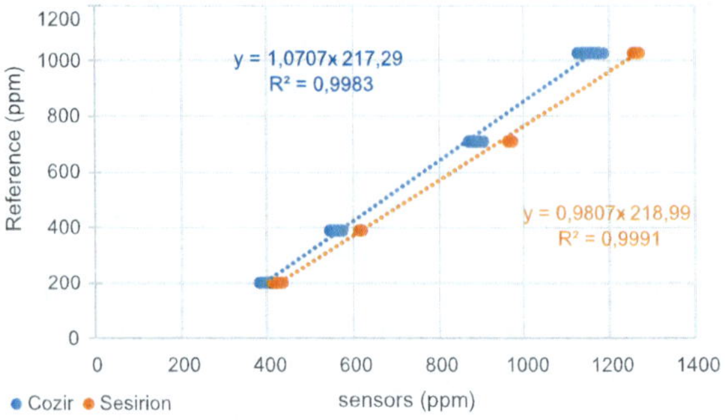


**Fig. 3** Performance comparison COZIR-A versus Sensirion

the concentration of $CO_2$ in the ambient air inside the closed calibration box. This step takes approximately 5 minutes until the sensors provide constant readings. In the second stage, the $CO_2$ concentration is removed until it reaches zero. To do this, the vacuum pump is turned on, and the air that is present inside the container is forced to pass through a glass gas scrubber containing a solution of calcium hydroxide $Ca(OH)_2$ and sodium hydroxide (NaOH). In this reaction, calcium carbonate ($CaCO_3$) is obtained as a product. The air is then passed through a trap bottle and then through a glass U-shaped tube containing silica gel. In this way, a closed system is formed, in which the $CO_2$ contained in the calibration box is gradually eliminated.

Finally, in the third stage it is necessary to generate a known quantity of the gas. The quantitative production of $CO_2$ is carried out using affordable and low-cost materials such as syringes and three-way stopcocks. Under ambient conditions (30 °C, 70% relative humidity, and an average atmospheric pressure of 1000 hPa), volumes of 1.2, 2.0, and 3.0 mL are injected into the calibration box in order to obtain concentrations of 191.39, 510.37, and 829.35 ppm of $CO_2$ respectively. Knowing the reference concentrations and the response of the sensor according to the gas injections, a relation is established by applying the mathematical model that best fits. The mathematical equations are used to correct the sensor readings (Fig. 3).

### 2.4 *Measurements Campaign*

The HZS-GARP-AQ-04 (installed in the back seat of an electric bicycle) was deployed in the city of Santa Clara to test its correct operation. Routes included some of the main roads in the city (Fig. 4), samples were taken at peak travel times both in the morning (7:00–8:00 AM) and in the afternoon (4:30–5:50 PM). The average vehicle speed was around 15 km/h with minimum speeds of 9 km/h and maximum speeds of 18 km/h according to GPS data. Momentary stops, of around two minutes, were made at strategic points in the city. As in previous works where air quality is measured using bicycles, the sampling frequency of this system was set at 20 s [8]. Therefore, at a speed of 15 km/h and with this sample rate it is possible to take a reading approximately every 80 m.

## 3 Results

Figure 5 shows the concentrations measured along the route made in the city of Santa Clara for the gases: $NO_2$, $SO_2$, and $O_3$ in the afternoon peak hours. The measured values confirm previous studies carried out throughout the country [5], which shows a low quality of the air we breathe in the most populated cities. It is worth noting that despite the high concentrations detected, they do not exceed the concentrations established by the Cuban standard NC 1020 [21], which addresses: air quality-pollutants-maximum admissible concentrations and guide values in habitable

**Fig. 4** Road traveled in the city of Santa Clara during sampling

areas. This standard establishes maximum exposure values in periods of one hour for: $SO_2 = 97$ ppb, $NO_2 = 87$ ppb and $O_3 = 77$ ppb. The concentration of ozone ($O_3$) exceeds the recommended values.

The behavior of the suspended particulate material in sizes $\leq 10\,\mu m$ and $\leq 2.5\,\mu m$ followed a distribution as shown in Fig. 5. This distribution also follows values measured in other Cuban cities with high vehicular traffic [6]. The NC 1020 standard establishes maximum exposure values of $PM_{10} = 200\,\mu g/m^3$ and $PM_{2.5} = 100\,\mu g/m^3$.

## 4 Conclusions

The main innovative aspect of this study is that lower-cost monitoring systems based on open-source hardware and software and LCS can be used in real applications by regular persons after proper calibration of the sensors included in the system.

The use of low-cost calibration procedures supported by regular equipment and proper filtering allows the creation of a product that is as cheap as possible but capable of delivering reliable measurements.

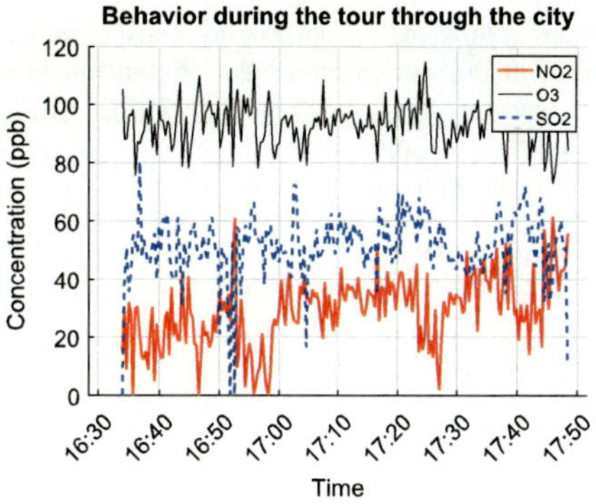


Afternoon gas concentration

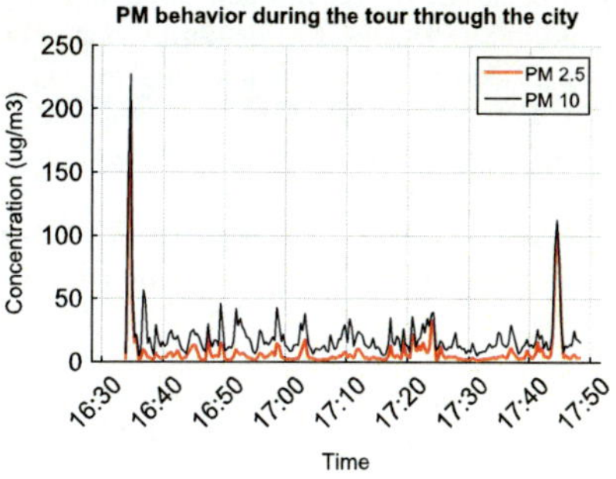


Particulate matter in the afternoon

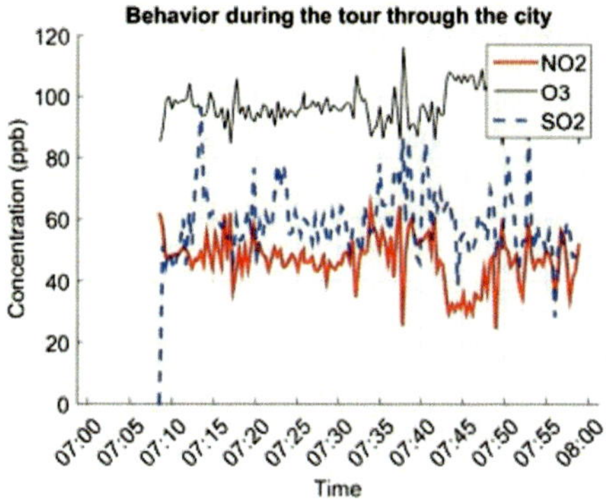


Gas concentration in the morning

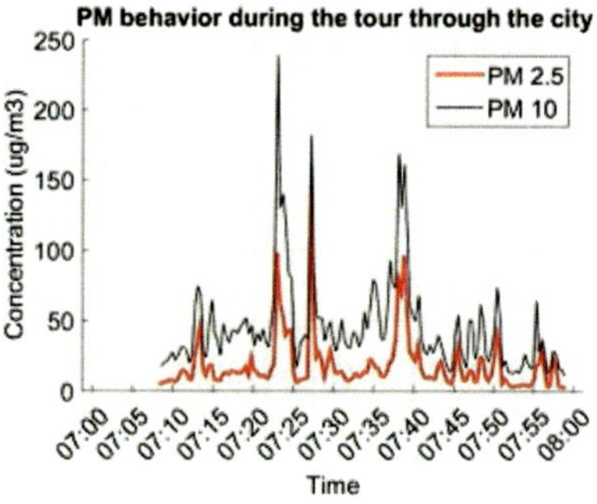


Particulate matter in the morning

**Fig. 5** Behavior of particulate material and several pollutants in the routes

The measurements in the city of Santa Clara show that the concentration of the measured parameters can be excessive for human health in some areas during peak hours.

# References

1. Artíñano, B., Narros, A., Díaz, E., Gómez, F.J., Borge, R.: Low-cost sensors for urban air quality monitoring: preliminary laboratory and in-field tests within the tecnaire-cm project. In: 2019 5th Experiment International Conference (exp. at'19), pp. 462–466. IEEE (2019)
2. Baldelli, A.: Evaluation of a low-cost multi-channel monitor for indoor air quality through a novel, low-cost, and reproducible platform. Meas.: Sensors **17**, 100059 (2021)
3. Concas, F., Mineraud, J., Lagerspetz, E., Varjonen, S., Liu, X., Puolamäki, K., Nurmi, P., Tarkoma, S.: Low-cost outdoor air quality monitoring and sensor calibration: a survey and critical analysis. ACM Trans. Sensor Netw. (TOSN) **17**(2), 1–44 (2021)

4. Cuéllar Luna, L., Maldonado Cantillo, G., Suárez Tamayo, S., del Puerto Rodríguez, A., Romero Placeres, M.: Estratificación del riesgo de muerte por enfermedades no transmisibles asociado a la contaminación ambiental en cuba. Medisan **25**(6), 1394–1410 (2021)
5. Cuesta-Santos, O., González-Jaime, Y., Sosa-Pérez, C., López-Lee, R., Bolufé-Torres, J., Reyes-Hernández, F.: La calidad del aire en la habana. actualidad. Revista Cubana de Meteorología **25**(3) (2019)
6. Cuesta-Santos, O., López-Lee, R., González-Jaime, Y., Sosa-Pérez, C., Bolufé-Torres, J., García-Valdés, E., Reyes-Hernández, F.: Comportamiento del material particulado en la habana. valores experimentales. Revista Cubana de Meteorología **26** (2020)
7. DeSouza, P., Anjomshoaa, A., Duarte, F., Kahn, R., Kumar, P., Ratti, C.: Air quality monitoring using mobile low-cost sensors mounted on trash-trucks: Methods development and lessons learned. Sustain. Cities Soc. **60**, 102239 (2020)
8. Gómez-Suárez, J., Arroyo, P., Alfonso, R., Suárez, J.I., Pinilla-Gil, E., Lozano, J.: A novel bike-mounted sensing device with cloud connectivity for dynamic air-quality monitoring by urban cyclists. Sensors **22**(3), 1272 (2022)
9. Idrees, Z., Zou, Z., Zheng, L.: Edge computing based iot architecture for low cost air pollution monitoring systems: a comprehensive system analysis, design considerations and development. Sensors **18**(9), 3021 (2018)
10. Kelechi, A.H., Alsharif, M.H., Agbaetuo, C., Ubadike, O., Aligbe, A., Uthansakul, P., Kannadasan, R., Aly, A.A.: Design of a low-cost air quality monitoring system using arduino and thingspeak. Comput. Mater. Contin **70**, 151–169 (2022)
11. Komarudin, U., Subarli, H., Najmurrokhman, A., et al.: Design and implementation of a low-cost air quality measurement instrumentation using intemet-of-things platform and cloud-based messaging service. In: 2021 1st International Conference on Electronic and Electrical Engineering and Intelligent System (ICE3IS), pp. 170–175. IEEE (2021)
12. Kotseva, I.: Engineering design-based learning in integrative stem education. Sci. Eng. Educ. **4**(1), 62–67 (2019)
13. Lim, C.C., Kim, H., Vilcassim, M.R., Thurston, G.D., Gordon, T., Chen, L.C., Lee, K., Heimbinder, M., Kim, S.Y.: Mapping urban air quality using mobile sampling with low-cost sensors and machine learning in Seoul, South Korea. Environ. Int. **131**, 105022 (2019)
14. Liu, X., Jayaratne, R., Thai, P., Kuhn, T., Zing, I., Christensen, B., Lamont, R., Dunbabin, M., Zhu, S., Gao, J., et al.: Low-cost sensors as an alternative for long-term air quality monitoring. Environ. Res. **185**, 109438 (2020)
15. Manisalidis, I., Stavropoulou, E., Stavropoulos, A., Bezirtzoglou, E.: Environmental and health impacts of air pollution: a review. Frontiers in Public Health, p. 14 (2020)
16. Mannucci, P.M., Franchini, M.: Health effects of ambient air pollution in developing countries. Int. J. Environ. Res. Public Health **14**(9), 1048 (2017)
17. Marques, G., Miranda, N., Kumar Bhoi, A., Garcia-Zapirain, B., Hamrioui, S., de la Torre Díez, I.: Internet of things and enhanced living environments: measuring and mapping air quality using cyber-physical systems and mobile computing technologies. Sensors **20**(3), 720 (2020)
18. Martinez, A., Rodríguez-Hernández, E., Hernandez, L., Schalm, O.: Design of a low-cost system for the measurement of variables associated with air quality (preprint). IEEE Embedded Systems Letters (ESL) (2022)
19. Meghji, J., Mortimer, K., Agusti, A., Allwood, B.W., Asher, I., Bateman, E.D., Bissell, K., Bolton, C.E., Bush, A., Celli, B., et al.: Improving lung health in low-income and middle-income countries: from challenges to solutions. The Lancet **397**(10277), 928–940 (2021)
20. Morawska, L., Thai, P.K., Liu, X., Asumadu-Sakyi, A., Ayoko, G., Bartonova, A., Bedini, A., Chai, F., Christensen, B., Dunbabin, M., et al.: Applications of low-cost sensing technologies for air quality monitoring and exposure assessment: how far have they gone? Environ. Int. **116**, 286–299 (2018)
21. Oficina.Nacional.de.Normalización: Nc: 1020 (2014): Calidad del aire - contaminantes - concentraciones máximas admisibles y valores guías en zonas habitables. (2014)
22. Park, Y.M., Sousan, S., Streuber, D., Zhao, K.: Geoair-a novel portable, gps-enabled, low-cost air-pollution sensor: design strategies to facilitate citizen science research and geospatial assessments of personal exposure. Sensors **21**(11), 3761 (2021)

23. Rai, A.C., Kumar, P., Pilla, F., Skouloudis, A.N., Di Sabatino, S., Ratti, C., Yasar, A., Rickerby, D.: End-user perspective of low-cost sensors for outdoor air pollution monitoring. Sci. Total Environ. **607**, 691–705 (2017)
24. Rodríguez-Hernández, E.: Diseño de un sistema de medición y monitoreo de variables asociadas a calidad del aire. Msc., Universidad Central "Marta Abreu" de Las Villas (2020)
25. Rodríguez-Hernández, E., Schalm, O., Martínez, A.: Development of a low-cost measuring system for the monitoring of environmental parameters that affect air quality for human health. ITEGAM-JETIA **6**(22), 22–27 (2020)
26. Sá, J.P., Alvim-Ferraz, M.C.M., Martins, F.G., Sousa, S.I.: Application of the low-cost sensing technology for indoor air quality monitoring: a review. Environ. Technol. & Innov. 102551 (2022)
27. Sendra, S., Garcia-Navas, J.L., Romero-Diaz, P., Lloret, J.: Collaborative lora-based sensor network for pollution monitoring in smart cities. In: 2019 Fourth International Conference on Fog and Mobile Edge Computing (FMEC), pp. 318–323. IEEE (2019)
28. Turner, M.C., Andersen, Z.J., Baccarelli, A., Diver, W.R., Gapstur, S.M., Pope III, C.A., Prada, D., Samet, J., Thurston, G., Cohen, A.: Outdoor air pollution and cancer: an overview of the current evidence and public health recommendations. CA: Cancer J. Clini. **70**(6), 460–479 (2020)
29. Wang, S., Ma, Y., Wang, Z., Wang, L., Chi, X., Ding, A., Yao, M., Li, Y., Li, Q., Wu, M., et al.: Mobile monitoring of urban air quality at high spatial resolution by low-cost sensors: impacts of covid-19 pandemic lockdown. Atmos. Chem. Phys. **21**(9), 7199–7215 (2021)
30. Zhang, H., Srinivasan, R., Ganesan, V.: Low cost, multi-pollutant sensing system using raspberry pi for indoor air quality monitoring. Sustainability **13**(1), 370 (2021)

# “José Martí” International Airport Runway Lights System Control and Monitoring

**Dayra Ivette Saurí Vázquez and Ariel Omar Cepero Díaz**

**Abstract** This paper presents the realization of the remote control and monitoring of the “José Martí” International Airport runway lights system. This system is conformed by 20 lights circuits connected into 2 substations. There is a PLC in each substation that communicates by Modbus TCP protocol with 2 redundant industrial computers located in the air traffic control tower (ACT) and with a computer located in the aerodrome maintenance personnel’s office. In the ACT it is executed a software to control remotely the activation, at the brightness level required for landing and takeoff operations, of the constant current regulators (CCR) that feed each circuit of the runway lights system, as well as, to display the status of the system. In the aerodrome maintenance personnel’s computer, it is executed a software to monitor the operative status of the CCR and to record historically the status events of the regulators, both principal and reserves, for each circuit, as well as to display a synoptic representation of the circuits that are operating. This software for monitoring with maintenance assistance objective is a novel element for the “José Martí” International Airport, that allows a rapid response to breakdowns, providing better guidance to the technicians for the maintenance of the runway lights system, and avoided to acquire a costly control and monitoring system that is offered by the regulators manufacturers themselves. It is also able to work with various CCR models.

**Keywords** PLC · HMI · CCR (Current Constant Regulator) · ACT · PAPI · Runway edge lights · Low approaching lights · High approaching lights

---

D. I. Saurí Vázquez (✉) · A. O. Cepero Díaz
José Martí” International Airport, Havana, Cuba
e-mail: dayra.sauri@hav.ecasa.avianet.cu

A. O. Cepero Díaz
e-mail: ariel.cepero@ecasa.avianet.cu



O. Llanes-Santiago (ed.), *Proceedings of 19th Latin American Control Congress (LACC 2022)*, Studies in Systems, Decision and Control 464,
https://doi.org/10.1007/978-3-031-26361-3_26

# 1 Introduction

The runway lights system of an aerodrome consists of a system of lights that offer a visual help to the pilots of the aircraft in the operations of approach to the aerodrome, landing, takeoff and movement along the runway or taxiways (TW). This system can be divided into subsystems of approach lights, runway edge and threshold lights, precision approach path indicator lights (PAPI) and taxiway lights; obeying international standards that specify its configuration, color and intensity [4, 5]. The lights of each subsystem are grouped into circuits and the current that circulates through them is kept constant with the use of constant current regulators (CCR), which can be operated manually or remotely, providing different normalized levels of electrical intensity. The operation of this system is generally controlled and monitored remotely from the air traffic control tower (ACT), which remain in contact as well with the aircraft pilots [3, 7].

The remote control and monitoring of the runway lights system at the "José Martí" International Airport (AIJM, from the initials in Spanish) presented limitations, because it did not take into account confirmation signals from the regulators for the surveillance and control of this process and, on the other hand, it did not record failure events or abnormal conditions that constitute alarms and that can be used to carry out maintenance on regulators and runway lights system. Therefore, it became necessary to develop a new control and monitoring system that would allow a better management of the actions carried out in the light signaling systems for guiding the operations. This new control and monitoring system has two visual interfaces, developed in high-level programming language C# in Microsoft Visual Studio [6], one for interaction with air traffic control operators and another for interaction with runway lights system maintenance operators. In addition, it was necessary to make changes to the connection of the control signals between the CCR and the control panel in each of the substations, located at both heads of the runway.

## *1.1 MCR³ Constant Current Regulator*

The MCR³ is a constant current regulator controlled by microprocessor based on consecutively spliced thyristors that adjust the current supply for the output transformer, produced and commercialized by the Belgian company ADB Safegate [1, 2]. A digital regulator determines the trip angle of the thyristors to adjust the output current to a reference value, based on the selected brightness level. It can have optional printed circuit boards for ground fault and lamp failure detection, and different remote control modes, as well as built-in circuit selectors. The operating logic is carried out through the interconnection between the different boards, which are presented below:

- Local Master Circuit (LMC), PCB 1513, for control of communication through the local bus. This card is related to the multiway interface, the JBUS control and software configuration with the computer.

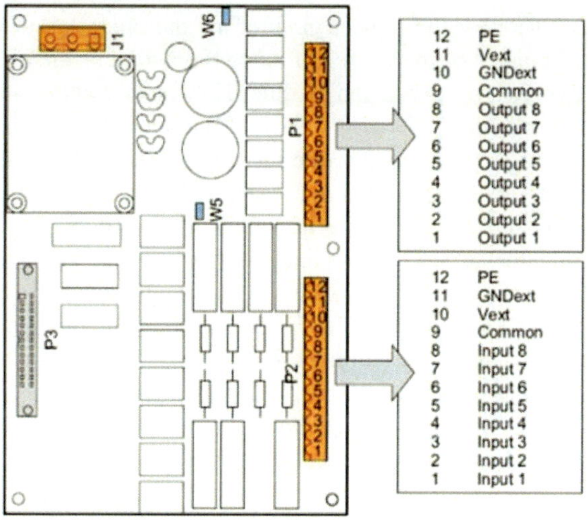


**Fig. 1** MW PCB 1486

- Thyristor Block Module (TBM), PCB 1517, produces the trigger signals of the gate of these to obtain the proper conduction angle. It also has overcurrent protection and asymmetric voltage output monitoring, as well as monitoring at the power transformer input.
- Current Control Logic (CCL), PCB 1516, compares the current demanded and obtained for its adjustment by means of another conduction angle requested from TBM, shows the value of the measurement obtained and cuts the current in case of an overcurrent.

The user interface, UI PCB 1507, indicates the output voltage (Vout), the input voltage (Vin), the nominal current (In), the output current (Iout), the operation mode (remote or local) of the regulator, the earth fault (EFD) and the lamp fault (LFD) if it has this option. The multi-wire (MW) interface, PCB1486, shown in Fig. 1, provides control and response indication signals for basic remote control functions between the CCR and the remote control system. Each multiwire card has 8 inputs and 8 outputs terminals, managing the signals between the regulator and the remote control system. It can use an internal or an external source of power supply of 24 or 48 V DC, and the configuration of the inputs/outputs is set by default from the software. The input control signals go to connector P2 and output confirmation signals go to connector P1.

## 2 Control and Monitoring of the Runway Lights System at the "José Martí" International Airport

In the ACT of the AIJM there are two Magelis HMIPPH6A2701 industrial computers in redundancy with a basic Supervisory Control and Data Acquisition (SCADA) software for the remote monitoring and operation of the runway lights circuits. In the control panel of each of the two substations there is a TM221CE40R PLC and a TM3DQ16R expansion module. They receive at the entrances the signals confirming the activation of every CCR and the activation of the taxiways from an auxiliary panel. The outputs activate the relay coils, which establish the 5 brightness levels necessary for the respective systems with a voltage of 48 V DC to the input connector P2 of the multiwire interface PCB1486 of the CCR. The PLC and their expansion modules, as well as the industrial computers, are produced by Schneider Electric. The programming software used to develop the logic of the PLC is the EcoStruxure Machine Expert Basic [8]. The status of the input signals and the orders of the output signals are stored in memory words of the PLC for being read and written by the industrial computers through an Ethernet network using the Modbus TCP protocol. This network is achieved through a switch located in the control panel of each of the substations, connecting each substation with the ACT. The Fig. 2 shows a diagram of the connection between the ACT, the substations and the runway lights system different circuits.

As part of the design for the new control and monitoring of the runway lights system, the previous electrical panel of both substations had to be modified to ensure the confirmation of turn on of all the CCR, both the principal ones and the reserve ones. The confirmation of turn on from the CCR that supplies the taxiway circuits is combined with the confirmation of short circuit from the auxiliary panel to determine with certainty the turn on or turn off of each taxiway circuit. Due to the reserve CCR for the runway edge can be used in any of the 2 substations, is necessary the confirmation signal of turn on of the reserve CCR of both substations.

For sending the activation at the required brightness level command actions to the regulators of the different circuits, visualizing the confirmation of turn on and keeping a record of the events on interruptions and restorations of each regulator in service, whether principal or reserve, two software were developed that meet all

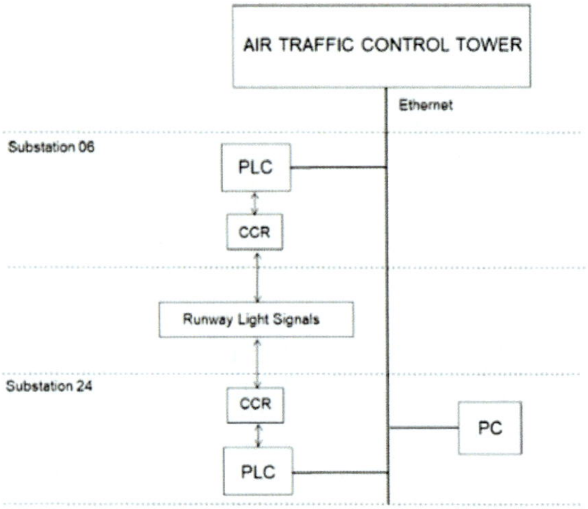


**Fig. 2** Communication diagram between the ACT and the substations for the control of the CCR

these requirements: a runway lights system remote control software and a software for monitoring the operation of the CCR. The first one is executed on the industrial computers at the ACT and allows the remote turning on, at the brightness level required for landing and takeoff operations, of the CCR that feed each circuit of the runway lights system, as well as displays the operational status of the different circuits. The second software runs on a computer at the aerodrome maintenance personnel's office and allows recording of regulators status events, both principals and reserves, for each circuit, as well as displaying the operational status of the different circuits too. With this history of events they are considered, the fault condition or restoration of the regulators, the entry into service of the regulators, specifying the connected circuit, and the active user control so that all actions carried out by the operator are in evidence. This allows a rapid response to breakdowns, providing better guidance to the technicians and personnel of the maintenance of the runway lights system.

The control software interface, presented in Figs. 3 and 4, is divided into two sections. In the upper section there is a visual representation of the circuits that are turn on by headend, depending on the communication with the substations. This communication is evidenced by the status of the buttons: green if there is communication with the substation and red if communication has been lost. In the case of this last option, the text to configure the IP of the substation's PLC that has no connection appears enabled. The lower section is subdivided into four sections: the first one shows the output brightness level of the CCR of each circuit; the second one groups the turn on buttons for the 5 levels of brightness and the turn off button of the corresponding circuit; the third one groups the particular on/off buttons for the taxiway circuits, which, unlike the other circuits, always turn on at brightness level 3; and the fourth column has two buttons for simultaneous on and off brightness level 3 of the PAPI, Runway edge lights (RL), Low approaching lights (LA) and High approaching lights (HA) circuits, two buttons with identical use for the taxiway circuits and a fifth button to set the headend in which all command actions will be executed. The same circuits are controlled in both headend: PAPI, RL and LA, and the HA circuit is only controlled from Substation 06 (RWY06). As for the taxiways, in SEE06 the turn on order is given for four taxiways and in Substation 24 (RWY24) for nine taxiways.

The CCR operation monitoring software interface consists of three windows. The initial window for user control, the main window to know the status of the regulators, circuits and recent events, and the event history window. The initial window for user control is offered in two modes: as an administrator (Fig. 5a) and as an operator (Fig. 5b). The administrator has the options to add or delete users and change the password of these, in addition to go to the main window. The operator has only the option to go to the main window.

In the main window, presented in Fig. 6, there is a synoptic diagram of the airport runway, where the circuits that are operating can be observed, as well as the information of the regulator in service for each circuit, the brightness level of each one, the most recent events associated with the status of each regulator, the user who have logged in, an access to the event history window and the existence or not of a

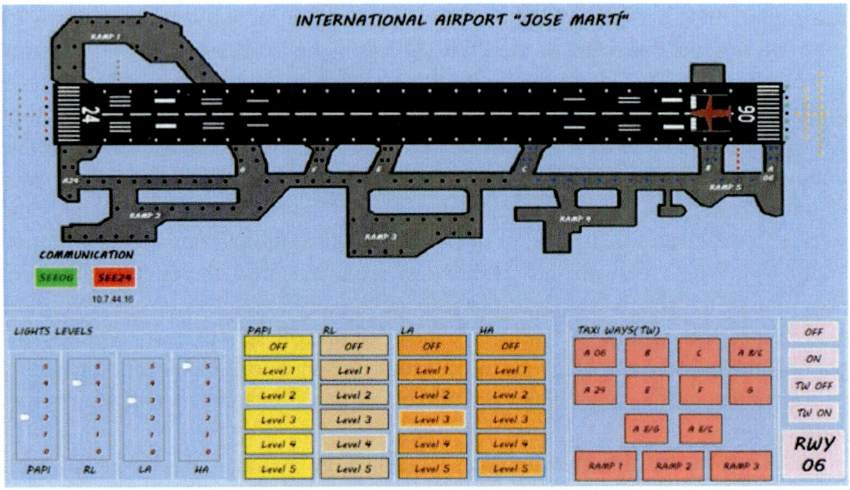


**Fig. 3** Control action of turn on and turn off of the circuits from the RWY06

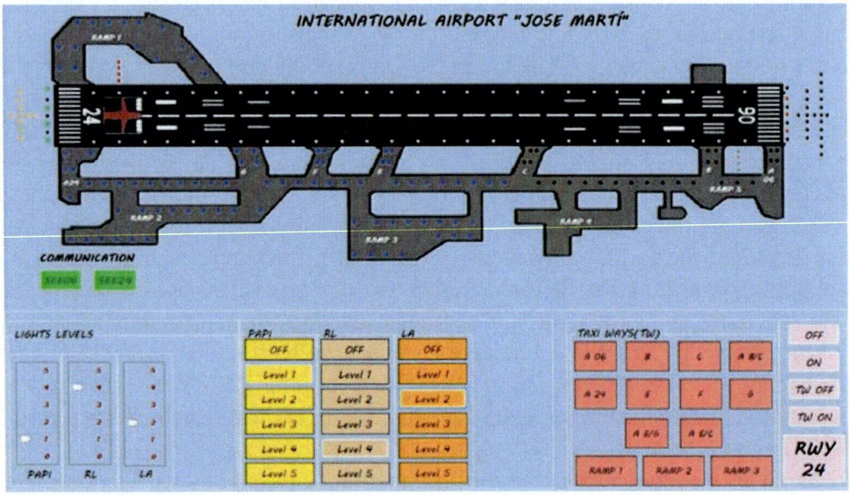


**Fig. 4** Control action of turn on and turn off of the circuits from the RWY24

connection with each headend. In case of losing communication with a substation due to the change of the IP address, the respective buttons are enabled to access an IP addresses configuration window (Fig. 7) where indicate the IP values of the PLC to restore communication with the substations.

In the main window, the status of the regulators is represented with the following legend:

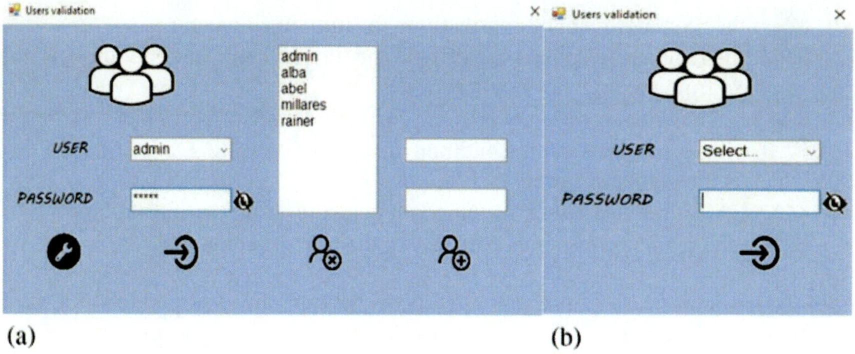


**Fig. 5** User validation window for **a** the administrator and **b** the operators

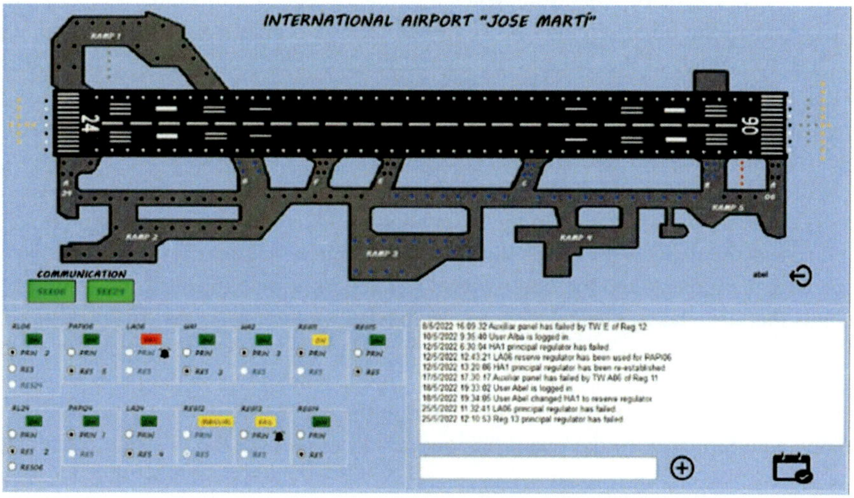


**Fig. 6** Main window

**Fig. 7** IP addresses configuration

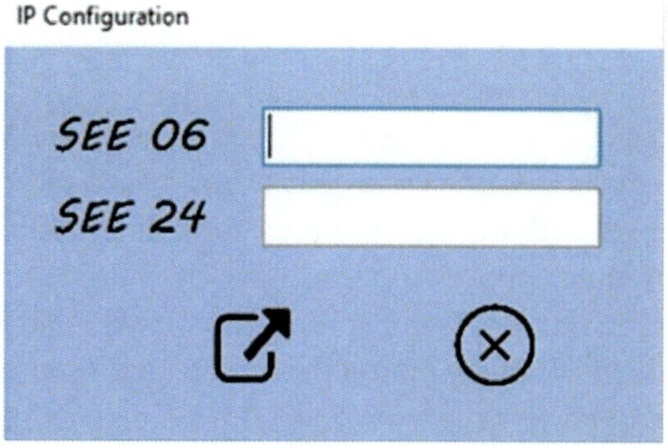

- ON if the regulator responds correctly to the turn on command.
- ON if the regulator responds correctly to the turn on command, but there is no response from the taxiway auxiliary panel.
- FAIL if the regulator does not respond correctly to the turn on order.
- FAIL if the previous fault condition disappears, awaiting confirmation of which is the regulator in service.
- MANUAL if the regulator was turned on manually from the substation.
- OFF if the regulator responds correctly to the turn off command.

In the particular case of taxiway circuits, as they are always operated at a fixed brightness level, it is not necessary to display said brightness level value on the regulator in service. In the rest of the circuits, the brightness level reached is shown next to the regulator in service. There are, in both substations, circuits that share the same reserve regulator. Therefore, when the reserve regulator of a circuit is selected, the option in the other circuit who share the reserve regulator is disabled, making it clear for which of the circuits the shared reserve regulator is in service. In this window there is also a text box to add or edit a comment to the event selected from the list, with the last ten events that have taken place. This allows to add clarity to the event that happened, as well as the possible causes or solutions.

To consult the events that have taken place during a selected month and year, press the event history button in the main window, located in the lower right corner, and access the window shown in Fig. 8. In this window there are 2 drop-down lists: one with the months and the other with the years, to be able to select the history of events that have taken place during a selected date (month and year) by clicking on the magnifying glass. These events are displayed in the list box, with the date, time and text format of the corresponding event. In this window, as in the main window, there is a text box to add or edit a comment to the event selected from the list. This allows to add clarity to the event that happened, as well as the possible causes or solutions.

## 3 Conclusions

The runway lights control system of Cuban airports is controlled locally or with archaic remote control systems, based on relay logic and copper wired communication, with the exception of some international airports, where similar more modern remote control solutions have been implemented, based on PLC, visualization on PC or Human–Machine Interface (HMI) and Ethernet communication between these devices. The solution presented here offers, as a novel element, the incorporation of software for monitoring with objectives of maintenance assistance for the regulators. It also represents a step towards technological sovereignty, since it is avoided to acquire costly control and monitoring systems that are offered by the CCR manufacturers themselves for airports. This software allows working with various CCR

**Fig. 8** Window to consult the events that have taken place an specific month and year

models, as it is independent of the regulator model installed in the runway lights control system.

The given solution presents a simple and intuitive interface for air traffic control operators and runway lights control system maintenance operators, who have shown their satisfaction in using this solution. It has been possible to reduce the time in the detection and solution of regulator failures and the historical record of the events has made it possible to take timely decisions for the purchase of spare parts. In the immediate future, this solution will be generalized to the rest of the Cuban airports.

## References

1. ADB.: Regulador de corriente constante controlado por microprocesador Tipo MCR$^3$ 2, 5 kVA a 30 kVA. Manual de Instalación. 1st edn. (2013a)
2. ADB.: Regulador de corriente constante controlado por microprocesador Tipo MCR$^3$ 2, 5 kVA a 30 kVA. Manual de Mantenimiento. 1st edn. (2013b)
3. Chacón García, A., Bustos Rodríguez, S., Velásquez Velásquez, Á.M.: Control y monitoreo de sistemas de iluminación aeroportuaria: simulación computacional para la pista norte del aeropuerto el dorado. Visión Electrónica **3**(1), 31–45 (2009). https://doi.org/10.14483/22484728.689
4. IACC.: Regulaciones Aeronáuticas Cubanas, RAC 14: Aeródromos y Helipuertos. Aeródromos, 5a edn. Instituto de Aeronáutica Civil de Cuba (2015)
5. ICAO.: Annex 14 to the Convention on International Civil Aviation. Aerodromes, vol. I. Aerodrome Design and Operations, 8th edn. International Civil Aviation Organization (2018)
6. Corporation, M.: Microsoft visual studio community 2019. Version **16**(2), 3 (2019)
7. Rivera Castañeda, D.A.: Sistema de monitoreo y control para los sistemas de luces de pista de los aeropuertos La Mina y Puerto Bolivar. Universidad Distrital Francisco José de Caldas, Colombia, Tesis de Grado (2019)
8. Schneider Electric.: EcoStruxure Machine Expert Basic. Version 1.0 (2019)

# Automatic System to Characterization of Creatinine by Optic Measurement '*On Line*' in Wastes Samples During Hemodialysis

**Abel Calle-Herranz, Carmenchu Regueiro-Busoch, Carmen Brígida Busoch-Morlán, and Angel Regueiro-Gómez**

**Abstract** **Introduction**: The renal diseases have a high social and economic impact due to the incapacity that produce in the population and the high costs of renal function replacement therapies. The measurement of uremic toxins such as creatinine is a first factor to determine if the patient should be submitted to the hemodialysis process (3.5–4.5 h per therapeutic session during alternate days). In this process, the clinical specialist acquires a sample of blood plasma (centrifuged blood) and mixing this with a chemical reagent to detect the toxin concentration with help of an optical instrument (spectrophotometer). Nowadays, serious limitations exist with this process because it is not continuous during the clinical treatment. **Objective**: The main objective of this research was related with the development of "On-line" measurement channel for the characterization of uremic toxins (creatinine) using the dialysis liquid (waste liquid) during the hemodialysis process of patient. **Methods**: This investigation verified experimentally validity of creatinine measurement "On line" with a new opto-electronic differential system related with the absorbance of biological samples (waste liquid) during the hemodialysis process. **Results**: The characterization of the optic system and their conditioning channel was carried out, and with this new device, creatinine concentrations were measured in plasma and in dialysis liquid with similar results. **Conclusions**: The new alternative proposal of the system of creatinine measurement in the dialysis liquid (waste samples) allowed to characterize of this toxin and it can be generalized in the commercially available systems.

A. Calle-Herranz (✉)
Centro Provincial de Electromedicina, La Habana, Habana, Cuba
e-mail: abel.calle@infomed.sld.cu

C. Regueiro-Busoch · C. B. Busoch-Morlán
Department of Automatic and Computation, School of Automatic and Biomedical Engineering, Universidad Tecnológica de La Habana José Antonio Echeverría (Cujae), Habana, Cuba

A. Regueiro-Gómez
Department of Bioengineering, School of Automatic and Biomedical Engineering, Universidad Tecnológica de La Habana José Antonio Echeverría (Cujae), Habana, Cuba



O. Llanes-Santiago (ed.), *Proceedings of 19th Latin American Control Congress (LACC 2022)*, Studies in Systems, Decision and Control 464,
https://doi.org/10.1007/978-3-031-26361-3_27

**Keywords** Hemodialysis · Creatinine '*on line*' detection · Dialysis's wastes samples

# 1 Introduction

The renal diseases have a social and economic high impact due to the high costs of the therapies of substitution of the renal function such as the hemodialysis (average cost >$60′000.00 USD per patient annually in USA and UK) or the kidney transplant ($150′000.00 USD) [1]. The renal failure can take place slowly in a period of months or years product of lesions in the kidney (Fig. 1 Left) that they are not detected generally, since initially they don't present symptoms detectable, that which goes developing the Disease Renal Chronicle (DRC) [2].

There are diverse the methods that they help to check the correct operation of the kidneys; but very few those that offer total accuracy. When the kidneys stop to work completely due to different causes (diabetes, high blood pressure, glomerulonephritis, polycystic kidney disease, urinary tract infections, drugs and toxins, kidney stones, accident and traumas and big loss of blood), the patient should undergo systematic dialysis at least three times per week or to receive a transplant (this last very delicate and complex option for the limitations of the donors' compatibility) [3]. The most common treatment is the hemodialysis, which allows that the blood goes by an extracorporeal circuit (Fig. 1 *Center*) which remove of waste products such as creatinine, urea and excess water from the blood, and once clean, the blood returns to the fellow's body [4].

In the process of cleaning several toxins are discarded, among which are the urea and the creatinine [5, 6], this last of clinical great interest in their mensuration during the hemodialysis process [7]; but nowadays it presents limitations due to the technological complexity for their measurement and the existence of some limitations of reagents, all that which limit an appropriate pursuit of the therapeutic treatment in the patients during the therapy [8]. To obtain an appropriate dialysis (Fig. 1 *Right*),

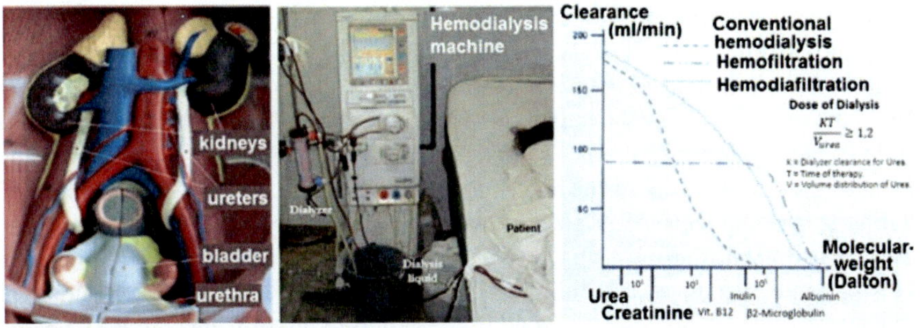


**Fig. 1** Components of urine system, example of hemodialysis machine and relation of clearance versus methods of dialysis

some aspects should be observed or controlled, such as the dialysis dose (DD) should contemplate:

$$DD = \frac{KT}{Vurea} \tag{1}$$

Besides an appropriate dialysis dose, a good clearing of other solutes of molecular bigger weight, the time and dialysis frequency should avoid arterial hypertension or medication necessity to avoid adverse symptoms intra or derived extra-dialysis of the procedure, the parameters of the metabolism calcium and phosphorus should be controlled, potassium and bicarbonate should be inside limits normal pre- and post-dialysis, the patients should be well fed:

$$nPCR > 1\frac{\frac{g}{kg}}{day} \tag{2}$$

Nowadays the measurement of creatinine is complex. A sample of blood is extracted of the patient, and this sample is moved to the clinical laboratory, where the sample is centrifuged, separating the plasma and join it with a specific reactive. The concentration of creatinine is measured with a spectrophotometer. The ideal process would be the continuous measurement of creatinine during the hemodialysis (typically among 3–4 h, three times per week for each patient), so that the hemodialysis machine controls the optimal dose on the patient, but nowadays this possibility on line is not possible in the commercial modern systems [9, 10]. This work shows a new alternative for the creatinine characterization in the waste's liquid of the hemodialysis process with the use of optic detection of sample's turbidity in continue mode.

## 2 Materials and Method

Figure 2 (*Left*) shows a new creatinine meter introduced in hemodialysis machine. This module is compound by an optic system (stimulation with LED and differential detection) and the channel of measurement and control (Fig. 2 *Right*).

First, the system is verified and it wait the reception of the command 'ON' to begin the process of signal acquisition. The system opens the valves ($Ev_1$ and $Ev_2$) and fills the carriage with a waste sample. If the operation is related with the cleaning of sensor, the system opens the valves and it permits the pass of sterilization liquid during some minutes synchronized with this process in the hemodialysis machine. The measurement during 3 min of the stored sample (each data is acquired each 10 s), permits that a new algorithm detects the changes of creatinine concentration. The average values obtained are send to the Control Unit of Concentrate (CUC) to modify the dose of dialysate 'On line' during the hemodialysis of patient.

An initial experiment in the laboratory was determining the power of different LEDs to used. We selected the interval from 250 to 900 nm of wavelength, including

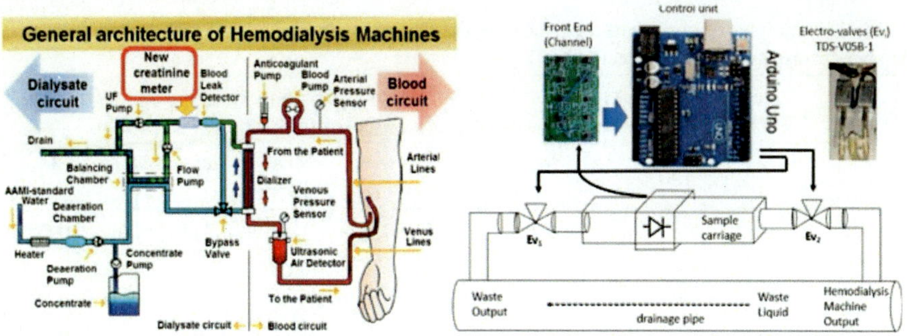


**Fig. 2** New system to measurement of creatinine (*Left*: Representation of hemodialysis process, *Right*: Simplified creatinine meter)

the typical interval used by commercial spectrophotometers in the Clinical Laboratory of hospital institutions, and it was necessary to carry out a sweeping of different wavelength for a better analysis of the used samples. To each selected LED, we were measured the emitted powers with an optic-meter and during the experiments, their polarization current was controlled in three typical values: 10, 8 and 5 mA. For the receiving circuit, two similar detectors were used in the optic differential system, one was placed appropriately to receive directly the signal across the sample carriage and another detector was employed to establish a reference in any work atmosphere. During the developed research in the laboratory a data acquisition card (NI-USB1208FS) has been used (connection USB 2.0). The interface was developed in LabView (V 17) that allows the good visualization of the acquired data, with the option of selecting the quantity of channels and the sampling frequency with a maximum limit of 50 kS/s and 12 bits of resolution.

**Protocol for the preparation of the experimental samples**

With the designed system we carried out a series of readings to the samples (10 ml of reagent produced by HELFA Diagnostic, which is equal to have 1 ml of hydroxide of sodium plus 9 ml of picric acid and creatinine as reference solution) applying this in the kinetic method, which is based on carrying out the readings of the prepared samples each 10 s, to different longitudes of wave. The analyzed samples are conformed for: Sample 1: 1 ml of reagent +100 μl creatinine, Sample 2: 1 ml of reagent +100 μl creatinine +100 μl of $H_2O$, Sample 3: 1 ml of reagent +100 μl creatinine +200 μl of $H_2O$, y Sample 4: 1 ml of reagent +100 μl creatinine + 300 μl of $H_2O$. During the process of the investigation a protocol was developed as established the norms for the work with these systems, getting ready the samples according to that regulated in the development of the kinetic method of Jaffe (the handling of samples and concentration of used reagents), accepted by the regulatory authority.

**Table 1** Examples of answer of the measurement channel with optic filters (references with well-known absorbances for the selected LED)

| Filter | Green LED (10 mA) Vo (Volts) | | | | | Red LED (5 mA) Vo (Volts) | | | | |
|---|---|---|---|---|---|---|---|---|---|---|
| | Meas 1 | Meas 2 | Meas 3 | Mean | Std. | Meas 1 | Meas 2 | Meas 3 | Mean | Std. |
| Without filter | 4.44 | 4.55 | 4.48 | 4.49 | 0.045 | 0.736 | 0.733 | 0.748 | 0.739 | 0.006 |
| 0 | 4.25 | 4.44 | 4.39 | 4.36 | 0.080 | 0.726 | 0.734 | 0.734 | 0.731 | 0.004 |
| 1 | 3.73 | 3.92 | 3.89 | 3.85 | 0.083 | 0.641 | 0.642 | 0.642 | 0.642 | 0.000 |
| 2 | 3.53 | 3.63 | 3.62 | 3.59 | 0.045 | 0.598 | 0.598 | 0.588 | 0.595 | 0.005 |
| 3 | 3.09 | 3.19 | 3.19 | 3.16 | 0.047 | 0.590 | 0.588 | 0.581 | 0.586 | 0.004 |
| 4 | 2.78 | 2.82 | 2.83 | 2.81 | 0.022 | 0.530 | 0.524 | 0.521 | 0.525 | 0.004 |
| 5 | 2.41 | 2.48 | 2.46 | 2.45 | 0.029 | 0.534 | 0.532 | 0.530 | 0.532 | 0.002 |
| 6 | 2.20 | 2.22 | 2.23 | 2.22 | 0.012 | 0.521 | 0.520 | 0.519 | 0.520 | 0.001 |
| 7 | 2.13 | 2.14 | 2.14 | 2.14 | 0.005 | 0.514 | 0.514 | 0.515 | 0.514 | 0.000 |
| 8 | 2.11 | 2.12 | 2.11 | 2.11 | 0.005 | 0.512 | 0.511 | 0.520 | 0.514 | 0.004 |

## 3 Results and Discussion

### *3.1 Characterization of the Optic Transmitters*

For the characterization of the optic transmitters a set of filters (Reference System for verification in Clinical Laboratories) was used, of which were known its absorbance value. The filters were placed between transmitter and the receiver, being obtained the answers of the system for the previously well-known wavelength (Table 1).

### *3.2 Experimental Results*

When the same samples are analyzed, in the same interval of time in the new development device and the spectrophotometer, similar results are obtained, that is to say, as the absorbance values increase, they diminish the values of tension, what would establish a relationship inversely proportional (Fig. 3).

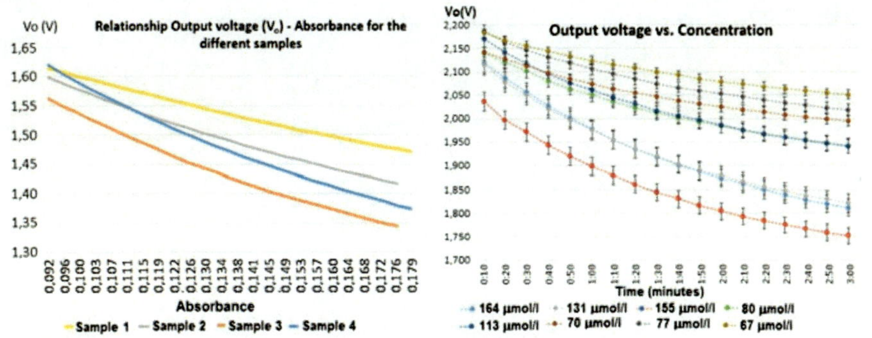


**Fig. 3** Examples of measurements (*Left*: Relationship between voltage output and the absorbance for different samples. *Right*: Real samples obtained of 8 patients' plasma)

## *3.3 Real Analysis of Samples of Clinical Laboratory*

For the study of real samples obtained in the unit of Clinical Laboratory, we carried out tests to plasma samples (8 patients), of which their respective creatinine concentrations were already known. The designed system offers solid results, because values are obtained with high repeatability for the wavelength used. Of each real sample, we took 100 µl and join it with a 1 ml of reagent, and analyze it with our new device proposed during 3 min, obtaining the values of voltage every 10 s, in a similar way to the kinetic method. Figure 3 (*Right*) shows the obtained results, when the creatinine concentrations increase, the output voltages diminish (related with the sample absorbance).

Nowadays the devices used in the laboratories of Institutions of Health are not able to determine values of concentrations below 60 µmol/l, and this situation during the hemodialysis therapies limits the process of evaluation of the creatinine quantity that the patient is expelling during the 3 or 4 h of this treatment.

## *3.4 Experimental Results with Wastes Samples*

For the generalization of the kinetic method developed previously, we applied the methodology on a group of wastes samples not controlled of different patients (samples obtained of hemodialysis wastes). The samples of all patients were analyzed in an interval of 30 min. The wastes were picked up fulfilling the pertinent measures to guarantee the biological security in the handling of the residuals and during the process of the patient's hemodialysis. We analyze in each interval of time, 100 µl of waste with 1 ml of reagent, measuring their value of voltage each 10 s, in a similar way to the kinetic method. The results in all patients were similar, with a descent of voltages in the system during the time of treatment, equivalent to an increase of detection of creatinine.

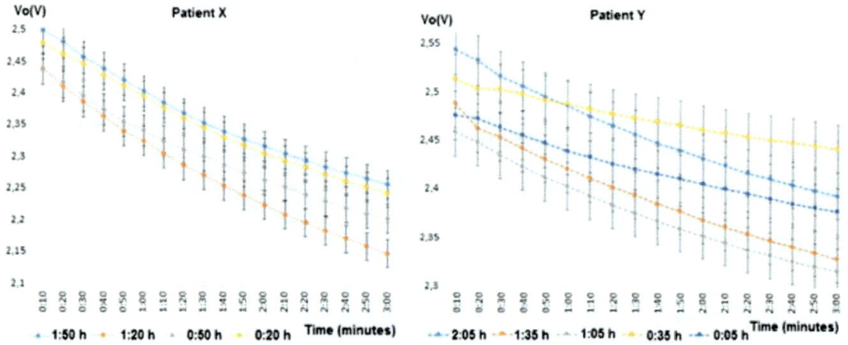


**Fig. 4** Experimental results with a new proposed system (*Left*: Patient X: weight: 120 kg, height: 180 cm, high Corporal Mass Index CMI = 37 kg/m$^2$ and half obesity, *Right*: Patient Y: weight: 60 kg, height: 165 cm and low CMI = 22 kg/m$^2$)

In the case of Patient X we began to take his wastes when the patient disposes of 1 h and 50 min to finish of hemodialysis process. Figure 4 (*Left*) shows the values of "On Line" voltages obtained and the behavior of the wastes samples was similar to the results obtained in the measurements carried out in the blood plasma. In Patient X, it is possible to observe the increase of output voltage values in the waste's samples acquired between 0:50 and 0:20 h (gray and yellow lines). In this time, the patient was lunched, and we observed the changes in the detection with an increase of his levels of the uremic toxins.

The same procedures were used with another Patient Y (Fig. 4 *Right*), but this dispose of 2 h and 5 min finish of hemodialysis process. The experimental results obtained and the behavior was similar to the measurement. In the Patient Y, it is possible to observe the increase of output voltage values in the samples acquired between 0:35 and 0:05 h (yellow and blue lines). In the previous interval (between 1:05 and 0:35 h), the patient lunched, and when his organism took the necessary nutrients of the ingested foods, an increase of the uremic toxins was appreciated in his blood and in the waste liquid.

The new proposed system detected the changes in the wastes samples of the patient during the process of hemodialysis, and it allows to activate the control of the mixture of the dialysate liquid (concentrate: water, ions and salts) to introduce in the hemodialysis machine with a continuous adjustment of the dose of dialysis during the treatment of patients.

The optic system had a good behavior (high linearity and robustness) with low dispersion and a little uncertainty ($< 0.5\%$) of the measurement values. The selection of green LED in the emission or stimulation of samples drove to the best results. Some authors have investigated using UV waves but their results depend on the control of the emission power, the complexity of the opto-electronic system and the reagent employed [11, 12].

The modern systems of hemodialysis in the country don't have this option of continuous measurement, because they include a protocol of characterization of toxins (mainly urea and creatinine) previous to the process of the patients' clinical treatment, with serious limitations in the existence of reagents and the transference process of blood samples acquired in the therapeutic rooms. With the new alternative proposed, the results in all patients analyzed were similar, with a descent of voltages output in the system (optical sensor) during the time of treatment, equivalent to a decrease in the creatinine detection.

The results of the research demonstrate that the method used is correct for the determination of creatinine in the waste liquid, however it's necessary increase the number of samples to improve the statistical analysis (accuracy and precision) as a dialysis dosimeter and to present a regulatory authority. On the other hand, it is convenient to comment that the new developed optical circuit (creatinine detector) can be added in a similar way to the system for optical detection of blood leak that already machines have, which facilitates the introduction of this improvement in available commercial systems.

## 4 Conclusions

The introduction of this new system in the hemodialysis machine permits the detection "On line" of uremic toxins (specially the creatinine concentration) and the adjust of the dose of dialysis in continuous mode, introducing a better result in this complex therapy.

**Acknowledgements** The authors appreciate the support offered by the project MORPHYs (PR-0435) development in the Department of Bioengineering (CEBIO), at Universidad Tecnológica de La Habana (Cujae).

## References

1. WHO.: World Health Statistics overview 2019: monitoring health for the SDGs (Sustainable Development Goals) (2019)
2. López-Heydeck, S.M., López-Arriaga, J.A., Montenegro-Morales, L.P., Cerecero-Aguirre, P., Vázquez-de Anda, G,F.: Laboratory analyses for early diagnosis of chronic kidney disease. Rev. Mex. Urol. **78**(1), 73–90 (2018). https://doi.org/10.24245/revmexurol.v78i1.1601
3. National Institute of Diabetes. Kidney disease of diabetes. 2 Dec 2015, Available from: http://www.niddk.nih.gov/health-information/health-topics/kidney-disease/kidney-disease-of-diabetes/Pages/facts.aspx
4. Caprov, N.: Demuestran la eficacia de un riñón artificial portátil (2016), Available from: https://facebook.com/miradaprofesional
5. Lewin, R.N.J.: What is the normal range for a creatinine blood test? (2020), Available from: https://www.medicalnewstoday.com/articles/
6. Mayo Clinic Staff.: Creatinine test (2018), Available from: https://www.facebook.com/sharer.php?u=https://www.mayoclinic.org/tests-procedures/creatinine-test/about/pac-20384646

7. Yamada, S., Yamamoto, S., Fukuma, S., Nakano, T., Tsuruva, K., Inaba, M.: Geriatric Nutritional Risk Index (GNRI) and creatinine index equally predict the risk of mortality in hemodialysis patients: J-DOPPS. Sci. Rep. **10**(5756) (2020). https://doi.org/10.1038/s41598-020-62720-6
8. Badell-Moore, A., Bacallao-Méndez, R., Mañalich-Comas, R.: Sobre el establecimiento de valores de referencia de la excreción de creatinina para la población cubana. Revista Cubana de Alimentación y Nutrición (RCAN) **25**(1) (Suplemento 1), pp. S28–S60 (2015). RNPS: 2221, ISSN: 1561-2929
9. Dalapati, P., Manik, N.B., Basu, A.N.: Optical characterization of the photodetector and light emitting diode over a wide range of temperature. Int. J. Adv. Res. Phys. Sci. **2**(6), 18–23 (2015)
10. Tomson, R., Uhlin, F., Fridolin, I.: Urea rebound assessment based in UV-absorbance in spent dialysate. ASAIO J. **60**(4), 459–465 (2014)
11. Stepanova, O., Konoplev, G., Gennady, Z., Kuznetsov, A., Frorip, A.: Analysis of solute kinetics during hemodialysis treatment by measuring the UV absorption of effluent dialysate at different wavelengths. IEEE, pp. 1–5. 978-1-7281-5761-0/20©2020
12. Paats, J., Adoberg, A., Arund, J., Fridolin, I., Lauri, K., Leis, L., et al.: Optical method and biochemical source for the assessment of the middle-molecule uremic Toxin_2-Microglobulin in spent dialysate. Toxins **13**, 255 (2021). https://doi.org/10.3390/toxins13040255

# Automatic System to Continuous and Differential Acquisition of Blood Pressure During Tilt Test

**C. Regueiro-Busoch, Y. J. Almeida-Pichardo, L. A. Noa-Lloren, C. B. Busoch-Morlán, and A. Regueiro-Gómez**

**Keywords** Tilt test · Blood pressure · Continuous and differential acquisition · Oscillometric method

## 1 Introduction

The cardiovascular illnesses constitute the first causes of death to international level [1], and it's necessary a systematic evaluation of the circulatory system. One option is related with the development of Tilt Test [2–4]. During the development of Tilt Test some physiological variables are measured, such as ECG, Arterial Blood Pressure and others (Fig. 1) [5].

The arterial blood pressure (ABP) constitutes a complex and important variable (Eq. 1) for the characterization of cardiovascular system behavior.

C. Regueiro-Busoch (✉) · C. B. Busoch-Morlán
Department of Automatic and Computation, School of Automatic and Biomedical Engineering, Universidad Tecnológica de La Habana José Antonio Echeverría (Cujae), Havana, Cuba
e-mail: carmenchu@automatica.cujae.edu.cu

C. B. Busoch-Morlán
e-mail: cbusoch@automatica.cujae.edu.cu

Y. J. Almeida-Pichardo · L. A. Noa-Lloren · A. Regueiro-Gómez
Department of Bioengineering, School of Automatic and Biomedical Engineering, Universidad Tecnológica de La Habana José Antonio Echeverría (Cujae), Havana, Cuba
e-mail: luisnoa9612@nauta.cu

A. Regueiro-Gómez
e-mail: regueiro@automatica.cujae.edu.cu



O. Llanes-Santiago (ed.), *Proceedings of 19th Latin American Control Congress (LACC 2022)*, Studies in Systems, Decision and Control 464,
https://doi.org/10.1007/978-3-031-26361-3_28

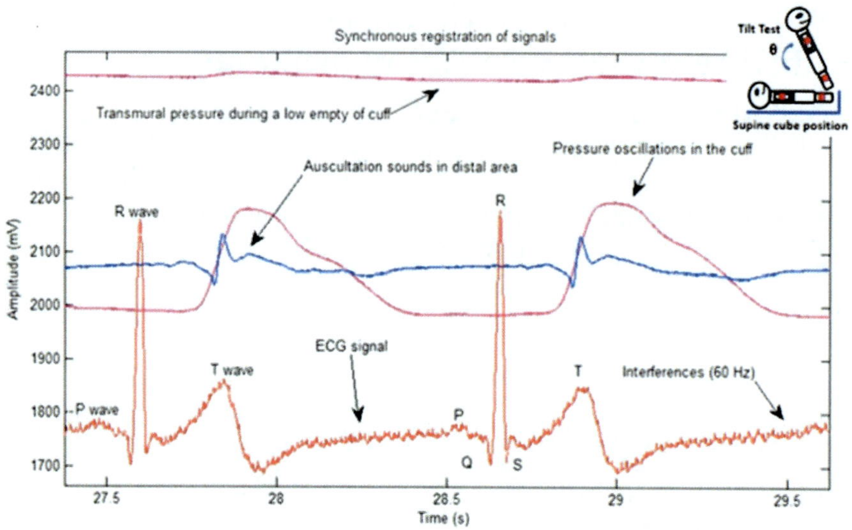


**Fig. 1** Example of physiological variables measurement during Tilt Test

$$p_m = \frac{1}{T} \cdot \int_0^T p(t)dt \tag{1}$$

For specific conditions during the systole if possible to determine the dependence of the pressure in the blood vessels, starting from the flow and the hemodynamic resistance of vessel (Eq. 2), k is the elasticity coefficient of the blood vessels and it establishes the proportionality among the pressure and the volume, and Rho is the hemodynamic resistance.

$$p = p_s e^{\frac{-t}{kR_{ho}}} \tag{2}$$

Numerous noninvasive methods are used in the clinical practical for the measurement of blood pressure, among them the most used are the auscultation of sounds (Korotkoff sounds), and the oscillometric method [6]. In general, these methods need the combined employment of an inflated cuff, which should be adjusted to the anatomy of the measurement area (typically the extremities), the inflated cuff limits the continuous record and none of these methods permits the identification directly of three clinical values: maximal, minimum and mean of blood pressure. Other interesting methods are the plethysmography methods (impedance or optical method), which permit the acquisition in continuous mode of pulse wave using the cuff placed in the finger or another specific area, with some limitations in the adjust of sensors (electrodes or optical sensors) during the process of acquisition [7].

During the Tilt test, it's necessary to obtain the blood pressure records in continuous mode to facilitate the study of different cardiovascular pathologies (vasovagal syndrome and others). Typically, these cardiovascular studies have an approximate duration between 60 and 80 min, time during which the inclination bed varies their position (between 0 and 70 grades). When the clinical specialist uses the classical methods (auscultatory or oscillometric method), it's necessary to stop the acquisition in a minimum interval of five minutes between samples, because the patients need to reestablish their circulation in their occluded members (extremities) during the filled/empty cycle of cuff. The photoplethysmography (optics or electric) shows limitations for the development of Tilt Test, due to the adjustment and compensation carried out, when the sensor is placed on the subject's finger [8–10].

## 2 Materials and Method

This research proposes a new alternative to the differential and continuous acquisition of arterial blood pressure during the Tilt Test. Figure 2 shows the new biomedical system designed to the acquisition of arterial blood pressure during the development of Tilt Test. It's composed for three blocks: Front End, Conversion & Control Unit (ARDUINO UNO) and Personal Computer (Visual Interface in LabVIEW).

Two different cuffs are used to measurement the arterial blood pressure in the higher and lower member (arm and leg). An analog multiplexer (IC-4066) is used

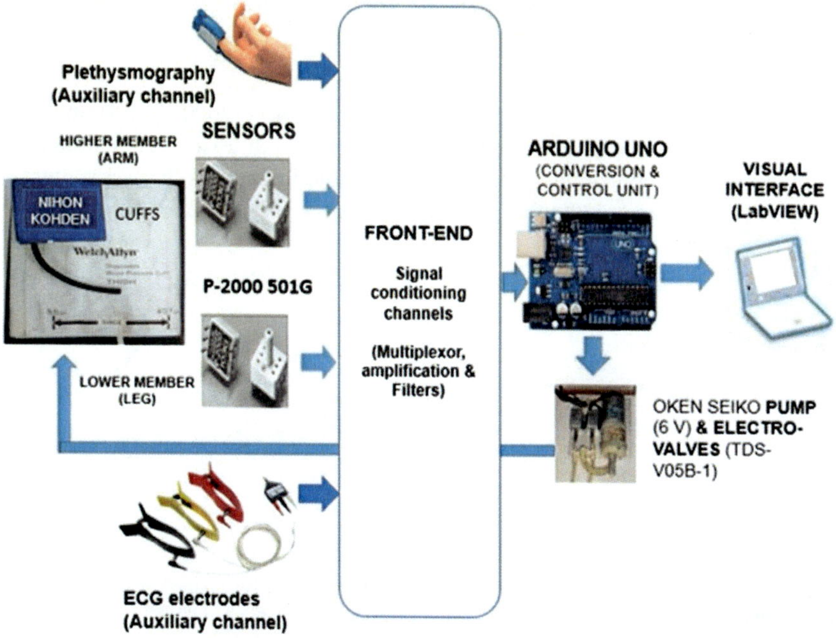


**Fig. 2** New system for continuous and differential measurement of ABP during a Tilt Test

to connect both sensor arrays (P2000-501G, COPAL Electronics, Japan), and the functional blocks in the front end permit the amplification (IC-AD620) and filtering (IC-LM324) of pressure signals obtained in mode continuous during the movement of bed (inclination).

The conversion and control unit, based in the use of a Kit ARDUINO UNO, permits the digitalization of signals with a good resolution (10 bits and error <0.5 mmHg), and this unit send the information obtained to a main computer via USB, using an interface development in LabVIEW for the control, visualization and spectral-time analysis of records. In a first reading cycle, the system detects the fellow's mean pressure in bed position, this value is taken as reference for the one filled later of the cuffs during the differential continuous records of the pressures in the individual's extremities (arm and leg).

If the specialist detects some problem in the fellow during the inclination test, he can stop the operation of the system with the help of a general pause button (STOP). During the development of research, two auxiliary channels (ECG and Photo-plethysmography) are used to obtain the temporal references of the physiological variables.

# 3 Results and Discussion

## 3.1 *Characterization of Front-End in a New System to ABP Acquisition During TT*

Figure 3 (Left) shows the high linearity of sensor arrays (P-2000- 501G) obtained in laboratory. The small differences in the obtained sensibilities are compensated by the program in the application, so the lineal answer is similar in both cases. Figure 3 (Right) shows the bandwidth of acquisition channel (3 Hz < f < 2500 Hz).

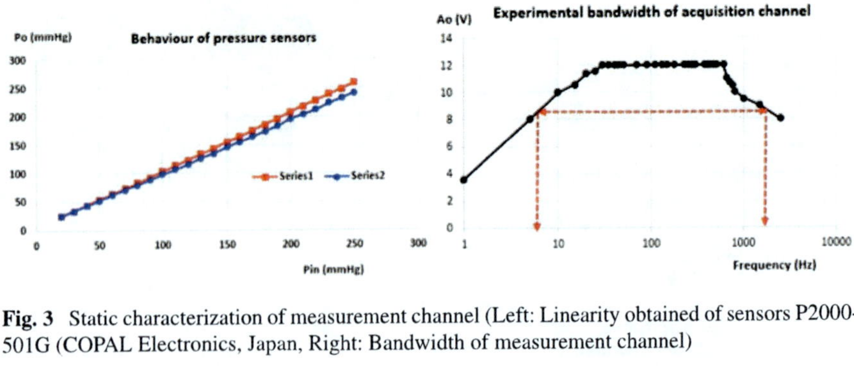


**Fig. 3** Static characterization of measurement channel (Left: Linearity obtained of sensors P2000-501G (COPAL Electronics, Japan, Right: Bandwidth of measurement channel)

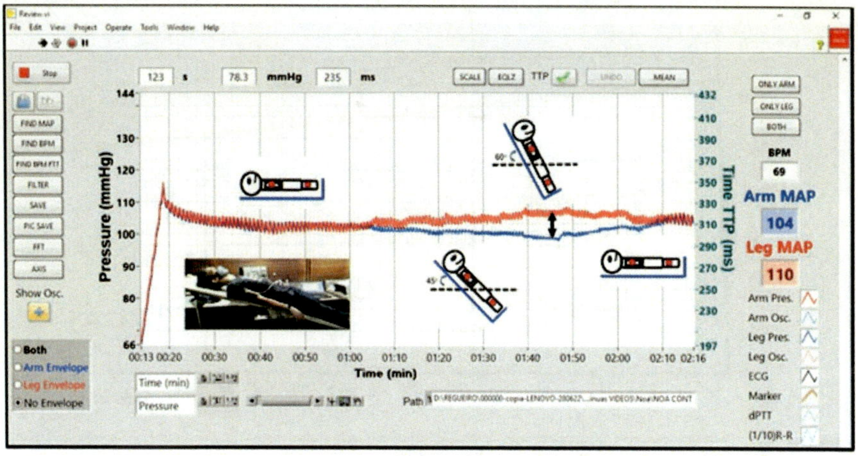


**Fig. 4** Example of blood pressure recording during Tilt Test in the Instrumentation & Bioengineering laboratory

## 3.2 *Interface Development in LabVIEW to Control and Display of Obtained Results*

Figure 4 shows the interface development in LabVIEW tool to control of KIT ARDUINO and one example of recorded blood pressure. Initialing the user can select the address of record files, the sampling frequency and the port of connection. During experimental phase in laboratory, 18 supposedly healthy voluntaries (12 man and 6 women with ages between 18 and 60 years) were measured with good results. The criteria for inclusion of the subjects in the study were their written informed consent. The figure shows the recording cycle, first in supine position, and later in different positions during the development of Tilt test.

In the record is possible to appreciate the behavior of the continuous differential record of blood pressure, existing a difference between the arm and leg pressures in different positions. Figure 5 shows two records obtained with the bigger dispersion among all the studied subjects, it is observed that the mean pressure (section I) depend on the anatomical and functional characteristics of subjects (personalized measurement) and all results are in the clinical interval accepted by the Universal Standard AAMI / ESH / ISO (ISO 81060 2:2018) for automatic systems. In the second section during the inclination of the table, similar variability is appreciated, when the inclination increases, the difference of acquired pressure from the members (arm and leg) grow, which constitutes an important result that the clinical specialists can evaluate in their diagnosis. In the last section is appreciated that a recovery of the pressures exists returning at the departure level in the horizontal position (supine cubit position). This behavior was observed for all the acquired records in the database during the investigation, not appreciating differs between men and women.

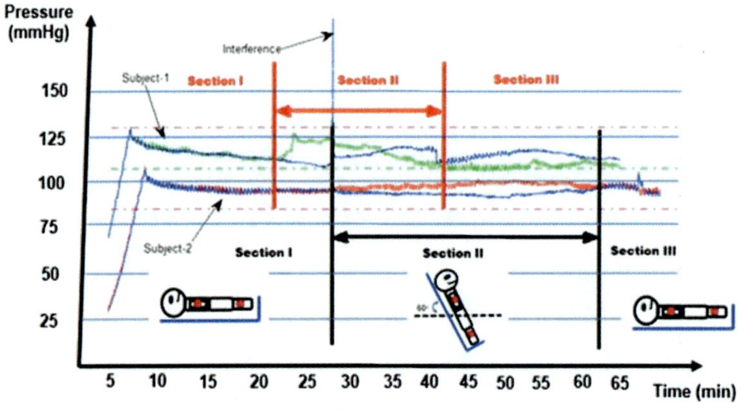


**Fig. 5** Examples of records in different subjects during a Tilt Test

The obtained results allow to analyze other clinical parameters of interest. Figure 6 shows the option of characterize the behavior of the pulse transit time (PTT) between different sections of the registers, for example, between the maximum of R wave and the maximum of Leg or Arm pressure ($PTT_A$ or $PTT_L$) or between the pressure maximums of the Leg and Arm ($PTT_{A\text{-}L}$).

Nowadays, we are working in the determination of relations between some cardiovascular illnesses (Ex.: Vagal Syncope and others) with the values obtained of differential MAP, and we are studying the behavior in the spectral domain of recorded signals. This new system allows to enlarge the number of physiologic variables to

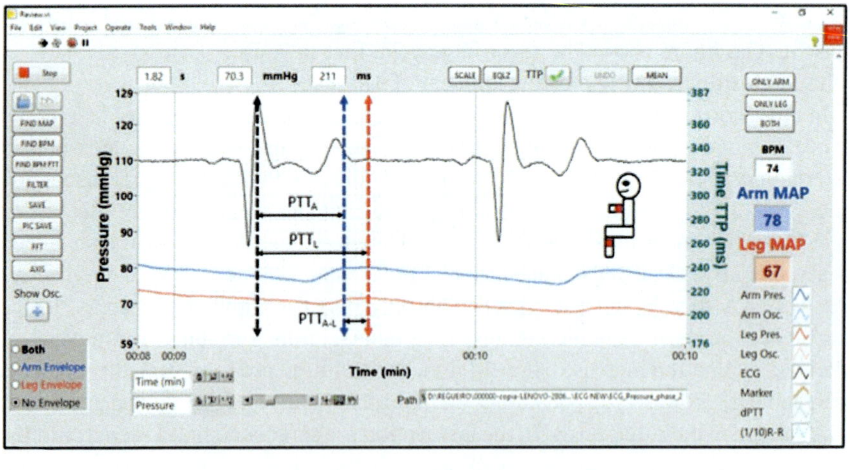


**Fig. 6** Example of record in a subject during a Tilt Test to characterize the Pulse Transit Time (PTT) between different areas (heart—arm, heart—leg and arm—leg)

acquire (ECG, breathing frequency, and others), and this allows to improve the diagnosis of the clinical specialists during the studies of the cardiovascular system from the submit patients to the Tilt Test.

## 4 Conclusions

The new system proposed permits the differential acquisition in mode continuous of the blood pressure variable during a Tilt Test with a good robustness and a low uncertainty. The acquisition channel shows a good response (linearity and bandwidth) and the developed interface is easy to use for the record and the spectral-time processing of the acquired variables.

**Acknowledgements** The authors appreciate the support offered by the project MORPHYs (PR-0435) development by Division of Instrumentation & Bioengineering at Universidad Tecnológica de La Habana (Cujae), for the development of this research (materials and instruments used for experiments).

## References

1. WHO.: World health statistics overview 2019: monitoring health for the SDGs, sustainable development goals, 1–28 (2019)
2. Altinsoy, M., Sutton, R., Kohno, R., Sakaguchi, S., Mears, R.K., Benditt, D.G.: Ambulatory ECG monitoring for syncope and collapse in United States, Europe, and Japan: the patients' viewpoint. J. Arrhythmia **00**, 1–8 (2021). https://doi.org/10.1002/joa3.12560
3. Buszko, K., Kujawski, S., Newton, J.L., Zalewski, P.: Hemodynamic response to the head-up Tilt Test in patients with syncope as a predictor of the test outcome: a meta-analysis approach. Front. Physiol. **10**, 184 (2019). https://doi.org/10.3389/fphys.2019.00184
4. Stewart, J.M., Shaban, M.A., Fialkoff, T., Tuma-Marcella, B., Visintainer, P., Terilli, C., et al.: Mechanisms of tilt-induced vasovagal syncope in healthy volunteers and postural tachycardia syndrome patients without past history of syncope. Physiol. Rep. **7**(13), e14148-e (2019). https://doi.org/10.14814/phy2.14148
5. Ninni, S., Kouakam, C., Szurhaj, W., Baille, G., Klug, D., Lacroix, D., et al.: Usefulness of head-up tilt test combined with video electroencephalogram to investigate recurrent unexplained atypical transient loss of consciousness. Arch. Cardiovasc. Dis. **112**(2), 82–94 (2019). https://doi.org/10.1016/j.acvd.2018.08.004
6. Meidert, A.S., Saugel, B.: Techniques for non-invasive monitoring of arterial blood pressure. Front. Med. **4**, 231 (2018). https://doi.org/10.3389/fmed.2017.00231
7. Lee, S., Chung, J., Bae, J., Cho, Y.J., Nam, K., Jeon, Y.: Continuous non-invasive arterial pressure monitoring (ClearSight System) and ankle blood pressure measurements as alternatives to conventional arm blood pressure. J. Clin. Med. **9**, 3615 (2020). https://doi.org/10.3390/jcm9113615, https://www.mdpi.com/journal/jcm
8. Bo, Y., et al.: Comparison between automated office blood pressure measurements and manual office blood pressure measurements—implications in individual patients: a systematic review and meta-analysis. Curr. Hypertens. Rep. **23**, 4 (2021). https://doi.org/10.1007/s11906-020-01118-1

9. Mol, A., Blom, M.E.C., van den Bosch, D.J., Van Wezel, R.J.A., Meskers, C.G., Maier, A.B.: Orthostatic blood pressure recovery measured using a sphygmomanometer is not associated with physical performance or number of falls in geriatric outpatients. Gerontology. Clinical Section: Brief Report, 2–5 (2021). https://doi.org/10.1159/000515658
10. Simonyan, M.A., Borovkova, E.I., Ishbulatov, Y.M., Skazkina, V.V., Karavaev, A.S., Shvartz, V.A., Kiselev, A.R.: Gender-related specificities of photoplethysmogram spectral assessment dynamics in healthy subjects during the passive tilt test. Russ. Open Med. J. **10**, e0115 (2021). https://doi.org/10.15275/rusomj.2021.0115
11. Regueiro-Busoch, C., Almeida-Pichardo Y. J., Noa-LLoren L. A., Busoch-Morlán, C., Regueiro-Gómez, A.: Automatic system to continuous and differential acquisition of Blood Pressure during Tilt Test. In: Proceedings of IFAC 2022. Havana, Cuba (2021)